ENVIRONMENTAL SCIENCE, ENGINEERING AND TECHNOLOGY SERIES

MYCORRHIZA

OCCURRENCE IN NATURAL AND RESTORED ENVIRONMENTS

ENVIRONMENTAL SCIENCE, ENGINEERING AND TECHNOLOGY SERIES

Additional books in these series can be found on Nova's website under the Series tab.

Additional E-books in these series can be found on Nova's website under the E-books tab.

ENVIRONMENTAL SCIENCE, ENGINEERING AND TECHNOLOGY SERIES

MYCORRHIZA

OCCURRENCE IN NATURAL AND RESTORED ENVIRONMENTS

MARCELA PAGANO
EDITOR

Nova Science Publishers, Inc.
New York

Copyright © 2012 by Nova Science Publishers, Inc.

All rights reserved. No part of this book may be reproduced, stored in a retrieval system or transmitted in any form or by any means: electronic, electrostatic, magnetic, tape, mechanical photocopying, recording or otherwise without the written permission of the Publisher.

For permission to use material from this book please contact us:
Telephone 631-231-7269; Fax 631-231-8175
Web Site: http://www.novapublishers.com

NOTICE TO THE READER

The Publisher has taken reasonable care in the preparation of this book, but makes no expressed or implied warranty of any kind and assumes no responsibility for any errors or omissions. No liability is assumed for incidental or consequential damages in connection with or arising out of information contained in this book. The Publisher shall not be liable for any special, consequential, or exemplary damages resulting, in whole or in part, from the readers' use of, or reliance upon, this material. Any parts of this book based on government reports are so indicated and copyright is claimed for those parts to the extent applicable to compilations of such works.

Independent verification should be sought for any data, advice or recommendations contained in this book. In addition, no responsibility is assumed by the publisher for any injury and/or damage to persons or property arising from any methods, products, instructions, ideas or otherwise contained in this publication.

This publication is designed to provide accurate and authoritative information with regard to the subject matter covered herein. It is sold with the clear understanding that the Publisher is not engaged in rendering legal or any other professional services. If legal or any other expert assistance is required, the services of a competent person should be sought. FROM A DECLARATION OF PARTICIPANTS JOINTLY ADOPTED BY A COMMITTEE OF THE AMERICAN BAR ASSOCIATION AND A COMMITTEE OF PUBLISHERS.

Additional color graphics may be available in the e-book version of this book.

Library of Congress Cataloging-in-Publication Data

Mycorrhiza : occurrence and role in natural and restored environments /
editor, Marcela Pagano.
   p. cm. -- (Environmental research advances)
Includes bibliographical references and index.
ISBN 978-1-61209-226-3 (hardcover : alk. paper)
1. Mycorrhizas. I. Pagano, Marcela.
QK604.2.M92M928 2011
631.4'6--dc22
        2010048376

Published by Nova Science Publishers, Inc. † New York

CONTENTS

Preface		**vii**
Chapter 1	Arbuscular Mycorrhizas in Natural Environments: An Overview *Marcela C. Pagano*	**1**
Chapter 2	Large-scale Diversity Patterns in Spore Communities of Arbuscular Mycorrhizal Fungi *Javier Álvarez-Sánchez, Nancy C. Johnson, Anita Antoninka, V. Bala Chaudhary, Matthew K. Lau, Suzanne M. Owen, Irene Sánchez-Gallen, Patricia Guadarrama and Silvia Castillo*	**29**
Chapter 3	Arbuscular Mycorrhizas and their Importance for Tropical Forest Formation in Brazil *Waldemar Zangaro*	**49**
Chapter 4	Arbuscular Mycorrhizal Fungi in the Amazon Region *Clara Patricia Peña-Venegas*	**75**
Chapter 5	Ocurrence of Mycorrhizas in Highland Fields *Marcela C. Pagano and Marta Noemi Cabello*	**87**
Chapter 6	Co-ocurrence of Arbuscular Mycorrhizas and Dark Septate Endophytes in Pteridophytes from a Valdivian Temperate Rainforest in Patagonia, Argentina *Natalia Fernández, Sonia Fontenla and María Inés Messuti*	**99**
Chapter 7	Mycorrhizas, an Important Component in Semiarid Sites *Marcela C. Pagano, Maria Rita Scotti and Marta Noemi Cabello*	**127**
Chapter 8	Mycorrhizal Status and Responsiveness of Early Successional Communities from Chaquean Region in Central Argentina *Carlos Urcelay, Paula A. Tecco, Marisela Pérez, Gabriel Grilli, M. Silvana Longo and Romina Battistella*	**147**

Chapter 9	Measuring and Estimating Ectomycorrhizal Fungal Diversity: A Continuous Challenge *Ornella Comandini, Andrea C. Rinaldi and Thomas W. Kuyper*	**165**
Chapter 10	Mycorrhizal Diversity in Native and Exotic Willows (Salix humboldtiana and S. alba) in Argentina *Mónica A. Lugo, Alejandra G. Becerra, Eduardo R. Nouhra and Ana C. Ochoa*	**201**
Chapter 11	Tree Species Composition and Diversity in Brazilian Freshwater Floodplains *Florian Wittmann*	**223**
Chapter 12	Arbuscular Mycorrhizae in Aquatic Plants, India *K. P. Radhika, James D'Souza and B. F. Rodrigues*	**265**
Chapter 13	Importance of Arbuscular Mycorrhizal Fungi for Recovery of Riparian Sites in Southern Brazil *Sidney Luiz Stürmer, Andressa Franzoi Sgrott, Felipe Luiz Braghirolli, Alexandre Uhlmann and Rosete Pescador*	**275**
Chapter 14	Mycorrhizas in Natural and Restored Riparian Zones *Marcela C. Pagano and Marta Noemi Cabello*	**291**
Index		**317**

PREFACE

This new book explores the links between mycorrhizas in the aquatic and riparian ecosystems, presenting research that points to the important role of arbuscular mycorrhizal in these environments. The volume seeks to spur further development of links for understanding AM function in the different soil systems, both in natural and restored ecosystems.

Chapter 1 - Arbuscular mycorrhizal fungi (AMF) are distributed worldwide in terrestrial temperate, tropical and arctic regions, as well as under waterlogged and aquatic conditions. The root system of most of the terrestrial higher plants growing under natural conditions presents mycorrhizal symbiosis. Most of the economically important agronomic plant crops as well as commercial fruit and economically important forest trees form mycorrhizas. However, little is known about plants from natural habitats, in spite of the higher biological activity in soils under natural conditions in comparison to cultivated or disturbed soils. Given their relevant role in ecosystems, this chapter will review AMF symbiosis in natural environments. Even though reviews have already been published recently, more reports of mycorrhizal status of plants are needed, as well as further studies of mycorrhizal fungal populations in relation to soil fertility and functionality. Moreover, there is little information about the reliable use of microbial parameters and use of biological potential for better plant nutrient management. Studies on mycorrhizas in natural environments are crucial in comparing AMF communities between undisturbed and disturbed ecosystems. The occurrence of different types of mycorrhizal fungi from various plants and from a wide range of soils and association of fungi with different types of host plants have already been recorded; however, the existing knowledge needs to be organized into a possible structure and a system of influencing factors on AMF needs to be developed. In this chapter, the presence of glomalin in soils and mycorrhizal fungi potential are also briefly outlined, drawing on results of research on natural environments, mostly in Brazil.

Chapter 2 - Surprising little is known about the factors controlling Arbuscular Mycorrhizal (AM) fungal diversity and distribution patterns. A better understanding of these factors is necessary before mycorrhizas can be effectively managed for their benefits in ecosystem restoration and agriculture. The goal of this chapter is to examine the relationships between AM fungal diversity, plant diversity and latitude across a variety of vegetation types and disturbance regimes. The authors created a large database by compiling seven distinct datasets from across North America to test the hypotheses that 1) diversity of AM fungal communities should be positively related to plant diversity; 2) AM fungal diversity should be higher in low latitudes than in high latitudes; and 3) disturbance and land use should influence

AM fungal diversity. The database was composed of 523 samples collected by research groups from Mexico and United States of America with eight different vegetation types and land uses. Abundance of 121 AM fungi taxa as well as data on geographic location, vegetation type, and disturbance history were included in the database. Contrary to their expectations, species richness and evenness of AM fungal spore communities were not correlated with plant richness or latitude. The influence of disturbance on AM fungal species diversity varied with climate. Their findings indicate that the factors controlling community diversity differ for plants and mycorrhizal fungi. Disturbance influences AM fungal spore diversity, and the outcome of these effects varies among ecosystems.

Chapter 3 - The importance of the arbuscular mycorrhizal (AM) fungus for establishment, survival and growth of the tropical native woody species belonging to different ecological groups of succession from Brazilian tropical forests is discussed. The response to AM inoculation and the degree of AM root colonization of the plant species decreased with the advance among successional ecological groups. High response to inoculation and high colonization among early successional woody species (pioneer and early secondary) emphasize the importance of AM association for the initial growth of these woody species, which are evolved in the initial tropical forest structuring. The late successional woody species (late secondary and climax), which dominates in the mature forests, showed absence or low both responsiveness to inoculation and degree of root colonization, indicating limitations to use of the AM fungus as strategies for mineral acquisition in early growth stages. In this chapter, some experimental results of AM inoculation in tropical native woody species from different ecological groups will be presented and discussed, especially in relation to the woody species development, the vegetal succession and the importance of AM association on the reclamation of degraded areas.

Chapter 4 - The Amazonia is the highest continuous tropical rain forest ecosystem with an area of approximately 7.413.827 Km^2. Seventy percent of the soils of the Amazon basin are Oxisols and Ultisols. They are characterized by being old, highly weathered and leached, with low pH (3-5.5), poor cation exchange capacity and limited availability of phosphorus content. Due to these particular conditions, plants have developed strategies to supply nutrient limitations such as mycorrhizas. Mycorrhizal community in the Amazon region is dominated by Arbuscular mycorrhizal fungi (AMF). *Glomus* is the dominant genus. AMF occupies the first 30cm of the topsoil in which colonize superficial fine roots of plants. This distribution is related to root distribution and nutrient availability into the soils. According to spore description high AMF diversity exists in the region. Low pH of the soil and soil phosphorus availability do not affect mycorrhizal colonization capacity. Effective AM symbiosis, quantified by the number of arbuscules, might be related to specific soil phosphorus ranges. More arbuscules and therefore a better capacity of phosphorus mobilization to the plant has been observed in soils with available phosphorus concentrations between 15-20ppm. AM symbiosis occurs across the region from natural to anthropic ecosystems. AMF diversity and spore abundance change according to soil intervention. Highest diversity and low spore numbers occur in natural forest ecosystems. No distinguishable patterns exist to anthropic environments. Number of spores, arbuscular mycorrhizal fungi diversity, and % colonization varies depending on soil management, vegetation composition and physical and chemical conditions of the soil.

Chapter 5 - The southeastern Brazilian highlands are centers of endemism and diversity; however, outcrop plant communities still lack systematic studies. Research in these areas, which are subject to mining, and where metal-tolerant plant species may exist, is essential to successful restoration. The aim of this review is to explore current information on the occurrence of mycorrhizas in outcrops, and to speculate about their mutualistic interactions. Root colonization in the studied Brazilian plant species as well as the arbuscular mycorrhizal (AM) spore diversity found in their rhizospheres are illustrated. The relevant findings are emphasized, such as the presence of vesicles and auxilliary cells in most roots. Occurrence of different types of AM fungi from six outcrop plants belonging to 6 families and 6 genera are presented. The associations of the fungi with different types of host plants as well as the soil properties are discussed. The mycorrhizal status of two angiosperms is reported for the first time. Highland field's plant species contains natural AM fungal species richness that can be affected by mining, and this supplies important information for restoration programs. Moreover, AM dependent plants in natural and restored highland field ecosystem, like *Eremanthus incanus*, can present higher potential for regeneration in habitats subject to disturbance. Research directions needed to increase understanding of mycorrhizal associations in these environments are indicated. The chapter discusses benefits and problems encountered, in order to highlight the need for a continual and integrated study of the highland ecosystems.

Chapter 6 - Plant roots are colonized by several fungi species, including those that form arbuscular mycorrhizas (AM). The contribution of AM to plant productivity, plant nutrition and soil composition is crucial, and in natural ecosystems plant diversity and community structure are strongly influenced by these organisms. Despite the importance of pteridophytes (lycophytes and ferns) in the origin and evolution of vascular plants, little is known about AM in this group of plants. The aim of this chapter is to describe the occurrence of AM in different species of pteridophytes present in the Valdivian temperate rainforest of Puerto Blest (Patagonia, Argentina), and to discuss different factors that might determine not only the mycorrhizal status but also the development of different AM morphologies in these plants (*Arum, Paris* or *Intermediated*-type). Sporophytes were sampled by random walk from terrestrial, epipetric and epiphytic habitats. The samples were stained and the percentage of root length colonized by these symbiotic fungi was quantified in all of them. A total of 21 species belonging to 10 families and 12 genera were found. Arbuscular mycorrhizas were present in 47.6 % of these species. Fifty percent of the families included only consistently mycorrhizal species (Blechnaceae, Dryopteridaceae, Gleicheniaceae, Lophosoriaceae and Pteridaceae), 20 % facultatively mycorrhizal species (Equisetaceae and Lycopodiaceae) and 30 % non-mycorrhizal species (Aspleniaceae, Grammitidaceae and Hymenophyllaceae). It was observed that while facultative species have an *Intermetiate*-type AM, consistently mycorrhizal species were only colonized by *Paris*-type associations. In addition to AM, another group of fungi were present within the roots of all the species considered in this study. They corresponded to dark septate endophytes (DSE) and were characterized by regularly septate hyphae and microsclerotia. The colonization pattern of DSE is described and its importance as possible "mycorrhizal associations" is discussed.

Chapter 7 - Tropical dry deciduous forests are under extreme climatic and edaphic environmental conditions, generally presenting shallow soils with low water availability. Fabaceae, Myrtaceae, and Meliaceae are the most common families with highest species richness in tropical dry forests, where plant communities are supposed to be dominated by mycorrhizal plants; however, the ecological traits of few plant species has been studied and verified as mycotrophic. The aim of this review is to explore the current information on the occurrence of mycorrhizas in the semiarid of Brazil, and to speculate about the benefit of symbiosis in semiarid sites. The diversity of arbuscular mycorrhizal fungi (AMF), analyzed based on spore morphology, revealed higher species richness and at least eighteen AM fungal known species present in the natural dry environments. The AMF spore diversity is illustrated, including those AMF which cannot be described at species level. The study of these areas, which commonly present sandy soils with low fertility, is essential to a successful restoration. For some regions, mixed forest used for land revegetation presents an AMF spore community composition more similar to that of the preserved sites. The spore community of the scrub vegetation usually present in disturbed sites is somewhat different. The semiarid's plant species contains natural AMF fungal species richness that can be affected by human activities, and AM dependent plants in natural ecosystems, may present higher potential for regeneration in habitats subjected to disturbance. The authors end with research directions that are needed to increase understanding of mycorrhizal associations in these environments. The problems encountered are discussed in this chapter, in order to highlight the need for a continual and integrated study of the semiarid ecosystems.

Chapter 8 - Theoretical models suggest that early successional communities are dominated by non-mycorrhizal plant species and thus, assume that mycorrhiza interactions do not play an important role in structuring these plant communities. Here the authors test these classic models gathering data from Chaquean region in central Argentina. The evidence shows that, in contrast to model predictions, most of the studied ruderal species in these communities harbor arbuscular mycorrhizal fungi (AMF) and dark septate endophytes (DSE) in their roots. In addition, data from synthetic experiment show that AMF have antagonistic effects on three abundant ruderal fast-growing pioneer species (growing alone and interacting with each other) and promote evenness in this simplified plant community. The effects of AMF on ruderal pioneers observed here mirror those attributed to herbivores and pathogens, suggesting a previous unrecognised mechanism by which AMF might promote secondary succession. Altogether, the evidence suggests that AMF and DSE should be included in models that predict the effects of fungal root symbionts in early secondary succession communities. The evidence so far suggests that AMF-ruderal species symbiosis has potential implications in agroecosystems management, ecological restoration and phytoremediation.

Chapter 9 - Thousands of ectomycorrhizal (ECM) fungal species exist, but estimates of global species richness of ECM fungi differ widely. Many genera have been proposed as being ECM, but in a number of studies evidence for the hypothesized ECM habit is lacking. Progress in estimating ECM species richness has therefore been slow. Recently, the authors have published evidence for the ECM habit of fungal species and for the identification of the mycobiont(s) in specific ECM associations, using published and web-based mycorrhiza literature. The identification methods considered were morpho-anatomical characterization of naturally occurring ECMs, pure culture synthesis, and molecular identification. In addition, stable isotope data of C and N, and phylogenetic information were also considered as relevant criteria to assess ECM habit. Their survey indicated that for 343 fungal genera an ECM status has been alleged, and for about two thirds (236 genera) of these supportive evidence of ECM status exists or can at least be entertained as the more reasonable hypothesis. On the basis of their literature search the authors conservatively estimated ECM species richness around 7750 species. However, on the basis of estimates of knowns and unknowns in macromycete diversity, they suggested that a final estimate or ECM species richness between 20,000 and 25,000 would be more realistic. Recent updates, taking into consideration evidence that became available after their study was released, have not changed figures substantially (234 genera, 7950 species), confirming that current knowledge of ECM fungal diversity, as supported by experimental evidence, is only partly complete, and that inclusion of many fungal genera in this trophic and ecological category is not verified at this stage. Care must thus be used when compiling lists of ECM and saprotrophic species on the basis of published information only. They reflect on the status of the various sources of evidence. They also discuss interesting avenues of future research, including a wider assessment and understanding of the ubiquity and diversity of the secondary root-associated fungi (root endophytes, especially Ectomycorrhiza-Associated Ascomycetes).

Chapter 10 - Mycorrhizal morphology and diversity vary not only within each particular host family of plants, but also with soil characteristics, nutrients availability, spatial-temporal conditions, micro-habitats, and host's age. Ectomycorrhizae (ECM) and arbuscular mycorrhizae (AM) provide nutritional benefits to their hosts, besides their effects in soil aggregation, soil pollutants sequestration and host's interconnection by hyphal network that allows nutrients transport, seedlings establishment and conservation of forest ecosystems. Furthermore, certain hosts species such as *Salix* spp. in the Salicaceae are involved in dual associations with ECM and Glomeromycota fungal symbionts. *Salix*, presents a wide distribution in South America, native *Salix humboldtiana* populations as well us the introduced species are mostly located in riparian ecosystems, or in temporarily flooded areas. Riparian zones have an important role regulating the movement of material and water between soil-river systems. Mycorrhizal diversity and colonization was studied in riparian populations of native *S. humboldtiana* Willd. and of exotic *S. alba* L. in semiarid riparian environments. Differences in ECM morphotypes and mycorrhizal colonization were found in both species. Effects of soil physical-chemical features on ECM diversity are discussed and ECM morphotypes are illustrated.

Chapter 11 - Neotropical freshwater floodplain forests are important landscape units because they consist of both flood-resistant species and of immigrants from the adjacent uplands, thereby concentrating large part of regional floristic biodiversity. They substantially contribute to the food webs of the terrestrial and the aquatic fauna, and fulfil a variety of ecologically and economically important functions. The inundation of roots and aboveground organs reduces oxygen availability to trees and is widely considered as a potential stress factor. Inundation is thus a powerful factor selecting the occurrence and distribution of tree species, which in turn influences species distribution and richness of most Neotropical floodplain forests.

Floodplain forests occur in all tropical biomes of Brazil. Most of its tree species are strongly zoned along the flooding gradient and associated habitat disturbance, thus reflecting the degree of adaptations developed to the unfavourable site conditions. Floodplains mostly are characterized by subsets of species of the adjacent uplands. Few flood-tolerant tree species are widely distributed, with occurrence in nearly all Brazilian biomes. However, these species are rarely restricted to flooded habitats. This evidences that these species not only tolerate flooding – but that they are generally tolerant to a wide range of stressful conditions.

Most floodplain forests are highly diverse, but lack endemic tree species – with exception of Amazonian large-river floodplains. The lack of endemism within flooded forests of the Cerrado and Atlantic forest domain can be traced to both, intense species exchange between floodplains and adjacent uplands and high connectivity by river systems that act as migration routes over huge geographic distance. The occurrence of severe droughts and wet periods during Tertiary and Quaternary climate change repeatedly interrupted evolutional processes of the flood-adapted tree floras. By contrast, elevated degrees of endemism in Amazonian large-river floodplains indicate comparatively s environmental conditions over large part of the Amazon basin since at least the early Paleocene.

Chapter 12 - Plant root fungal interactions called mycorrhizae are found in approximately 90% of all vascular plants. Hydrophytes were regarded as nonmycorrhizal until a decade ago as arbuscular mycorrhizal (AM) fungi require oxygen to thrive and it had been assumed AM fungi have little significance in wetland ecosystems. In India, studies are scarce on mycorrhizal association in wetland habitats viz., aquatic and marshy lands, khazan lands, and mangroves ecosystem. The first report of mycorrhizal colonization in aquatic plants was recorded in modified leaves of *Salvinia cucullata*. Recent study indicated that 16 of 20 plant species from aquatic and marshy habitats in Goa exhibited AM fungal root colonization. A study on occurrence of AM fungal association in five subtropical ponds of marshy habitats from Shillong revealed that nonmycorrhizal plants exhibited poor plant growth and biomass. The presence of AM fungal association in 24 plant species of sedges (representing six genera) was confirmed from different vegetation types of Western Ghats. In Goa, occurrence of AM fungal association in 28 plant species belonging to 16 families from khazan land ecosystem indicated adaptability to salt stress conditions in plant growth and survival. Mycorrhizal association in mangroves ecosystem in the Ganges river estuary indicated tolerance of mycorrhizae to various types of physical and chemical stresses in soil. This paper reviews the work carried out on diversity and role of AM fungi in wetland ecosystems from India.

Chapter 13 - The endomycorrhizal association established between arbuscular mycorrhizal fungi (Glomeromycetes) and plant roots are known to have a pivotal role on plant survival and nutrition. In this chapter, the authors convey that the triad riparian forest-carbon sequestration-arbuscular mycorrhizal fungi should be considered as environmental friendly strategy to recovery riparian sites. Greenhouse and field experiments carried out in soils occurring at the Itajai valley basin have demonstrated the importance of this association for growth and carbon allocation on plants species used to revegetation of riparian sites. Inoculation of pioneer species with selected fungal isolates have shown that more carbon is allocated to roots when compared to non-inoculated plants or plants inoculated with indigenous fungi. For *Schinus terebinthifolius* they have evidence at field conditions that inoculation with *Scutellospora heterogama*, *Glomus clarum* and *Acaulospora koskei* increased significantly the quantity of carbon stored in individual plants, calculated using allometric equations. Riparian sites are habitats that harbors a great diversity of glomeromycetes and 42 species has been registered in works developed in the USA and Brazil, most of them in the genera *Glomus* and *Acaulospora*. It is also noticeable that all members of *Paraglomus* (*P. occultum*, *P. brasilianum*, *P. laccatum*) and *Archaeospora trappei* are commonly found in riparian sites. Considering the role of arbuscular mycorrhizal fungi on growth and carbon accumulation of pioneer plants and the high diversity of glomeromycetes in riparian soils, the inoculation of plants to be used in revegetation process of riparian sites should be viewed as a environmental friendly strategy to recovery these areas and improve carbon sequestration.

Chapter 14 - The riparian vegetation, used for cattle and agriculture practices, has long been negatively influenced by human activities, especially domestic sewage and mining activity. Only recently the river basins are considered as part of the landscape and have begun to be studied for their environmental management. Moreover, a study of interactions among the functional groups of microorganisms in riparian regions is essential to a successful restoration, and to improve restoration programs. The purpose of this review is to explore the current information on the occurrence of mycorrhizas in riparian ecosystems in Brazil, and to speculate about the role of symbiosis. This chapter discusses arbuscular mycorrhizal (AM) root colonization, drawing on results of research of native and invasive plant species in Brazil. As expected, the studies revealed that most native plant species in riparian areas show mycotrophy. AM spore diversity found in rhizospheric soils is illustrated, and relevant findings are emphasized. As generally found, legumes present a higher AM colonization, being the *Arum*-type the most commonly observed. A high AMF spore richness in the rhizospheric soils of most plant species in riparian zones was observed. In general, AM fungal species belongs to eight genera, which are related to the sites and regions studied; however, *Glomus* and *Scutellospora* spores were common. Some terrestrial pteridophytes showed presence of AM structures; however, some ferns show dominant extraradical hyphae whereas their mycorrizal status remains uncertain. The restored areas presented higher AM species richness than the degraded areas. The AM symbiosis is a common and important component in the riparian vegetation, and should be included in future restoration programs. The benefits and problems encountered are discussed in this chapter.

In: Mycorrhiza: Occurrence in Natural and Restored...
Editor: Marcela Pagano

ISBN: 978-1-61209-226-3
© 2012 Nova Science Publishers, Inc.

Chapter 1

ARBUSCULAR MYCORRHIZAS IN NATURAL ENVIRONMENTS: AN OVERVIEW

Marcela C. Pagano[*]
Federal University of Ceará, Fortaleza, Ceará, Brazil

ABSTRACT

Arbuscular mycorrhizal fungi (AMF) are distributed worldwide in terrestrial temperate, tropical and arctic regions, as well as under waterlogged and aquatic conditions. The root system of most of the terrestrial higher plants growing under natural conditions presents mycorrhizal symbiosis. Most of the economically important agronomic plant crops as well as commercial fruit and economically important forest trees form mycorrhizas. However, little is known about plants from natural habitats, in spite of the higher biological activity in soils under natural conditions in comparison to cultivated or disturbed soils. Given their relevant role in ecosystems, this chapter will review AMF symbiosis in natural environments. Even though reviews have already been published recently, more reports of mycorrhizal status of plants are needed, as well as further studies of mycorrhizal fungal populations in relation to soil fertility and functionality. Moreover, there is little information about the reliable use of microbial parameters and use of biological potential for better plant nutrient management. Studies on mycorrhizas in natural environments are crucial in comparing AMF communities between undisturbed and disturbed ecosystems. The occurrence of different types of mycorrhizal fungi from various plants and from a wide range of soils and association of fungi with different types of host plants have already been recorded; however, the existing knowledge needs to be organized into a possible structure and a system of influencing factors on AMF needs to be developed. In this chapter, the presence of glomalin in soils and mycorrhizal fungi potential are also briefly outlined, drawing on results of research on natural environments, mostly in Brazil.

Keywords: Mycorrhizal symbioses – Natural environments – Soil glomalin – waterlogged and aquatic conditions

---

[*] Federal University of Ceará, Fortaleza, Ceará, Brazil
[*] marpagano@gmail.com; pagano@netuno.lcc.ufmg.br

INTRODUCTION

In natural ecosystems arbuscular mycorrhizas (AM) are obligatory root symbionts, typically found as mixed communities with multiple species colonizing any given plant root. Since the functionality of the symbiosis is highly variable and dependent upon the identity of the AM and host species involved (Johnson et al. 1997, Klironomos 2003), the composition of AM species colonizing a given plant has important implications for its fitness. AMF species composition may be an important determinant of plant community composition (van der Heijden et al. 1998), and therefore, changes of AM community composition could have important implications for ecosystem function (Rillig 2004, Mummey et al. 2009).

In 2002, Brundrett summarized the coevolution of mycorrhizal fungi and roots based on paleobotanical and morphological studies, as well as phylogenetics. Seminal works, such as those by van der Heijden and Sanders (2003) and Smith and Read (2008), are important to update the emergent information about the mycorrhizal symbioses and their ecology.

Wang and Qui (2006) summarized data for substantial numbers of flowering plants, through a compilation of a checklist of mycorrhizal occurrence among 3,617 species (263 families) of land plants, and found that about 92% of land plant species and families were colonized by mycorrhizas. They showed the status of plants species from natural as well as from anthropogenic environments, such as agrosystems from all the continents; however, due to time limitation and unavailability of some literature, 659 of 961 references were used. In a more recent review, Brundrett (2009) also showed data on mycorrhizas of plants in natural habitats. The mycorrhizal status of aquatic and land plants covering most major habitats and geographic regions of the world are shown in this work. However, Brundrett (2009) highlighted the gaps in existing knowledge due to: poorly sampled habitats, un-sampled plant families, families with complex root strategies, as well as the need to study the mycorrhizal colonization linked to plant diversity, ecology or physiology at the ecosystem scale. Moreover, he pointed out the need for corrections as regards the status of families or genera published in earlier studies.

Additionally, Brundrett (2009) compiled data from 125 published papers and used them to estimate rates of errors for diagnosis of AM, EM and NM roots as well as the frequency of use of different definitions of AM. He found that near 200 families are yet to be sampled and the majority of these are very small (families with 15 species). He remarked that most of these unallocated families are likely to contain AM plants as they are sister to, or nested within, clades known to predominantly contain AM plants. He summarized data on mycorrhizas of plants in natural habitats from 128 publications, covering most major habitats and geographic regions in the world.

It has been showed that mycorrhizal fungi are ubiquitous soil inhabitants, occurring virtually worldwide in temperate, tropical and arctic regions, except for under waterlogged conditions; however, reports on AM in some macrophytes from lakes and streams (Beck-Nielsen and Madsen 2001, Kai and Zhiwei 2006), from rivers (Marins et al. 2009), from river estuaries (Sengupta and Chaudhuri 2002), in marshy plants (Radhika and Rodrigues 2007) and in riparian zones (Guevara and López 2007, Yang et al. 2008, Pagano et al. 2008b,c,d,e, 2009a) have increasingly pointed out their importance to these ecosystems.

Most non-mycorrhizal (NM) plants are marginalized in wet, saline, dry, disturbed, or cold habitats or extremely infertile soils, where plant productivity is low and inoculum of AM could be scarce (Brundrett 1991). Moreover, Wang and Qiu (2006) have showed that arbuscular mycorrhiza is the predominant and ancestral type of mycorrhiza in land plants, and ectomycorrhiza independently evolved from AM many times through parallel evolution.

Furthermore, since soil disturbance by human activities decreases nitrogen-fixing bacteria and mycorrhizal fungi usually associated with roots (Cooke and Lefor 1990), the environmental benefits of AM in restoration of the natural cover, which is an important aspect of environmental management, is increasingly studied.

The purpose of this review is: (1) to summarize existing literature in order to designate the distribution of AM, or to point conflicting information; (2) to update the most recent reports of diversity of plants with AM in natural environments. Data showing mycorrhizal benefits in natural ecosystems are also here compiled.

SOIL CONDITIONS AND MYCORRHIZAS

Soil heterogeneity is especially important in tropical regions, where it has been estimated that the distribution of more than 30% of tree species is restricted by soil properties (Clark et al. 1999). The mineral nutrient content of soils is known to modulate intrinsic attributes in plants such as growth rate (Poorter 1989) and morphology (Schreeg et al. 2005). Furthermore, the plant-associated interactions such as intra- and interspecific competition (Casper and Jackson 1997), herbivory (Boege and Dirzo 2004), and mycorrhizal colonization (Treseder and Vitousek 2001) are also affected.

Generally speaking AM live under poor and very unstable soil conditions and environmental factors greatly affect their diversity and their root colonization. Thus the status of a plant species depends on the growth environment. However, there is variation among the colonization of plant species in response to environmental variables. Siqueira and Saggin Júnior (2001) showed large differences in AM colonization, responsiveness to inoculation, mycorrhizal dependency and efficiency of phosphorus uptake among Brazilian tree species, which were also influenced by available soil P in solution. However, in Brazil, for nearly all native tree species, there is a generalized lack of studies involving evaluations of growth and responses to soil amendment and nutrient application under field conditions (Resende et al. 2005).

Low soil fertility and limited P availability favor the symbiotic association because the AM fungi may increase the P supply for plants. The hypothesis that plants with coarsely branched roots and with few or no root hairs are expected to be more dependent on mycorrhiza than plants with finely branched roots is broadly accepted, but not extensively demonstrated (Siqueira and Saggin Júnior 2001).

Jakobsen et al. (2003) stressed the great potential for detailed studies of nutrient transport in mycorrhizae formed in natural ecosystems in order to understand the complex effect on plant community. Many natural ecosystems in sandy soils support highly mycotrophic systems, and can be used as models for microcosmos experiments (Read 2003).

AM can live in slightly alkaline soil (pH 7.5-8.5) and some can even tolerate high pH condition (Yang et al. 2008); they are especially efficient in the uptake of inorganic phosphorus (Jakobsen et al. 2003). Moreover, reports have shown that roots of plants growing in saline-sodic soil conditions in natural grasslands from Argentina were highly colonized by AM (Garcia and Mendoza 2007).

Research in the last ten years has altered the way we think about the function of mycorrhizas in natural ecosystems; hence much more emphasis is now placed on the development and operation of the external mycelium and its role in linking plants together in mycorrhizal guilds (Smith and Read 2008).

AM FUNGI AND PLANT SUCCESSION

Overall, little is known about the relation between AM and native plant succession in natural ecosystems (Janos 1980; Zangaro Filho 1997; Siqueira et al. 1998), there being conflicting information.

In Brazil the native vegetation is predominantly AM (Stürmer and de Souza 2009) and there is a great need for more detailed studies with each tree species (Resende et al. 2005). Some authors (Carvalho, 1994; Lorenzi, 1998 a, b) still disagree as to the ecological classification of many species. The choice of species that can be more apt for planting in a site with a given set of characteristics will be more precise as more knowledge of ecology is gathered (Resende et al. 2005).

Zangaro et al. (2008) showed that mycorrhizal root colonization and spore density were positively correlated with fine root length, root hair length, root hair incidence, and negatively correlated with fine root diameter, in a tropical forest in southern Brazil (see Chapter 3, in this book). They also suggested that slow-growing species are less able to maintain AM fungi due to their low growth in relative high soil fertility and shading. However, as regards slow-growing plant species, it is known that they are more adapted to low natural fertility soils (Marschner 1991, Lambers and Poorter 1992).

Additionally, studies of Siqueira et al. (1998) reported that pioneer species from southeast Brazil showed high susceptibility to infection and colonization of AM fungi. Zangaro et al. (2000) showed that mycorrhizal dependence is higher in early successional species, and much lower in late successional species from south Brazil. Siqueira and Saggin-Júnior (2001) and Zangaro et al. (2005, 2007) verified that AMF root colonization is essential for both nutrient acquisition and plant survival in tropical seedlings of early successional Brazilian woody species, with apparent root morphology for high uptake capacity, but not able to ensure adequate nutrition to maintain their inherent fast growth rate.

On the other hand, other researchers, such as Janos (1980a,b, 1995), suggested that the dependence on mycorrhizal association is higher in late successional species in the tropical forests of Costa Rica, and that early successional species are non-mycotrophic or facultatively mycotrophic.

Therefore, the literature presents controversial aspects concerning the behavior of tree species in terms of AM colonization with the advance of sucessional group. To establish criteria for classification of native species it is important to consider rationalization of fertilizer use and cost reduction, as well as the benefits of this symbiosis.

ROOT COLONIZATION IN NATURE

Mycorrhizal symbiosis is almost universally present in terrestrial plants in nature (Read 2003), but some aquatic plants also present AM colonization (Marins et al. 2009), and nowadays this is being investigated in more detail (Table 1). Moreover, studies of AM ecology in natural ecosystems have great potential to lead to improvements in agriculture (Koide and Mosse 2004).

AMF can associate with multiple hosts (Molina et al. 1992) and the hyphae from the roots of established plant frequently colonize new roots of adjacent plants (Read et al. 1985). The same species of AM may develop differentially in different plant species; however, further research is needed regarding this aspect (Jakobsen et al. 2003). Commonly, the first AM taxa to invade a root is frequently the most abundant colonizer (Hart and Reader 2002).

There has been a marked advance on the survey of plant species and their AM symbionts in natural environments (Kessler et al. 2009). In this chapter, due to space restrictions and unavailability of some of the literature, 16 references were selected for review (Table 1). It should be pointed out that some reports include root colonization, indicating the morphologies of the AM present, and also the spore diversity found in the rhizosphere of the plant species studied; however, most reports do not show all these characteristics, which could be very useful to draw more conclusions about patterns of root colonization and AM species associated with plant species in different regions along the planet.

It can be observed that variations in occurrence of fungal structures provide information about the fungi in relation to nutrient transfer and plant growth (Jakobsen et al. 2003); root colonization declines with host ontogeny since older roots are higher suberized, and factors that alter root volume (root age and season) influence AM colonization (Reviewed by Egerton-Waburton et al. 2003).

Details of Brazilian native species colonization by AM are given by Pagano and Cabello (Chap. 5, this Vol.) and Pagano et al. (Chap.7, this Vol.). They reported the predominance of *Arum*-type colonization, and the presence of significant AM morphological structures, such as intraradical and extraradical hyphae, vesicles and arbuscules. The Glomineae-type of colonization, present in the majority of species evaluated by Pagano, is in line with the presence of *Glomus* spores in their rhizospheres. Interestingly, none or few DSE were observed in root of plants in natural environments. Practically only some ferns present the *Paris* type (Table 1).

Studies of AM colonization are important for seedling production and preparation of technologies for successful restoration or afforestation, because of the fact that vegetal species exhibit different AM dependency (Siqueira and Saggin-Junior 2001). Moreover, the literature shows that different plant species often harbor quite distinct AM fungal communities (Eom et al. 2000).

According to rDNA identification in roots, natural ecosystems present a higher AMF richness in tropical forests > grasslands > temperate forests (Öpik et al. 2006). Molecular identification directly in host plant roots is important to select efficient AM isolates; nevertheless, further analysis is necessary.

Table 1. Updated list of some families and genera/species of arbuscular mycorrhizal plants in natural habitats; R = Reference

Order	Family	Species or genus	Habit	Ecology/ Habitat	AM type	PC	Spores genus or species reported	Country/ Site	R
Pteridphytes	Blechnaceae	Blechnum polypoides	terrestrial	Riparian Forest	Paris	ar, eh, h, iv	A. scrobiculata, Glomus macrocarpum	Brazil (BR)	1
		Blechnum penna-marina	terrestrial	CM	Paris	v,c,h	NI	AR	8
	Dryopteridaceae	Onoclea sensibilis	terrestrial	facultative wetland, fen	NI	a,v,	NI	New York State (USA)	7
	Thelypteriaceae	Thelypteris palustris	terrestrial	facultative wetland, fen	NI	a,v,	NI	NYS	7
	Ophioglossaceae	Botrychium virginianum	terrestrial	mixed deciduous woodland	NI	h	Glomus, Scutellospora†	Hungary	9
		Botrychium crenulatum, B. lanceolatum	terrestrial		-		Glomus	Argentina (AR)	10
	Aspleniaceae	Asplenium salignum	epiphytic	rain forest (RF)	NI	h	NI	Indonesia, Malaysia (IM)	13
	Cibotiaceae	Cibotium barometz	terrestrial	RF	NI	h	NI	IM	13
	Cyatheaceae	Cyathea lurida	terrestrial	RF	NI	h	NI	IM	13
		Sphaeropteris moluccana	terrestrial	RF	NI	h	NI	IM	13
	Davalliaceae	Wibelia denticulata	epiphytic	RF	NI	h	NI	IM	13
	Dennstaedtiaceae	Microlepia puberula	terrestrial	RF	NI	h	NI	IM	13
	Dipteridaceae	Cheiropleuria bicuspis	terrestrial	RF	NI	h	NI	IM	13
		Dipteris conjugata	terrestrial	RF	NI	h	NI	IM	13
	Dryopteridaceae	Bolbitis heteroclita	terrestrial	RF	NI	h	NI	IM	13
		Lomagramma lomarioides	terrestrial	RF	NI	h	NI	IM	13
		Pleocnemia hemiteliiformis	terrestrial	RF	NI	h	NI	IM	13
		Teratophyllum aculeatum	terrestrial	RF	NI	h	NI	IM	13
		Polystichum montevidense	terrestrial	CM	Paris	v,h	NI	AR	8
	Gleicheniaceae	Diplopterygium longissimum	terrestrial	RF	NI	h	NI	IM	13
		Dicranopteris flexuosa*	terrestrial	Riparian zone	NI	eh, ac	Acaulospora, Dentiscutata biornata, Glomus macrocarpum	BR	1

Family	Species	Habit	Vegetation			Fungi		
Hymenophyllaceae	*Abrodictyum obscurum*	terrestrial	RF	NI	h	NI	IM	13
	Abrodictyum pluma	terrestrial	RF	NI	h	NI	IM	13
	Crepidomanes minutum	epiphytic	RF	NI	h	NI	IM	13
	Hymenophyllum penangianum	epiphytic	RF	NI	h	NI	IM	13
	Hymenophyllum serrulatum	epiphytic	RF	NI	h	NI	IM	13
Lindsaeaceae	*Lindsaea bouillodii, L. ensifolia, L. parasitica*	terrestrial	RF	NI	h	NI	IM	13
	Sphenomeris chinensis	terrestrial	RF	NI	h	NI	IM	13
	Tapeinidium luzonicum, T.pinnatum	terrestrial	RF	NI	h	NI	IM	13
Lycopodiaceae	*Lycopodiella cernua*	terrestrial	RF	NI	h	NI	IM	13
	Huperzia saururus	Terrestrial	CM	*Arum*	a,v,h	NI	AR	8
Lygodiaceae	*Lygodium microphyllum*	terrestrial	RF	NI	h	NI	IM	13
Marattiaceae	*Angiopteris evecta*	terrestrial	RF	NI	h	NI	IM	13
	Ptisana silvatica	terrestrial	RF	NI	h	NI	IM	13
Oleandraceae	*Oleandra neriiformis*	terrestrial	RF	NI	h	NI	IM	13
Polypodiaceae	*Taenitis dimorpha*	terrestrial	RF	NI	h	NI	IM	13
	Polypodium bryopodum	Epiphytic or rupestral	CM	*Paris*	v,h,c	NI	AR	8
Pteridaceae	*Haplopteris angustifolia, H. scolopendrina*	epiphytic	RF	NI	h	NI	IM	13
	Pityrogramma calomelanos	terrestrial	RF	NI	h	NI	IM	13
	Pityrogramma trifoliata	terrestrial	Riparian zone	I	rf	*Acaulospora bireticulata, A. scrobiculata , A. excavata , A. laevis, A. spinosa, Glomus, Scutellospora*	BR	1
	Pteris oppositipinnata	terrestrial	RF	NI		NI	IM	13
	Argyrochosma nívea, Adiantum raddianum	fern	Chaco Serrano woodland (CSW)	*Paris*	h	NI	AR	15
Schizaeaceae	*Anemia tomentosa*	shrub	CSW	*Paris*	h	NI	AR	15
Selaginellaceae	*Selaginella furcillifolia, S. intermedia, S. minutifolia, S. plana, S. roxburghii, S. Stipulata*	terrestrial	RF	*NI*	h	NI	IM	13

Table 1 (Continued)

Order	Family	Species or genus	Habit	Ecology/ Habitat	AM type	PC	Spores genus or species reported	Country/ Site	R
	Tectariaceae	*Pleocnemia conjugata*	terrestrial	RF	NI		NI	IM	13
		Pteridrys syrmatica	terrestrial	RF	NI		NI	IM	13
		Tectaria crenata, T. griffithii	terrestrial	RF	NI		NI	IM	13
	Thelypteridaceae	*Christella parasitica*	saxicolous	RF	NI		NI	IM	13
		Coryphopteris viscosa	terrestrial	RF	NI		NI	IM	13
		Mesophlebion crassifolium, M. persquamiferum, M. trichopodum	terrestrial	RF	NI		NI	IM	13
		Metathelypteris dayi	terrestrial	RF	NI		NI	IM	13
		Sphaerostephanos appendiculatus, S. heterocarpus, S. larutensis, S. Penniger, S. Polycarpus	terrestrial	RF	NI		NI	IM	13
		Thelypteris dentata	terrestrial	Riparian zone	*Arum-Paris*		*Acaulospora, Glomus macrocarpum*	BR	1
	Woodsiaceae	*Deparia petersenii*	terrestrial	RF	NI		NI	IM	13
		Diplazium asperum, D. bantamense, D. cordifolium, D. malaccense, D. procumbens, D. sorzogonense, D. Tomentosum	terrestrial	RF	NI		NI	IM	13
Gymno sperms	Equisetaceae	*Equisetum giganteum*	terrestrial	CSW	*Paris*	h	NI	AR	15
	Ephedraceae	*Ephedra triandra*	herb	CSW	*I*	h	NI	AR	15
Monocots	Araceae	*Symplocarpus foetidus*	herb	obligate wetland, fen	NI	a,v	NI	NYS	7
	Cyperaceae	*Bulbostylis*	herbs	Highland field	ND	ar, iv.h	*Acaulospora, Dentiscutata biornata, Glomus*	BR	1
		Carex hystericina	herbs	obligate wetland	NI	v.h	*NI*	NYS	7
		Carex sterilis	herbs	Obligate wetland	NI	a,v	*NI*	NYS	7

	Species		Habitat					
	Eriophorum viridcarinatum	herbs	Obligate wetland	NI	a,v	NI	NYS	7
	Cyperus entrerianus	Herb	CSW	NI	h	NI	AR	15
	Cladium jamaicense	graminoid	wetland	NI	h	NI	EEUU	4
Alliaceae	*Nothoscordum gracile*	herb	CSW	NI	h	NI	AR	15
Poaceae	*Paspalum repens*	macrophyte	Aquatic	*Paris*	h	*Acaulospora aff. laevis*	BR	2
	Agrostis stolonifera	grass	facultative wetland	NI	a,v	NI	NYS	7
	Bromus ciliatus	Grass	facultative wetland	NI	a,v	NI	NYS	7
	Calamagrostis canadensis	Grass	facultative wetland	NI	a,v	NI	NYS	7
	Glyceria striata	grass	obligate wetland	NI	a,v	NI	NYS	7
	Phragmites communis	perennial herbage	Riparian forest	NI	v,h	NI	China	5
	Chloris	grass	CSW	*Paris*	h	NI	AR	15
	Cortaderia selloana, Cenchrus incertus	grass	CSW	*Arum*	h	NI	AR	15
	Urochloa plantaginea	herb	Riparian sites	*Arum*	h	NI	BR	1
	Deyeuxia hieronymi	grass	CM	*Paris-Arum*	v,h,c	NI	AR	8
	Festuca tucumanica, Poa sturckertii	grass	CM	*Paris*	v,h,c	NI	AR	8
	Schismus arabicus	annual	fog-free Pacific coastal desert (PCD)	NI	h	*Glomus*	Chile (CHI)	16
Juncaceae	*Juncus brachycephaulus*	graminoid	obligate wetland	NI	a,v	NI	NYS	7
	Juncus tenuis	graminoid	facultative wetland	NI	a,v	NI	NYS	7
Iridaceae	*Iris versicolor*	herb	obligate wetland	NI	a,v	NI	NYS	7
	Sisyrinchium chilense	herb	CSW	*Arum*	h	NI	AR	15
	Sisyrinchium graminifolium	Geophyte	fog-free Pacific coastal desert (PCD)	NI	a,v,h	*Glomus*	Chile (CHI)	16

Table 1 (Continued)

Order	Family	Species or genus	Habit	Ecology/ Habitat	AM type	PC	Spores genus or species reported	Country/ Site	R
	Pontederiaceae	*Eichhornia crassipes*	macrophyte	Aquatic	*Paris*	h	NI	BR	2
		Pontederia cordata	macrophyte	Aquatic	*Paris*	h	NI	BR	2
	Nymphaeaceae	*Nuphar spp*	macrophyte	Aquatic	NI		NI	Florida wetland (EEUU)	4
	Arecaceae	*Sabal mexicana*	palm	Riparian	ND	h	NI	México	3
		Trithrinax campestris	palm	CSW	NI	h	NI	AR	15
	Typhaceae	*Typha spp.*	macrophyte	Aquatic	NI		NI	EEUU	4
		Typha latifolia	macrophyte	Aquatic		a,v,h	NI	NYS	7
		Panicum hemitomon	grass	wetland	NI	a,v,c,h	NI	EEUU	4
		Sparganium emersum	amphibious plant	intermittent aquatic habitats		h	NI	SW Slovenia (Europe)	14
	Bromeliaceae	*Deuterocohnia longipetala*	terrestrial	arid environment	NI		NI	AR	6
		Dyckia floribunda, D. velascana	herb	arid environment	NI		NI	AR	6
		Lotus glaber, Stenotaphrum secundatum	herb	natural grassland	NI	h, ar, v	NI	AR	10
		Puya spathacea	bromeliad	CSW	*Arum*	h	NI	AR	15
Dicots	Aizoaceae	*Mesembryanthemum cristallinum*	Annual, CAM	PCD	NI	a,v,h	*Gigaspora albida, Glomus aggregatum, Scutellospora*	Chile (CHI)	16
	Amaranthaceae	*Gomphrena pulchella*	herb	CSW	NI	h	NI	AR	15
	Amaryllidaceae	*Hippeastrum bagnoldii*	geophyte	PCD	NI	a,h	*Glomus*	Chile (CHI)	16
		Leontochir ovallei	geophyte	PCD	NI	a,v,h	*Glomus, Scutellospora*	Chile (CHI)	16
		Rhodophiala	geophyte	PCD	NI	v,h	*Glomus, Scutellospora*	Chile (CHI)	16
	Asteraceae	*Eremanthus incanus*	tree	Highland field		ac, iv, rh,	*Gigaspora margarita, Dentiscutata biornata, A. spinosa*	BR	1

Family	Species	Growth form	Habitat	Type	Codes	Fungi	Location	Ref
	Doellingeria umbellata	herb	facw, fen	NI	a,v	NI	NYS	7
	Euthamia graminifolia	herb	Fac, fen	NI	a,v	NI	NYS	7
	Packera aurea	herb	Facw, fen	NI	a, v, h	NI	NYS	7
	Solidago patula, S. Rugosa, S. uliginosa	herb	obligate wetland, fen	NI	a,v,h	NI	NYS	7
	Symphyotrichum boreale, S. puniceum	herb	Wetland, fen	NI	a,v,h	NI	NYS	7
	Achyrocline satureoides, A. tomentosa, Baccharis articulata, Eupatorium arnottianum, Flourensia campestris, F. oolepis, Grindelia cabrerae, Heterothalamus alienus, Stevia satureiifolia, Vernonia mollissima, V. nudiflora	shrub	CSW	NI	h	NI	AR	15
	Chaptalia sinuata, Gaillardia megapotamica, Tagetes minuta, Trichocline reptans, T. Sinuata, Xanthium cavanillesii, Zinnia peruviana	herb	CSW	NI	h	NI	AR	15
	Achyrocline satureoides	shrub	CM	*Paris*	v,c,h	NI	AR	8
	Gamochaeta americana, Hypochaeris radicata	herb	CM	*Paris*	v,c,h	NI	AR	8
	Chaetanthera linearis	annual	PCD	NI	v,h	*Glomus*	Chile (CHI)	16
	Encelia canescens	Shrub, C3	PCD	NI	a,v,h	*Glomus*	Chile (CHI)	16
	Helenium aromaticum	annual	PCD	NI	a,h	*Glomus*	Chile (CHI)	16
	Perityle emoryi	Chamaephyte	PCD	NI	a,h	*Gigaspora albida, Glomus*	Chile (CHI)	16
Berberidaceae	*Berberis ruscifolia*	shrub	CSW	NI	h	NI	AR	15
	Berberis hieronymi	shrub	CM	Paris	v,c,h	NI	AR	8
Boraginaceae	*Cryptantha glomerata*	annual	PCD	NI	a,h	*Glomus*	Chile (CHI)	16
	Tiquilia litoralis	Hemicrypto-phyte	PCD	NI	a,h	*Glomus, Scutellospora*	Chile (CHI)	16
Campanulaceae	*Lobelia siphilitica*	herb	Facw	NI	v,h	NI	NYS	7
Euphorbiaceae	*Croton sarcopetalus*	shrub	CSW	*Arum*	h	NI	AR	15
Chenopodiaceae	*Atriplex repanada*	Shrub, C4	PCD	NI	a,v,h	NI	Chile (CHI)	16
	Acalypha communis, Euphorbia portulacoides, Euphorbia portulacoides	Herb	CSW	*Arum*	h	NI	AR	15

Table 1 (Continued)

Order	Family	*Species or genus*	Habit	Ecology/ Habitat	AM type	PC	Spores genus or species reported	Country/ Site	R
	Gentianaceae	*Gentianella achalensis*	herb	CM	*Paris*	v,h	NI	AR	8
	Malvaceae	*Pavonia viscosa*	herb	Higland field	*Arum*	ar, rh, iv, ov	NI	BR	1
		Gaya parviflora	herb	CSW	*Arum*	h	NI	AR	15
		Abutilon grandifolium, Mediolastrum gilliesii	shrub	CSW	I	h	NI	AR	15
		Sphaeralcea cordobensis	Shrub	CSW	*Paris*	h	NI	AR	15
		Cristaria glaucophylla	Shrub	PCD	NI	a,v,h	*Gigaspora albida, Glomus*	Chile (CHI)	16
	Melastomataceae	*Tibouchina multiflora*	arboreal	Higland field	*Arum*	ac, ov	*Acaulospora, Gigaspora*	BR	1
	Eriocaulaceae	*Paepalanthus bromelioides*	perennial herbaceous	Higland field	ND	ac, c, ov	*Gigaspora margarita, Dentiscutata biornata, A. spinosa*	BR	1
	Hydrocharitaceae	*Limnobium variegatum*	macrophyte	Aquatic	*Paris*	h	*Acaulospora aff. laevis, Acaulospora delicata, Glomus luteum*	BR	2
	Apiaceae	*Hydrocotyle americana*	herb	obligate wetland, fen	NI	a,v	NI	NYS	7
		Sium latifolium	amphibious plant	intermittent aquatic habitats		h	NI	SW Slovenia (Europe)	14
		Eryngium horridum	herb	CSW	NI	h	NI	AR	15
		Eryngium agavifolium	herb	Córdoba mountains (CM)	*Paris*	ar, v, c	NI	AR	8
		Oreomyrrhis andicola	herb	CM	*Arum*	ar, v,h	NI	AR	8
	Acanthaceae	*Hygrophyla cf. costata*	macrophyte	Aquatic	*Paris*			BR	2
		Stenandrium dulce, Dicliptera squarrosa	herb	CSW	NI	h	NI	AR	15

Haloragaceae	*Myriophylum brasiliense*	macrophyte	Aquatic	*Paris*	h	*Acaulospora delicata, Acaulospora longula, Acaulospora aff. myriocarpa*	BR	2
Malphigiaceae	*Janusia guaranitica*	vine	CSW	*Arum*	h	NI	AR	15
Polygonaceae	*Polygonum acuminatum*	macrophyte	Aquatic	*Paris*	h	*Acaulospora scrobiculata, Glomus claroideum, Glomus aff. invermaium*	BR	2
	Polygonum ferrugineum	macrophyte	Aquatic	*Paris*	h	*Glomus lamellosum, Glomus luteum*	BR	2
	Polygonum stelligerum	macrophyte	Aquatic	*Paris*	h	NI	BR	2
	Ruprechtia apetala	tree	CSW	*Paris*	h	NI	AR	15
Elaeagnaceae	*Elaeagmus angustifólia*	tree	Riparian forest	NI	v	NI	China	5
Tamaricaceae	*Tamarix*	shrub	Riparian forest	NI	a,v,h	NI	China	5
Fabaceae	*Alhagi sparsifolia*	shrub	Riparian forest	NI	v,h	NI	China	5
	Adesmia filifolia	Annual, C3	PCD	NI	a,h	*Glomus*	Chile (CHI)	16
	Adesmia tenella	Annual, C3	PCD	NI	h	*Glomus*	Chile (CHI)	16
	Halimodendrom halodendron	shrub	Riparian forest	NI	v,h	NI	China	5
	Glycyrrhiza inflata	perennial herb	Riparian forest	NI	v,h	NI	China	5
	Adesmia filifolia	Annual, C3	PCD	NI	a,h	*Glomus*	Chile (CHI)	16
	Inga edulis	tree	Riparian sites	Arum	ar, ov	*A. scrobiculata , A. Excavata, A. spinosa, Fuscutata rubra, Dentiscutata cerradensis, Racocetra gregaria,*	BR	1

Table 1 (Continued)

Order	Family	Species or genus	Habit	Ecology/ Habitat	AM type	PC	Spores genus or species reported	Country/ Site	R
		Inga edulis	tree	Riparian sites	Arum	ar, ov	A. scrobiculata , A. Excavata, A. spinosa, Fuscutata rubra, Dentiscutata cerradensis, Racocetra gregaria,	BR	1
		Erythrina speciosa	tree	Riparian sites	Arum	ar, ov, iv	A. scrobiculata, Dentiscutata cerradensis, Glomus taiwanense, Glomus constrictum	BR	1
		Plathymenia reticulata	tree	dry forest	Arum	v,a,h	A. scrobiculata, A rhemii, Gigaspora margarita,Glomus brohultii	BR	1
		Centrosema coriaceum	climber	Higland field	Arum	ar, c, v, ac, eh	A. spinosa, Gigaspora margarita, Scutellospora	BR	1
		Mimosa bimucronata	tree	Riparian sites	Arum		Acaulospora scrobiculata, Acaulospora aff bireticulata, A. spinosa, Dentiscutata cf cerradensis, Glomus	BR	1
		Centrolobium tomentosum	tree	Riparian sites	Arum	ar, v, h	Acaulospora scrobiculata,A. mellea, Acaulospora aff bireticulata, A. spinosa, Dentiscutata cf cerradensis, Glomus	BR	1

Family	Species	Habit	Site	Type	Structures	Fungi	Country	Ref
	Centrolobium tomentosum	tree	Riparian sites	Arum	ar, v, h	Acaulospora scrobiculata, A. mellea, Acaulospora aff bireticulata, A. spinosa, Dentiscutata cf cerradensis, Glomus	BR	1
	Acacia atramentária, Collaea argentina, Prosopis alba, P. nigra, Sophora linearifolia	shrub	CSW	*Arum*	h	NI	AR	15
	Acacia praecox, A. visco	tree	CSW	*Arum*	h	NI	AR	15
	Senna aphylla	shrub	CSW	*Arum-Paris*	h	NI	AR	15
Bignoniaceae	*Tabebuia heptaphylla*	tree	dry forest	*Arum*			BR	1
Solanaceae	*Lycium ruthenicum*	srhub	riparian forest	NI	v,h	NI	China	5
	Cestrum parqui, Nicotiana glauca	shrub	CSW	I	h	NI	AR	15
	Nicotiana longiflora, Nierembergia linariaefolia, Solanum sisymbrifolium	herb	CSW	*Paris*	h	NI	AR	15
	Petunia axillaria, Salpichroa origanifolia	herb	CSW	*Arum*	h	NI	AR	15
	Solanum palinacanthum	herb	CSW	I	h	NI	AR	15
	Schizanthus candidus	Annual herb, C3	PCD	NI	a,v,h	*Glomus, Scutellospora*	Chile (CHI)	16
Apocynaceae	*Poacynum hendersonii*	herb	Riparian forest	NI	v,h	NI	China	5
	Aspidosperma quebracho-blanco	tree	CSW	NI	h	NI	AR	15
	Amblyopetalum coccineum	herb	CSW	NI	h	NI	AR	15
	Asclepias mellodora	herb	CSW	NI	h	NI	AR	15
Compositae	*Hexinia polydichotoma*	Herb	Riparian forest	NI	v	NI	China	5
	Karelinia caspica	herb	Riparian forest	NI	v,h	NI	China	5

Table 1 (Continued)

Order	Family	Species or genus	Habit	Ecology/ Habitat	AM type	PC	Spores genus or species reported	Country/ Site	R
		Calyptocarpus biaristatus	herb	Riparian sites	Arum	h	NI	BR	1
	Cannaceae	*Canna indica*	herb	CSW	*Arum*	h	NI	AR	15
	Celtidaceae	*Celtis tala*	tree	CSW	I	h	NI	AR	15
		Celtis pallida	shrub	CSW	*Paris*	h	NI	AR	15
	Convolvulaceae	*Dichondra sericea, D. microcalyx, Ipomoea purpurea*	herb	CSW	*Arum*	h	NI	AR	15
	Lauraceae	*Cryptocarya corrugata, Beilschmiedia collina*	tree	rain forest	NI	h	NI	Austrália (AUS) / northern Queensland	11
	Liliaceae	*Camassia biflora*	geophyte	PCD	NI	a,v,h	*Glomus, Scutellospora*	Chile (CHI)	16
	Myrtaceae	*Syzygium endophloium, S. wilsoni*	tree	rain forest	NI	h	NI	AUS	11
	Sapindaceae	*Castanospora alphandii*	tree	rain forest	NI	h	NI	AUS	11
	Sterculiaceae	*Franciscodendron laurifolium*	tree	rain forest	NI	h	NI	AUS	11
	Monimiaceae	*Doryphora aromatica*	tree	rain forest	NI	h	NI	AUS	11
		Artocarpus heterophyllus	tree	Madhupur Forest	NI	h	*Glomus, Entrophospora, Gigaspora*	Bangladesh	12
	Anacardiaceae	*Schinopsis brasiliensis*	tree	Dry forest	*Arum*	a,v,h	*Acaulospora scrobiculata, Acaulospora aff bireticulata, A. rhemii,, Dentiscutatas, Glomus*	BR	1
		Schinus fasciculata	shrub	CSW	NI	h	NI	AR	15
		Schinus areira	tree	CSW	NI	h	NI	AR	15

	Lithraea molleoides	tree	CSW	NI	h	NI	AR	15
	Schinopsis haenkeana	tree	CSW	NI	h	NI	AR	15
Lythraceae	*Heimia salicifolia*	shrub	CSW	*Paris*	h	NI	AR	15
	Cuphea glutinosa	herb	CSW	*Arum*	h	NI	AR	15
Buddlejaceae	*Buddleja cordobensis*		CSW				AR	15
Caprifoliaceae	*Viburnum dentatum*	Tree shrub	fac wet, fen	NI	a,v,h	NI	NYS	7
	Triadenum virginicum	herb	obl wet	NI	a,v,h	NI	NYS	7
Cornaceae	*Cornus sericea*	Tree shrub	fac wet	NI	a,v,h	NI	NYS	7
Grossulariaceae	*Ribes hirtellum*	shrub	obl wet	NI	a,v,h	NI	NYS	7
Lamiaceae	*Lycopus americanus, L. uniflorus*	herb	obl wet	NI	a,v,h,	NI	NYS	7
	Prunella vulgaris	herb	fac wet	NI	a,v,h	NI	NYS	7
	Lepechinia meyenii	herb	CM	*Paris-Arum*	h,c	NI	AR	8
	Satureja odora	shrub	CM	*Arum-Paris*	v,h	NI	AR	8
	Teucrium scordium	amphibious plant	intermittent aquatic habitats		a,h	NI	SW Slovenia (Europe)	14
	Hedeoma multiflora	herb	CSW	*Arum*	h	NI	AR	15
Nolanaceae	*Nolana paradoxa*	annual	PCD	NI	a,h	*Glomus aggregatum*	Chile (CHI)	16
Onagraceae	*Oenothera affinis*	herb	CSW	*Arum*	h	NI	AR	15
	Oenothera	annual	PCD	NI	h	*Glomus or Gigaspora albida, Scutellospora*	Chile (CHI)	16
	Camissonia dentata	annual	PCD	NI	h	*Glomus aggregatum, Glomus*	Chile (CHI)	16
Oxalidaceae	*Oxalis conorrhiza*	herb	CSW	I	h	NI	AR	15
Passifloraceae	*Passiflora caerulea*	vine	CSW	*Arum*	h	NI	AR	15
Primulaceae	*Lysimachia ciliata*	herb	fac wet	NI	a,v,h	NI	NYS	7
Plantaginaceae	*Plantago tomentosa*	herb	CSW	I	h	NI	AR	15
	Plantago australis	herb	CSW	*Arum*	h	NI	AR	15
	Plantago hispidula	annual	PCD	NI	h	*Glomus, Scutellospora*	Chile (CHI)	16
Portulacaceae	*Calandrinia grandiflora*	Annual, CAM	PCD	NI	h	*Glomus*	Chile (CHI)	16
	Philippiamra celosioides	Annual, CAM	PCD	NI	a,h	*Glomus*	Chile (CHI)	16

Table 1 (Continued)

Order	Family	Species or genus	Habit	Ecology/ Habitat	AM type	PC	Spores genus or species reported	Country/ Site	R
	Rhamnaceae	Colletia spinosissima	vine	CSW	Arum	h	NI	AR	15
	Ranunculaceae	Clematis virginiana	vine	Fac wet	NI	a,v,h	NI	NYS	7
		Thalictrum pubescens	herb	Facwet	NI	a,v, h	NI	NYS	7
		Clematis montevidensis	vine	CSW	Paris	h	NI	AR	15
		Thalictrum decipiens	herb	CSW	Paris	h	NI	AR	15
	Rosaceae	Alchemilla pinnata, Duchesnea indica	herb	CM	Arum	v,h,c	NI	AR	8
		Polylepis australis	Shrub-tree	CM	Arum	v,h,c	NI	AR	8
		Acaena myriophylla	Herb	CSW	Arum	h	NI	AR	15
		Kageneckia lanceolata	tree	CSW	Arum	h	NI	AR	15
		Dasiphora fruticosa	shrub	Facu	NI	a,v,h	NI	NYS	7
		Fragaria virginiana	herb	Facu	NI	a,v,h	NI	NYS	7
		Rubus pubescens	subsrhub	Facwet	NI	a,v,h	NI	NYS	7
	Rubiaceae	Galium labradoricum	herb	Obl	NI	a,v,h	NI	NYS	7
		Galium latoramosum, Borreria densiflora	herb	CSW	Paris	h	NI	AR	15
		Cruckshanksia	Hemicryptophyte, CAM	PCD	NI	A,v,h	Glomus, Gigaspora albida	Chile (CHI)	16
	Rutaceae	Fagara coco	tree	CSW	Arum-Paris	h	NI	AR	15
		Flindersia bourjotiana	tree	rain forest	NI	h	NI	AUS	11
	Salicaceae	Salix humboldtiana	tree	CSW	Arum	h	NI	AR	15
		Populus euphratica	tree	Riparian forest	NI	a,v,h	NI	China	5
	Santalaceae	Jodinia rhombifolia	tree	CSW	NI		NI	AR	15
		Quinchamalium excresens	annual	PCD	NI	A,v,h	Glomus, Gigaspora albida	Chile (CHI)	16
	Scrophulariaceae	Bartsia crenoloba	herb	CM	Arum-Paris	a,v,h	NI	AR	8
		Chelone glabra	herb	Obl	NI	a,v,h	NI	NYS	7
		Gratiola officinalis	amphibious plant	intermittent aquatic habitats		a,h		SW Slovenia (Europe)	14

Cactaceae	*Opuntia sulphurea, Gymnocalycium monvillei, Trichocereus candicans, Acanthocalycium spiniflorum*	cactaceae	CSW	*Arum*	h	NI		AR	15
Tecophilaeaceae	*Zephyra elegans*	geophyte	PCD	NI	A,v,h	*Glomus, Gigaspora albida*	Chile (CHI)	16
Turneraceae	*Turnera sidoides*	herb	CSW	*Arum*	h	NI		AR	15
Urticaceae	*Pilea pumila*	facw			a,v,h			EEUU	7
Verbenaceae	*Aloysia gratissima, Lippia turbinata*	shrub	CSW	*Arum*	h	NI		AR	15
	Glandularia dissecta, G. peruviana, Verbena litoralis	herb	CSW	*Arum*	h	NI		AR	15
Violaceae	*Viola cucullata*	herb	Facw	NI	a,v,h	NI		NYS	7

[†] Molecular identification. A, h and v correspond to arbuscules, hyphae and vesicles reported in the studies, respectively. Presence of AMF colonization is indicated by h. 1= Pagano (2009), Chapter 5, 7 and 16 (this book); 2= Marins et al. (2009); 3= Guevara and López (2007); 4= Ipsilantis and Sylvia (2007a,b); 5= Yang et al. (2008); 6= Lugo et al. (2009); 7= Weishampel and Bedford (2006); 8 = Menoyo et al. (2007); 9= Kovács et al. (2007); 10 = Garcia and Mendoza (2007); 11= Gehring and Connell (2006); 12 = Dhar and Mridha (2006); 13 = Kessler et al. (2009); 14 = Šraj-Kržič et al. (2009); 15 = Fracchia et al. (2009); 16 = Dhillion et al. (1995). Obl = obligate wetland, facu = facultative upland; facwet = facultative wetland; CM = Córdoba mountains; IM= Indonesia, Malaysia; RF= rain forest; CSW = Chaco Serrano woodland; PCD = fog-free Pacific coastal desert, Chile.

Recently, increasing reports on AM colonization of roots in plants of aquatic environments (Kai and Zhiwei 2006, Šraj-Kržič et al. 2006, Radhika and Rodrigues 2007, Marins et al. 2009) confirm their frequent occurrence in these environments. Other results recently reported, such as the functional importance of AM enhancing tolerance to water excess and improving nutrient acquisition in plants in exchange for oxygen and organic carbon in aquatic environments (Andersen and Andersen 2006, Fougnies et al. 2007), as well as spatial (Wolfe et al. 2007) and temporal (Bohrer et al. 2004, Escudero and Mendoza 2005) patterns in the presence of AM in wetland ecosystems, bring about new insights that merit further study.

AM Spore Distribution and Diversity

In spite of the fact that there is little information of AMF in natural ecosystems, a typical estimate of diversity on the basis of spore counts ranges between 5 – 20 different species in a community (Sanders et al. 1996).

Allen et al. (1995) found that some species such as *Glomus*, *Acaulospora* and *Scutellospora* have a broad geographic distribution. However, Johnson et al. (2009) stressed that little is known about the factors that control the distribution of AMF, pointing out an association of AM species (spores) composition with latitude, temperature and precipitation. They showed that *Acaulospora*, *Ambispora* and *Scutellospora* spores are more dominant at low latitudes, suggesting a relation with both biotic and abiotic factors.

A high spore number in soil can be found in natural grasslands (81 spores per gram of dry soil) (Garcia and Mendoza 2007), or the reverse (lower spore numbers as in some soils from Brazil), reflecting the variation in spore numbers occurring in the different natural ecosystems. For tropical rain forests in México more than 200 spores in 100 g dry soil were found (Guadarrama and Álvarez-Sánchez 1999). In the semiarid as well as in the cerrado vegetation types of Minas Gerais State, Pagano et al. (2008a,e, 2009a,b) showed *Glomus* as dominant, followed by *Acaulospora*, *Scutellospora* (sensu Walker and Sanders) and *Gigaspora*.

In the rhizospheres of native species, Pagano et al. (2009) showed that the AM sporulation increased in the dry season compared to the rainy period. Frequently, common AMF species were a brown *Glomus* species, *G. macrocarpum* and *Acaulospora scrobiculata*, also found by Picone (2000) for Nicaragua and Costa Rica. Details of spore numbers in some Brazilian ecosystems are given by Pagano and Cabello (Chap. 5, this Vol.) and Pagano et al. (Chap. 7, this Vol.).

AMF has also a role in soil aggregation throughout the production of a protein of increasing interest for the science, called glomalin, which occurrence is discussed in the next section.

Glomalin in Natural Environments

Glomalin is a protein produced by all AMF tested to date (Wright et al. 1996, 1998, Nichols 2003), and has been detected in numerous soils in large amounts (Wright and

Upadhyaya 1998, Nichols 2003). Glomalin has a role in soil aggregation representing a large pool of carbon and nitrogen (Miller and Jastrow 2002); however, knowledge of glomalin in soils is extremely limited, and the protein has not yet been considered in ecological and biogeochemical studies, although concentrations of glomalin in the soil seem to be responsive to global change factors (elevated atmospheric CO2) (Rillig et al. 1999).

In addition, glomalin lasts in the soil over 6 years, with a possible range of 6-42 years (Rillig et al. 2001), supporting the hypothesis that glomalin is relatively stable in soils, and that one-time sampling may be sufficient to study this response variable. However, the generality of this observation has yet to be tested in a wider range of ecosystems (Lutgen 2003).

Reports have shown that land-use type significantly affected glomalin concentrations, with native forest soils and afforested areas having the highest concentrations (Rillig et al. 2003, Pagano et al. 2009a). Various results show that glomalin may be useful as an indicator of land-use change effects on forest soils, but further studies are needed to demonstrate the generality of these patterns (Rillig et al. 2003).

In Brazil, reports on soil glomalin are continuously increasing, such as those from Oliveira et al. (2009) for revegetated dune areas in the Paraíba State; however, most of them are available only in conference abstracts or theses not easily retrievable for consultation.

CONCLUSIONS

In the introduction to this chapter, I briefly described that mycorrhizal fungi are distributed worldwide in terrestrial temperate, tropical and arctic regions, as well, they are found under waterlogged and aquatic conditions.

I have also presented an overview of the new reports of mycorrhizal status of plants highlighting the importance of mycorrrhizae as an essential component for establishment and sustainability of plant communities, as well as the need for further studies of the mycorrhizal fungal populations in relation to the soil fertility and functionality.

Throughout the chapter, I have provided some examples of the occurrence of different types of AM fungi from various plants and from a wide range of soils and association of the fungi with different types of host plants were recorded; however, we need to organize the existing knowledge into a possible structure and create a system of influencing factors on AMF.

Since this chapter is primarily concerned with AMF occurrence in natural environments, I have refrained from discussing the potential of AMF for restoration practices, especially for the increasing percent of plant species belonging to different categories of threat as reported by Fuchs and Haselwandter (2008). However, some native Brazilian species (*S. brasiliensis*, *M. urundeuva*), classified within the threatened or near threatened category of the official Brazilian endangered species list, reported as mycotrophic, are mentioned in this chapter.

Finally, I have provided some examples from many efforts made in recent years to accrue benefits of the presence of glomalin in soils and mycorrhizal fungi potential in Brazil. Nonetheless, further study in natural environments is required to achieve maximum benefits from these microorganisms and their associations, as well as to compare AMF communities between undisturbed and disturbed ecosystems.

REFERENCES

Allen, EB; Rincón E; Allen MF; Pérez-Jimenez A; Huante P. Disturbance and seasonal dynamics of mycorrhizae in a tropical deciduous forest in Mexico. *Biotropica,* 1998, 30, 261-274.

Andersen, FØ; Andersen, T. Effects of arbuscular mycorrhiza on biomass and nutrients in the aquatic plant *Littorella uniflora. Freshwater Biol.,* 2006, 51, 1623-1633.

Beck-Nielsen, D; Madsen, TV. Occurrence of vesicular-arbuscular mycorrhiza in aquatic macrophytes from lakes and streams. *Aquat. Bot.,* 2001, 71, 141-148.

Becerra, A; Cabello, M; Chiarini, F. Arbuscular mycorrhizal colonization of vascular plants from the Yungas forests, Argentina. *Ann. For. Sci.,* 2007, 64, 765-772.

Boege, K; Dirzo, R. Intraspecific variation in growth, defense and herbivory in *Dialium guianense* (Caesalpiniaceae) mediated by edaphic heterogeneity. *Plant Ecology,* 2004,175, 59-69.

Bohrer, K.E., Friese, C.F., Amon, J.P.: Seasonal dynamics of arbuscular mycorrhizal fungi in differing wetland habitats. *Mycorrhiza,* 2004,14, 329-337.

Brundrett, M; Kendrick, B. The roots and mycorrhizas of herbaceous woodland plants. I. Quantitative aspects of morphology. *New Phytol.,* 1990, 114, 457-468.

Carvalho, PER. *Espécies florestais brasileiras: Recomendações silviculturais, potencialidades e uso da madeira.* Embrapa, Brasília, Brazil, 1994 (in Portuguese).

Casper, BB; Jackson RB. Plant competition underground. *Ann. Rev. Ecolog. Syst.,* 1997, 28, 545-557.

Clark, DB; Palmer, MW; Clark, DA. Edaphic factors and the landscape-scale distributions of tropical rain forest trees. *Ecology,* 1999, 80, 2662-2675.

Cooke, JC; Lefor, MW. Comparison of Vesicular-Arbuscular Mycorrhizae in Plants from Disturbed and Adjacent Undisturbed Regions of a Coastal Salt Marsh in Clinton, Connecticut, USA. *Environmental Management,* 1990, 14, 131-137.

Dhar, PP; Mridha MAU. Biodiversity of arbuscular mycorrhizal fungi in different trees of madhupur forest, Bangladesh. *J. For. Res.,* 2006, 17, 201-205.

Dhillion, SS; Vidiella, PE; Aquilera, LE; Friese, CF; Leon, E; Armesto, JJ; Zak, JC. Mycorrhizal plants and fungi in the fog-free Pacific coastal desert of Chile. Mycorrhiza, 1995, 5,381-386.

Egerton-Warburton, LM; Kuo, J; Griffin, BJ; Lamont, BB. The effect of aluminium on the distribution of calcium, magnesium and phosphorus in mycorrhizal and non-mycorrhizal seedlings of *Eucalyptus rudis*: a cryo-microanalytical study. *Plant Soil,* 1993, 481-484.

Eom, AH; Hartnett, DC; Wilson, GWT. Host plant species effects on arbuscular mycorrhizal fungal communities in tallgrass prairie. *Oecologia,* 2000, 122, 435-444.

Escudero, V., Mendoza, R.: Seasonal variation of arbuscular mycorrhizal fungi in temperate grassland along a wide hydrologic gradient. *Mycorrhiza,* 2005, 15, 291-299.

Fougnies, L, Renciot, S, Muller, F, Plenchette, C, Prin, Y, de Faria, SM, Bouvet, JM, Sylla, SN, Dreyfus, B, Bâ, AM. Arbuscular mycorrhizal colonization and nodulation improve flooding tolerance in *Pterocarpus officinalis* Jacq. seedlings. *Mycorrhiza,* 2007, 17, 159-166.

Fracchia, S; Aranda, A; Gopar, A; Silvani,V; Fernandez, L; Godeas, A. Mycorrhizal status of plant species in the Chaco Serrano Woodland from central Argentina. *Mycorrhiza*, 2009, 19, 205-214.

García, IV; Mendoza, RE. Arbuscular mycorrhizal fungi and plant symbiosis in a saline-sodic soil. *Mycorrhiza*, 2007, 17, 167-174.

Gehring C A., Connell J H. Arbuscular mycorrhizal fungi in the tree seedlings of two Australian rain forests: occurrence, colonization, and relationships with plant performance. *Mycorrhiza*, 2006, 16, 89-98.

Guevara, R; López, JC. Quality of rooting environments and patterns of root colonization by arbuscular mycorrhizal fungi in strangler figs in a Mexican palmetto woodland. *Mycorrhiza*, 2007, 17, 589-596.

Guadarrama, P; Álvarez-Sánchez, FJ. Abundance of arbuscular mycorrhizal fungi spores in different environments in a tropical rain forest, Veracruz, Mexico. *Mycorrhiza*, 1999, 8, 267-270.

Ipsilantis I, Sylvia, DM. Interactions of assemblages of mycorrhizal fungi with two Florida wetland plants, *Applied Soil Ecology*, 2007, 35, 261-271.

Ipsilantis I, Sylvia, DM. Abundance of fungi and bacteria in a nutrient-impacted Florida wetland. *Applied Soil Ecology*, 2007, 35, 272-280.

Jakobsen, I; Smith, SE; Smith, FA. Function and diversity of Arbuscular mycorrhizae in carbon and mineral nutrition. In: van der Heijden MGA and Sanders IR, editors. *Mycorrhizal Ecology*. Berlin: Springer; 2003; 75-92.

Janos, DP. Vesicular-arbuscular mycorrhizae affect lowland tropical rain forest plant growth. *Ecology*, 1980a, 61,151-162

Janos, DP. Mycorrhizae influence tropical succession. *Biotropica*, 1980b, 12,56-64.

Johnson, NC; Graham, JH; Smith, FA. Functioning ofmycorrhizal associations along the mutualism–parasitism continuum. *New Phytologist*, 1997, 135, 575-585.

Johnson, NC; Álvarez Sánchez, FJ; Antoninka, A; Chaudhary, VB; Lau, MK. *Biogeography of arbuscular mycorrhizal fungi: Distribution and community composition patterns*. In: 6th International Conference on Mycorrhiza ICOM6, 2009, Belo Horizonte. Abstracts ICOM6. Viçosa: editors, 2009, 1, 11-12.

Kai, W; Zhiwei, Z. Occurrence of arbuscular mycorrhizas and dark septate fungi endophytes in hydrophytes from lakes and streams in Southwest China. *Int. Rev. Hydrob.*, 2006, 91, 29-37.

Kessler, M; Jonas, R; Cicuzza, D; Kluge, J; Piatek, K; Naks, P; Lehnert, M. A survey of the mycorrhization of Southeast Asian ferns and lycophytes. *Plant Biology*, 2009, 1-7, doi:10.1111/j.1438-8677.2009.00270.x

Klironomos, JN. Variation in plant response to native and exotic arbuscular mycorrhizal fungi. *Ecology*, 2003, 84, 2292-2301.

Koide, RT; Mosse, B. A history of research on arbuscular mycorrhiza. *Mycorrhiza*, 2004, 14, 145-163.

Kovács,GM; Balázs, T; Pénzes, Z. Molecular study of arbuscular mycorrhizal fungi colonizing the sporophyte of the eusporangiate rattlesnake fern (*Botrychium virginianum*, Ophioglossaceae). *Mycorrhiza*, 2007, 17,597-605.

Lambers, H; Poorter H. Inherent variations in growth rate between higher plants: A search for fisiological causes and ecological consequences. *Advances in Ecological Research,* 1992, 23,188-261.

Lorenzi, H. *Árvores brasileiras: Manual de identificação e cultivo de plantas arbóreas nativas do Brasil*, 2nd ed. Volume 1. Editora Plantarum, Nova Odessa, Brazil, 1998a. (in Portuguese).

Lorenzi, H. *Árvores brasileiras: Manual de identificação e cultivo de plantas arbóreas nativas do Brasil*, 2nd ed. Volume 2. Editora Plantarum, Nova Odessa, Brazil, 1998b. (in Portuguese).

Lugo, MA; Molina, MG; Crespo, EM. Arbuscular mycorrhizas and dark septate endophytes in bromeliads from South American arid environment. *Symbiosis*, 2009, 47, 17-21.

Lutgen, ER; Muir-Clairmont, D; Graham, J; Rillig, MC. Seasonality of arbuscular mycorrhizal hyphae and glomalin in a western Montana grassland. *Plant and soil*, 2003, 257, 71-83.

Marins, JF; Carrenho, R; Thomaz, SM. Occurrence and coexistence of arbuscular mycorrhizal fungi and dark septate fungi in aquatic macrophytes in a tropical river-floodplain system. *Aquatic Botany*, 2009, 91 13-19.

Marschner, H. Mechanisms of adaptation of plants to acid soils. *Plant and Soil*, 1991, 134,1-20.

Menoyo, E; Becerra, AG; Renison, D. Mycorrhizal associations in *Polylepis* woodlands of Central Argentina. *Can. J. Bot.*, 2007, 85, 526-531.

Molina, R; Massicotte, H; Trappe, JM. Specificity Phenomena in Mycorrhizal symbioses: community-ecological consequences and practical implications. In: *Mycorrhizal Functioning and Integrative Plant-Fungal Process*, Allen MF, editor. Chapman Hall, New York; 1992, 357-423.

Mummey, DL; Antunes, PM; Rillig, MC. Arbuscular mycorrhizal fungi pre-inoculant identity determines community composition in roots. *Soil Biology & Biochemistry*, 2009, 41, 1173-117.

Nichols, KA; Wright, SF. Contributions of fungi to soil organic matter in agroecosystems. In: Magdoff, F., Weil, R.R., editors. *Soil Organic Matter in Sustainable Agriculture*. CRC Press, Boca Raton, 2004, 179-198.

Oliveira, JRG; Souza, RG; Silva, FSB; Mendes, ASM; Yano-Melo, AM. Role of autoctone community of arbuscular mycorrhizal fungi (AMF) on the development of native plant species in revegetated restinga dunes from coastal region of Paraíba State. *Revista Brasil. Bot.*, 2009, 32, 663-670.

Öpik, M; Moora, M; Liira, J; Zobel, M. Composition of root-colonizing arbuscular mycorrhizal fungal communities in different ecosystems around the globe. *Journal of Ecology*, 2006, 94, 778-790.

Pagano, MC. *Characterization of Glomalean mycorrhizal fungi and its benefits on plant growth in a semi-arid region of Minas Gerais (Jaíba Project), Brazil*. PhD thesis. Belo Horizonte: Federal University of Minas Gerais; 2007.

Pagano, MC; Cabello, MN; Scotti, MR. Phosphorus response of three native Brazilian trees to inoculation with four arbuscular mycorrhizal fungi. *J. Agric. Technol.*, 2007, 3, 231-240.

Pagano, MC; Cabello, MN; Bellote, AF; Sa, NM; Scotti, MR. Intercropping system of tropical leguminous species and *Eucalyptus camaldulensis*, inoculated with rhizobia and/or mycorrhizal fungi in semiarid Brazil. *Agrofor. Syst.*, 2008a, 74, 231-242.

Pagano, MC; Marques, MS; Cabello, MN; Scotti, MR. Mycorrhizal associations in native species for the restoration of velhas river riparian forest, Brazil. In: FERTBIO 2008-Londrina, PR, Brasil. 15-19 Setembro, 2008b (in Portuguese).

Pagano, MC; Passos RV; Viana P; Cabello MN; Scotti MR Riparian forest restoration: arbuscular mycorrhizae in disturbed and undisturbed soils. In: VI Congreso Latinoamericano de Micología, Mar del Plata, Argentina, 10-13 Novembro de 2008c.

Pagano, MC; Marques, MS; Sobral, M; Scotti, MR. Screening for arbuscular mycorrhizal fungi for the revegetation of eroded riparian soils in Brazil. In: VI Congreso Latinoamericano de Micología, Mar del Plata, Argentina. 10-13 Novembro de 2008d.

Pagano, MC; Scotti, MR. Recuperação de mata ciliar degradada do Rio Sabará - Minas Gerais. VII Simpósio Nacional sobre Recuperação de Áreas Degradadas. Curitiba-PR, Brazil, 9 a 11 de outubro de 2008e, (in Portuguese).

Pagano, MC; Persiano, AIC; Cabello, MN; Scotti, MR. *Survey of arbuscular mycorrhizas in preserved and impacted riparian environments*. In: 6th International Conference on Mycorrhiza ICOM6, 2009a, Belo Horizonte. Abstracts ICOM6. Viçosa: editors, 2009a, 1, 59-60.

Pagano, MC; Scotti, MR; Cabello, MN. Effect of the inoculation and distribution of mycorrhizae in *Plathymenia reticulata* Benth under monoculture and mixed plantation in Brazil. *New Forests* 2009b, 38:197-214.

Pagano, MC; Cabello, MN; Scotti, MR. Agroforestry in dry forest, Brazil: mycorrhizal fungi potential. In: LR Kellymore, editor. *Handbook on Agroforestry: Management Practices and Environmental Impact*. New York, Nova Science Publishers, 2010, 367-387.

Picone, C. Diversity and abundance of arbuscular-mycorrhizal fungus spores in tropical forest and pasture. *Biotropica*, 2000, 32, 734-750.

Poorter, H. Interspecific variation in relative growth rate: on ecological causes and physiological consequences. In: Lambers H, Cambridge ML, Konings H, Pons TL, editors. *Causes and consequences of variation in growth rate and productivity of higher plants*. SPB Academic, The Hague, The Netherlands, 1989, 45-68.

Radhika, KP; Rodrigues, BF. Arbuscular mycorrhizae in association with aquatic and marshy plant species in Goa, India. *Aquat. Bot.*, 2007, 86, 291-294.

Read, DJ. The structure and function of the vegetative mycelium of mycorrhizal roots. In: Jennings DH and Rayner ADM, editors. *The ecology and Physiology of the fungal Mycelium*. Cambridge University Press; 1984; 215-240.

Read, DJ. Mycorrhizas in ecosystems. *Experientia,* 1991, 47,376-391.

Read, DJ. Towards Ecological Relevance – Progress and Pitfalls in the Path Towards an understanding of mycorrhizal functions in nature. In: van der Heijden MGA and Sanders IR, editors. *Mycorrhizal Ecology*. Berlin: Springer; 2003; 3–29.

Resende, AV; Furtini Neto, AE; Curi, N. Mineral Nutrition and Fertilization of Native Tree Species in Brazil: Research Progress and Suggestions for Management. *Journal of Sustainable Forestry*, 2005, 20, 45-81.

Rillig, MC; Ramsey, PW; Morris, S; Paul, EA. Glomalin, an arbuscular-mycorrhizal fungal soil protein, responds to land-use change. *Plant and Soil,* 2003, 253, 293-299.

Rillig, MC; Wright, SF; Nichols, KA; Schmidt, WF; Torn, MS. Large contribution of arbuscular mycorrhizal fungi to soil carbon pools in tropical soils. *Plant and Soil*, 2001, 233, 167-177.

Rillig, MC. Arbuscular mycorrhizae and terrestrial ecosystem processes. *Ecology Letters*, 2004, 7, 740-754.

Rizzini, CT. *Tratado de fitogeografia do Brasil: aspectos ecológicos, sociológicos e florísticos*. 2nd Edition. Rio de Janeiro: Âmbito Cultural Edições Ltda.; 1997 (in Portuguese).

Sanders, IR; Clapp, JI; Wiemken, A. The genetic diversity of arbuscular mycorrhizal fungi in natural ecosystems - a key to understanding the ecology and functioning of mycorrhizal symbiosis. *New Phytol.*, 1996, 133,123-134.

Schreeg, LA; Kobe, RK; Walters, MB. Tree seedling growth, survival, and morphology in response to landscape-level variation in soil resource availability in northern Michigan. *Can. J. For. Res.*, 2005, 35,263-273.

Sengupta, A; Chaudhuri S. Arbuscular mycorrhizal relations of mangrove plant community at the Ganges river estuary in India. *Mycorrhiza,* 2002, 12,169-174.

Siqueira, JO; Saggin-Júnior, OJ. Dependency on arbuscular mycorrhizal fungi and responsiveness of some Brazilian native woody species. *Mycorrhiza,* 2001, 11,245-255.

Siqueira, JO; Carneiro, MAC; Curi, N; Rosado, SCS; Davide, AC. Mycorrhizal colonization and mycotrophic growth of native woody species as related to sucessional groups in Southeastern Brazil. *For. Ecol. Manag.*, 1998,107,241-252.

Smith, SE; Read, DJ. *Mycorrhizal Symbiosis*. New York: Elsevier; 2008.

Šraj-Kržič, N; Pongrac, P; Klemenc, M; Kladnik, A; Regvar, M; Gaberščik, A. Mycorrhizal colonisation in plants from intermittent aquatic habitats. *Aquat. Bot.*, 2006, 85, 333-338.

Šraj-Kržič, N; Pongrac, P; Regvar, M; Gaberščik, A. Photon-harvesting efficiency and arbuscular mycorrhiza in amphibious plants. Photosynthetica, 2009, 47,1, 61-67.

Stürmer, SL; de Souza, FA. Community structure of arbuscular mycorrhizal fungi in Brazilian ecosystems: patterns, methodologies and perspectives. In: 6th International Conference on Mycorrhiza ICOM6, 2009, Belo Horizonte. Abstracts ICOM6. Viçosa: editors, 2009, 1, 11.

Trappe, JM. Phylogenetic and ecologic aspects of mycotrophy in the angiosperms from an evolutionary standpoint. In: Safir GR, editors. *Ecophysiology of VA mycorrhizal plants*. CRC, Boca Raton, 1987, 5-25.

Treseder, KK; Vitousek, PM. Effects of soil nutrient availability on investment in acquisition of N and P in Hawaiian rain forests. *Ecology*, 2001, 82, 946-954.

van der Heijden, MGA; Klironomos, JN; Ursic, M; Moutoglis, P; Streitwolf- Engel, R; Boller, T; Wiemken, A; Sanders, IR. Mycorrhizal fungal diversity determines plant biodiversity, ecosystem variability and productivity. *Nature*, 1998, 396, 69-72.

Wang, B; Qiu, YL. Phylogenetic distribution and evolution of mycorrhizas in land plants. *Mycorrhiza,* 2006, 16,299-363.

Weishampel, PA; Bedford BL. Wetland dicots and monocots differ in colonization by arbuscular mycorrhizal fungi and dark septate endophytes. *Mycorrhiza*, 2006, 16,495-502.

Wolfe, BE; Weishampel, PA; Klironomos, JN. Arbuscular mycorrhizal fungi and water table affect wetland plant community composition. *J. Ecol.*, 2007, 94, 905-914.

Wright, SF; Upadhyaya, A. A survey of soils for aggregate stability and glomalin, a glycoprotein produced by hyphae of arbuscular mycorrhizal fungi. *Plant and Soil,* 1998, 198, 97-107.

Wright, SF; Upadhyaya, A. Extraction of an abundant and unusual protein from soil and comparison with hyphal protein of arbuscular mycorrhizal fungi. *Soil Science,* 1996, 161, 575-586.

Yang, Y; Chen, Y; Li, W. Arbuscular mycorrhizal fungi infection in desert riparian forest and its environmental implications: A case study in the lower reach of Tarim River. *Progress in Natural Science,* 18, 2008, 983-991.

Zangaro, W; Nisizaki, SMA; Domingos JCB; Nakano EM. Micorriza arbuscular em espécies arbóreas nativas da bacia do Rio Tibagi, Paraná. *Cerne,* 2002, 8, 077-087 (in Portuguese).

Zangaro, W; Nishidate, FR; Vandresen, J; Andrade, G; Nogueira, MA. Root mycorrhizal colonization and plant responsiveness are related to root plasticity, soil fertility and successional status of native woody species in southern Brazil. *Journal of Tropical Ecology,* 2007, 23,53-62.

Zangaro, W; Assis, RL; Rostirola, LV; Souza, PB; Gonçalves, MC; Andrade, G; Nogueira, MA. Changes in arbuscular mycorrhizal associations and fine root traits in sites under different plant successional phases in southern Brazil. *Mycorrhiza,* 2008, 19, 37-45.

Zhao, X; Yu, T; Wang, Y; Yan, X. Effect of arbuscular mycorrhiza on the growth of *Camptotheca acuminata* seedlings. *Journal of Forestry Research,* 2006, 17,121-123.

In: Mycorrhiza: Occurrence in Natural and Restored...
Editor: Marcela Pagano

ISBN: 978-1-61209-226-3
© 2012 Nova Science Publishers, Inc.

Chapter 2

LARGE-SCALE DIVERSITY PATTERNS IN SPORE COMMUNITIES OF ARBUSCULAR MYCORRHIZAL FUNGI

Javier Álvarez-Sánchez[1,], Nancy C. Johnson[2], Anita Antoninka[2], V. Bala Chaudhary[2], Matthew K. Lau[2], Suzanne M. Owen[2], Irene Sánchez-Gallen[1], Patricia Guadarrama[1] and Silvia Castillo[1]*

[1]Facultad de Ciencias, Universidad Nacional Autónoma de México. Circuito Exterior, Ciudad Universitaria 04510 México
[2] Department of Biological Sciences, Northern Arizona University, Box 5640, Flagstaff, AZ, 86011 USA

ABSTRACT

Surprising little is known about the factors controlling Arbuscular Mycorrhizal (AM) fungal diversity and distribution patterns. A better understanding of these factors is necessary before mycorrhizas can be effectively managed for their benefits in ecosystem restoration and agriculture. The goal of this chapter is to examine the relationships between AM fungal diversity, plant diversity and latitude across a variety of vegetation types and disturbance regimes. We created a large database by compiling seven distinct datasets from across North America to test the hypotheses that 1) diversity of AM fungal communities should be positively related to plant diversity; 2) AM fungal diversity should be higher in low latitudes than in high latitudes; and 3) disturbance and land use should influence AM fungal diversity. The database was composed of 523 samples collected by research groups from Mexico and United States of America with eight different vegetation types and land uses. Abundance of 121 AM fungi taxa as well as data on geographic location, vegetation type, and disturbance history were included in the database. Contrary to our expectations, species richness and evenness of AM fungal spore communities were not correlated with plant richness or latitude. The influence of disturbance on AM fungal species diversity varied with climate. Our findings indicate

---

* javier.alvarez@ciencias.unam.mx

that the factors controlling community diversity differ for plants and mycorrhizal fungi. Disturbance influences AM fungal spore diversity, and the outcome of these effects varies among ecosystems.

Keywords: Arbuscular mycorrhizal fungi, disturbance, diversity, land use, North America.

INTRODUCTION

Arbuscular mycorrhizal (AM) fungi are ancient soil organisms belonging to the Phylum Glomeromycota (Schüßler et al. 2001). These fungi form obligate symbiotic relationships with plant roots in most terrestrial ecosystems (Trappe 1987, Harrison 1997, Camargo-Ricalde 2002). In this symbiosis, the fungi facilitate plant uptake of soil resources and reciprocally gain fixed carbon (Smith and Read 1997). Moreover, AM fungi can affect plants' competitive abilities (Guadarrama et al. 2004) and play a determinant role in regeneration and plant succession (Carey et al. 1992, Allsopp and Stock 1994, Kardol et al. 2006). Mycorrhizal hyphae contribute to soil aggregate formation, soil conservation and reestablishment of plants in disturbed areas (Bethlenfalvay 1992, Cuenca et al. 1998, Miller and Jastrow 2000). In this regard, AM fungi play key roles in plant community structure, and ecosystem processes.

Despite their wide distribution and importance in ecosystems, the factors that influence the geographic distribution of AM fungi in natural and managed ecosystems are still unclear, in large part because of the difficulty of studying these fungi in nature, and the complex factors that can affect their distribution. Chaudhary et al. (2008) proposed a model for Glomeromycotan biogeography that illustrates this complexity. In the early Devonian (400 million years ago), plants had endophytic associations resembling vesicular–arbuscular mycorrhizas (Brundrett 2002), and continental drift may have contributed to diversification and large scale distribution patterns of AM fungi (Allen et al. 1995, Chaudhary et al. 2008). At a more proximate timescale, abiotic forces such as climate and soil properties, combined with biotic forces such as host plant community composition and inter- and intra-specific interactions are likely to be important factors (Rillig et al. 2002, Pringle et al. 2009). Furthermore, Glomeromycotan distribution may be controlled by intrinsic properties of the fungi such as their dispersal ability (Warner et al. 1987) and speciation and extinction rates (Chaudhary et al. 2008). Finally, disturbance history is likely to be an important driver of Glomeromycotan biogeography (Egerton-Warburton et al. 2007, Alguacil et al. 2008).

Glomeromycota are obligate biotrophs, consequently, their distribution is ultimately dependent on the distribution of living plants. Despite this restriction, AM fungi are often reported to have low host specificity because in controlled experiments individual fungal taxa have been shown to colonize hundreds of different plant species (Trappe 1987, Allen et al. 1995, Sanders 2003). However, upon closer examination, AM fungi do exhibit some degree of host preference in natural systems (Bever et al. 2001, Helgason et al. 2002); certain taxa of AM fungi proliferate with certain plant species (e.g. Johnson et al. 1992, Bever et al. 1996, Helgason et al. 1998, Vandenkoornhuyse et al. 2002). Several studies show that AM fungi can influence plant diversity and productivity (van der Heijden et al. 1998, Hartnett and Wilson 1999). Moreover, it has been reported that individual species of AM fungi and different assemblages of fungal taxa have variable effects on plant productivity and community structure (Klironomos et al. 2000, Klironomos 2002). Consequently, mycorrhizal symbioses may influence how plant species coexist and how soil resources are distributed among co-occurring plant species (van der Heijden et al. 2003). It is clear that AM fungi are an important regulator of plant community structure, but what are the reciprocal effects of plants on the community structure of AM fungi?

Wardle et al. (2004) suggest that aboveground and belowground diversity is often linked, but studies show contradictory patterns in the relationship between plant diversity and mycorrhizal fungal diversity. Arbuscular mycorrhizal fungal diversity may have positive (Burrows and Pfleger 2002, Oehl et al. 2003, Chen et al. 2004, Börstler et al. 2006), negative (Pietikäinen et al. 2007, Antoninka in preparation), no relationship (Johnson and Wedin 1997), or scale dependent relationships (Landis et al. 2004) with plant diversity.

There is generally a negative relationship between diversity and latitude among many groups of macro and microorganisms (Rosenzweig 1995, Kaufman and Willig 1998, Willig et al. 2003), though exceptions exist (Hillebrand 2004, Hughes et al. 2006). Since tropical plant communities are generally much more diverse than temperate plant communities, one may expect that AM fungal diversity may also exhibit a negative relationship with latitude. Surprisingly little is known about latitudinal patterns in AM fungal distribution. Taxonomists have informally noted that AM fungal diversity appears to be greater in tropical locations, with a greater diversity of Gigasporaceae species in particular closer to the equator (Walker 1992, Herrera-Peraza et al. 2001). Formal comparisons of AM fungal diversity between tropical and temperate locations using field spore data (Chaudhary et al. in preparation) and meta-analysis (Öpik et al. 2008) have demonstrated higher diversity in the tropics (but see Stutz et al. 2000). Studies within temperate North America, have demonstrated latitudinal gradients in species distributions (Koske 1987, Allen et al. 1995), but a large-scale latitudinal gradient study of AM fungal species that spans temperate and tropical locations has never been conducted for AM fungi.

Disturbance can play a pivotal role in ecosystems from promoting species diversity (Odion and Sarr 2007) to reducing ecosystem function and diversity (Jiménez-Esquilina et al. 2007, Odion and Sarr 2007), depending on the type and severity of the disturbance. AM fungi can be sensitive to disturbances that alter soil characteristics, such as tillage, agricultural and land management practices. But, can these disturbances influence AM fungal diversity? Tillage practices can alter AM fungal community composition and reduce sporulation of certain AM species (Jansa et al. 2002, Alguacil et al. 2008). Nitrogen enrichment or fertilization can reduce or increase AM species richness and diversity depending on the amount of available phosphorus and types of available host-plants (Egerton-Warburton et al. 2007). As the amount of N-deposition increases, it can promote a shift in AM fungal species composition, favoring certain *Glomus* species and displacing 'larger' spore species (Egerton-Warburton and Allen 2000). Soil compaction has been shown to reduce or have no effect on AM fungal colonization (*reviewed in* Entry et al. 1996, Nadian et al. 1997). Management practices, such as forest fuel reduction, can reduce AM fungal propagule abundance (Korb et al. 2004), reduce AM species richness and alter community composition (Owen et al. 2009). Disturbances associated with exotic plant invasions can either inhibit or promote AM fungal growth (Pringle et al. 2009).

Disturbances associated with climate change can also influence the distribution of AM fungi. Simulated climate change, by warming can increase the hyphal length of AM fungi and root colonization (Rillig et al. 2002). CO_2 enrichment tends to increase AM fungal abundance while N deposition tends to decrease AM fungal abundance (*reviewed by* Treseder 2004).

Some AM fungi respond differently to soil disturbances, for example Hart and Reader (2004) found that species from the suborder Glomineae were much less resilient to soil disturbances than species from the suborder Gigasporineae, likely due to the traits of either colonizing plant roots mostly by hyphae or by spores.

Molecular methods are increasingly useful for identification of Glomeromycota in field collected samples (Börstler et al. 2006, Hempel et al. 2007), and these methods have been successfully applied to biogeographical studies (Oehl et al. 2003, Öpik et al. 2008). However the usefulness of molecular data to study large scale patterns of AM fungal diversity is limited because currently there are relatively few data sets available which use the same gene regions to identify AM fungi. Unlike molecular data, many data sets of AM fungal spore populations from a wide range of locations are currently available. Analysis of AM fungal spore populations in soils can provide a useful indicator of community composition and diversity (e.g. check Bever et al. 2001, check Landis et al. 2004) as long as the strengths and weaknesses of this method are recognized (Table 1). Spore community data does not provide a perfect measure of AM fungal diversity because not all AM fungi produce spores (Clapp et al. 1995). Also, there is no relationship between spore abundance and total fungal biomass in roots and soils; some taxa of AM fungi produce copious spores, while other taxa produce very few spores per unit biomass. Consequently cross-species comparison of spore abundance is not useful; however, within a single species of AM fungus, comparison of spore abundance across treatments or environmental gradients can be very informative (e.g. Johnson et al. 1991, 1992). Furthermore, large scale patterns in species richness and community composition of AM fungal spores across environmental gradients can be very useful (e.g. Egerton-Warburton et al. 2007).

This chapter reports the results of a large-scale study of the species richness, evenness and diversity of AM fungal spores across temperate and tropical latitudes and plant diversity and disturbance gradients in natural and managed lands. The goal of this chapter is to help elucidate the factors controlling the distribution and diversity of this important group of fungi by addressing three questions. First, what is the relationship between plant diversity and AM fungal diversity? Second, is AM fungal diversity greater at tropical latitudes than temperate latitudes? Finally, how does disturbance and land use influence AM fungal diversity?

Table 1. Strengths and weaknesses of spore analysis

Strengths	Weaknesses
Spore population data CAN	Spore population data CAN NOT
Elucidate occurrence of spore-forming AM fungi	Detect occurrence of AM fungi that do not sporulate
Inform ecological inferences *within* morpho-species of AM fungi	Inform ecological inferences *between* morpho-species of AM fungi
Measure reproductive allocation to spores	Measure total biomass of AM fungi
Reveal patterns in AM fungal community structure	Reveal AM fungal functioning

METHODS

We compiled seven distinct datasets to generate a large database composed of 523 samples from 8 sites in Mexico and United States of America (USA) (Figure 1). These sites included eight distinct vegetation types and land uses: tropical wet forest, tropical dry forest, temperate forest, pasture, grassland, shrubland, old field and cropland. More details of the study sites can be found in Table 2. A total of 121 AM fungal taxa were reported in the database, of which 76 were identified to species. Geographic location, disturbance history, climate and soil properties were included in the database. Although data were collected by different researchers and for different purposes, our collection techniques were similar enough to allow for useful comparisons.

Figure 1. Map of sites location, according to vegetation type.

Sample Collection and Spore Extraction

USA sites - Soils were collected using a pipe or soil sampler ranging in size from 2.5 to 5 cm in diameter and 15 to 20 cm deep. Soils were placed in plastic bags and stored frozen until they were processed. Spores were extracted from soils using a modified wet-sieving and density centrifugation technique (Gerdemann and Nicolson 1963, McKenney and Lindsey 1987). A 25 g to 30 g soil sample was placed in a 2 L bucket and vigorously filled with water. The soil suspension was poured through two nested sieves with 250 μm and 25 μm openings. Material collected on the 250 μm sieve was rinsed into a petri dish and spores were picked out using a fine forceps. Material collected on the 25 μm sieve was rinsed into (50 ml) centrifuge tubes and spun at 500 g for 3 minutes. The supernatant was poured away and the pellet re-suspended in 2M sucrose and spun at 500 g for 1.5 minutes. The sucrose supernatant was collected on the 25 μm sieve, and the sucrose residue was rinsed away using tap water. The rinsed material was suctioned through a gridded membrane filter and a subsample of spores on the filter were scraped off the membrane filter and mounted on microscope slides using polyvinyl-lacto-glycerol (PVLG). Spores were counted and identified morphologically using a compound microscope (200 - 1,000X). Spores were identified to species using original species descriptions (Schenck and Perez 1990) and on-line references of species descriptions (INVAM, http://invam.caf.wvu.edu and http://www.lrz-muenchen.de/~ schuessler/amphylo).

Mexican sites - Soils were collected and stored as USA sites. We applied the wet sieving and decantation technique by Gerdemann and Nicolson (1963), modified by Brundrett et al. (1996). Spores were separated from 100 g (50 g for Tropical Dry Forest and Temperate Forest) of soil by wet sieving and decanting (Gerdemann and Nicolson 1963). Each swirling soil suspension was poured through two mesh sieves, 700 and 45 μm. The sediment was suspended (in water) and sieved again. This process was done 3 times.

Table 2. General description of the study sites

Study site	General site description	Location (lat/long)	MAP and MAT	Vegetation type/ land use (and number of samples)	Distur-bance
Cedar Creek LTER in Minnesota, USA	2200 ha Natural Reserve 50 km. The BioCON experiment is described in Reich et al. (2001)	45°24′N, 93°11′W	800 mm; 6.7°C	(1) Old forest, abandoned agricultural fields and pristine savannah and grasslands (15) (2) Grasses *Agropyron repens*, *Bromus inermis*, *Koeleria cristata*, *Poa pratensis*, *Andropogon gerardii*, *Bouteloua gracilis*, *Schizachyrium scoparium*, *Sorghastrum nutans*; herbaceous *Achillea millefolium*, *Anemone cylindrica*, *Asclepias tuberosa*, *Solidago rigida*, and N-fixing legumes *Amorpha canescens*, *Lespedeza capitata*, *Lupinus perennis Petalostemum villosum* (186)	P, ES, LS
Grand Staircase-Escalante National Monument, Utah, USA	An area of 769,000 ha near the towns of Big Water, Cannonville, Escalante, and Boulder.	37°24′N, 111°41′W	296 mm; 10°C	Shrubland with *Artemisia tridentata* and *Artemisia filifolia* (216)	ES, LS, P

Pinyon-Juniper, Colorado, USA	Semi-arid, mid-latitude Steppe	37°49'N, 108°55'W	480 mm; 8.9°C	Shrubland with *Pinus edulis*, *Juniperus osteosperma*, *Quercus gambelii*, *Cercocarpus montanu*, *Elymus elymoide*, *Bouteloua gracilis* (70)	HD, P
Taylor Woods Ponderosa pine, Arizona, USA	Growing stock was established in 1962 and every 10 years since, study areas have been maintained at a range of growing stock levels within an even-aged management strategy	35°16'N, 111°43'W	560 mm; 6°C	Semi-arid montane conifer forest (21)	ES, LS, P
Temperate forest, Mexico	At southern Mexico City, Central Mexico	19°14' to 19°17' N, 99°19' to 99°16'W	1200; 1100 mm lower altitude; 7.1, 9.2 to12° C	Dominant species are *Pinus hartwegii* at the top, *Abies religiosa* in the middle and *Quercus rugosa* in the lower altitude (6)	LS, P
Tropical Rain forest, Veracruz, Mexico, northern site	Los Tuxtlas Biosphere Reserve, southeastern of Veracruz State	18°32' to 18°37' N, 95°02' to 95°08' W	4500 mm; 24°C	Tropical rain forest (7)	P
Tropical rain forest, Veracruz, Mexico, southern site	At southern Los Tuxtlas.	18°15' to 18°24' N, 94°41' to 94°56' W	2900 mm; 24°C	Tropical rain forest; four land uses were considered: mature forest, successional forest, grassland and maize crop (80)	ES, LS, P
Tropical dry forest, Oaxaca, Mexico	Southern Mexico, at Nizanda, Oaxaca.	16° 38' to 16° 83' N, 94° 00' to 95° 00' W	1000 mm; 25°C	Parcels were grouped in three different successional stages (64)	ES, LS

MAP: Mean Annual Precipitation, MAT: Mean Annual Temperature. HD: high level disturbance; ES: medium level disturbance, early successional; LS: low level disturbance, late successional; P: pristine.

Samples were also centrifuged according to the sucrose density gradients technique by Daniels and Skipper (1982). Soil caught in the small sieve was transferred with water into 50 ml centrifuge tubes and centrifuged (Solbat C-600, 24°C) for 4 min at 3,500 rpm. The supernatant liquid was carefully decanted; the pellet was re-suspended in a sucrose solution (440 g/L) and centrifuged at 3,500 rpm for 50 sec. The supernatant was sieved (45 μm) and washed thoroughly for at least one minute (Walker, unpublished data). A Tween solution (two drops of Tween in 100 ml of water) and 5% chlorine solution were added for one and five minutes, respectively. The solid material from the sieve was washed in a Petri dish. Spores were mounted with PVLG on permanent slides marked with a grid. Spores were counted at a compound microscope at magnification up to 10X. Spores were identified to species as USA sites.

Data Analysis

An Analysis of Variance of Multiple Regression (Zar 1999) was conducted for spore species richness, evenness, diversity (using the Shannon diversity index) and total abundance, with the predictors, plant species richness (data were natural log transformed), latitude and disturbance. We used the following disturbance categories: high disturbance, low (early successional), medium (late successional), and mature (pristine sites) (Table 2). Pristine refers to undisturbed sites and high refers to USA sites with severe soil disturbances (e.g., Pinyon-Juniper, Colorado, USA, due to long duration and high temperature fires). Because disturbance effects are different between Mexican (e.g., improved plant species responses) and USA (e.g., negative responses), we analyzed them separately.

RESULTS AND DISCUSSION

Table 3 gives the statistical results for each of the analyses discussed below.

Table 3. Results of the statistical analysis: F (P); R^2

	USA			Mexico		
	Plant species richness	Latitude	Disturbance level	Plant species richness	Latitude	Disturbance level
AM fungal spore species richness	2.02 (0.16); 0.01	0.46 (0.49); 0.001	24.54 (<0.0001); 0.19	5.02(0.03); 0.03	0.14 (0.71); 0.001	14.92 (<0.0001); 0.13
AM fungal spore species evenness	8.4 (0.004); 0.03	0.11 (0.75); 0.0003	13.76 (<0.0001); 0.11	6.15 (0.014); 0.03	0.06 (0.81); 0.0003	1.76 (0.17); 0.02
AM fungal diversity (H')	2.79 (0.10); 0.01	0.55 (0.46); 0.002	0.08 (0.97); 0.001	0.46 (0.5); 0.002	0.15 (0.70); 0.001	8.34 (0.0003); 0.08
AM fungal abundance	1.85 (0.17); 0.01	0.27 (0.60); 0.001	14.78 (<0.0001); 0.12	4.81 (0.03); 0.02		0.4 (0.67); 0.004

Relationship between AM Fungal and Plant Diversity

There was no consistent relationship between plant diversity and the diversity of Glomeromycotan spores across the full dataset. There was a weak negative relationship between the species richness of AM fungal spores and plant species richness (Figure 2a) which was driven by lower spore species richness in the Mexican sites with high plant diversity ($R^2 = 0.03$). Species evenness of AM fungal spores had no relationship with richness of the plant communities (Figure 2b). There was no also relationship between the Shannon diversity of AM fungal spores and plant species richness (Figure 2c).

Other studies have also shown there to be very little connection between the diversity of plant hosts and their associated AM fungi. In lowland evergreen forests and pastures in Nicaragua and Costa Rica, Picone (2000) found that most of the AM fungal species produced more spores in pasture, and local AM fungal species richness did not significantly decline following conversion of forest to pasture. An extreme example of uncoupled above and belowground diversity was observed in a study showing no loss of AM fungal diversity when a diverse Costa Rican dry forests was converted to monocultures of *Hyparrhenia rufa*, an introduced C-4 grass from Africa (Johnson and Wedin 1997). Lovelock and Ewel (2005) in a study in monocultures and polycultures of cacao in Costa Rica, found that richness and the mean spores' number were similar between land uses, but in the polyculture *Acaulospora spinosa* and *A. morrowiae*, were the most and least abundant, respectively; however, according to plant species composition, the abundance was inverse in the monocultures.

Several studies do show a positive relationship with AM fungal diversity and plant diversity (Burrows and Pfleger 2002, Oehl et al. 2003, Chen et al. 2004, Börstler et al. 2006). Most of these studies were experimental, meaning that diversity had been manipulated in the field after only few years of diversity treatments. It is possible that in the longer term other factors, such as soil chemistry or soil texture play a larger role in influencing the assemblage of AM fungal spores or that these factors explain the patterns in the short term as well. For example, the Oehl et al. (2003) study demonstrates an increase in AM fungal diversity when going from a mono-crop system to natural grassland, where soil disturbances like tillage and fertilizer use have also likely shaped the AM fungal communities. More research is needed to tease out the complex interactions among biotic and abiotic factors in shaping AM fungal community structure.

Latitudinal Patterns in AM Fungal Diversity

There was no relationship between AM fungal spore richness, evenness or diversity and latitude (Figure 3). Plant diversity does show a relationship with latitude, but as AM fungal diversity is not closely related to plant diversity, perhaps this result should not be surprising. This confirms the statement of Allen et al. (1995) that diversity of mycorrhizal fungi does not follow patterns of plant diversity. These authors suggested that because AM fungi are generalists, their diversity would be determined by physiological adaptations to the environment. Recently, for phytoplankton patterns, Barton et al. (2010) discuss about two other possible explanations for changes in diversity: "hot spots" of enhanced diversity reflecting lateral dispersal, and an enabled long exclusion time scales by weak seasonality in the tropics; it is interesting that we had high AM fungal diversity in USA, and in the tropical Mexico sites. The relative availability of soil phosphorus and nitrogen may be an important determinant of the structure and function of AM fungal communities (Johnson 2010).

Influence of Disturbance on AM Fungal Diversity

Disturbance had a strong influence on species richness of AM fungal spore communities (Table 3). In the USA, species richness progressively decreased with medium and high levels of disturbance; while species richness increased under low disturbance in the Mexican sites (Figure 4a). Evenness was highest in the high disturbance plots in the USA sites but not in the Mexican sites (Figure 4b, Table 3). In the Mexican sites, diversity was greatest in sites with low levels of disturbance, but diversity was not sensitive to disturbance in the USA sites, probably because the opposite responses of species richness and evenness canceled each other (Figure 4c, Table 3).

It is interesting to note that low levels of disturbance did not reduce species richness in either the Mexican or the USA sites. In fact, in the Mexican dataset, species richness and diversity were significantly greater in sites with low levels of disturbance. This suggests that low levels of aboveground disturbance do not harm, and may even encourage sporulation by AM fungi. Perhaps plant carbon allocated belowground to roots and AM fungal symbioses is increased by low levels of disturbance, and this may increase solar radiation to remaining vegetation. Low levels of disturbance did not stimulate spore richness in the USA sites. Perhaps this difference among the datasets occurs because in the more northern sites the vegetation is less light limited and more limited by cold temperatures and drought. There was an extremely high disturbance treatment in the USA data set, and under this treatment spore species richness was significantly reduced while evenness was increased.

This pattern found in areas of high disturbance was mostly driven by sampling slash pile burns, which are management treatments intended to reduce forest fuels. This method of fuels reduction often creates lasting soil disturbances from long duration, high severity fires. Slashes pile burns have been shown to reduce AM fungi abundance (Korb et al. 2004, Owen et al. 2009). This result suggests that a few hardy spore species became dominate after extreme disturbance.

Management Implications

We found that species richness and evenness of Glomeromycotan spore communities were not correlated with plant richness or latitude; however, they were sensitive to disturbance (Table 3). The observation that there was no relationship between the plant diversity and Glomeromycotan diversity supports the notion that AM fungi are ancient clonal organisms that have evolved to infect relatively more ephemeral host communities which may change through successional processes. Individual plants associate with multiple AM fungal taxa and vice versa. To be successful, AM fungal taxa must be adapted to provision the local plant community with belowground resources under the ambient soil, and environmental conditions. Over long periods of time, conditions of the abiotic environment may be less variable than the biotic environment. Consequently, AM fungal communities may have adapted to be relatively resistant to change following shifts in plant community composition. This could be important to the success of traditional methods of slash-and-burn agriculture if it means that diverse AM fungal communities that are well adapted to the local environment may be present for several years after the native vegetation has been replaced with crops.

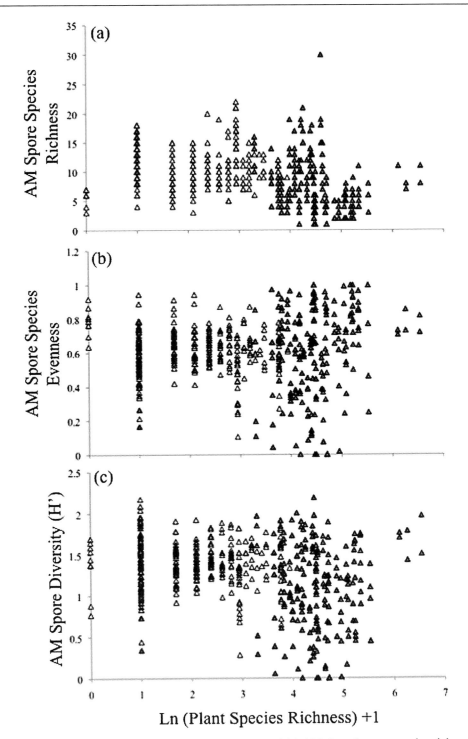

Figure 2. Relationships between plant species richness and (a) AM fungal spore species richness, (b) AM fungal spore species evenness and (c) AM fungal spore diversity. Mexican sites are indicated as gray triangles, and USA sites as open triangles.

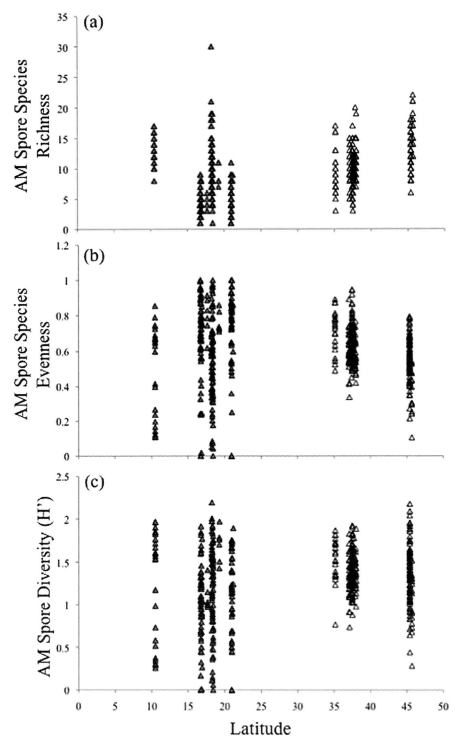

Figure 3. Relationships between latitude and (a) AM fungal spore species richness, (b) AM fungal spore species evenness and (c) AM fungal spore diversity. Mexican sites are indicated as gray triangles, and USA sites as open triangles.

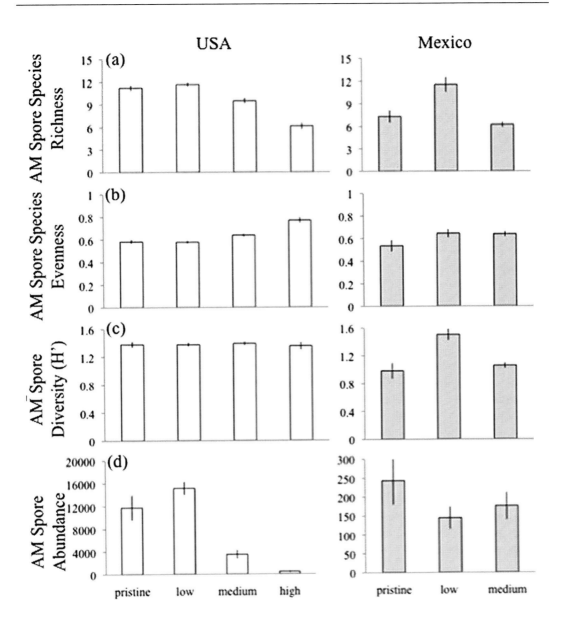

Figure 4. (a) AM fungal species richness, (b) evenness, and (c) diversity by disturbance ranking for sites in USA and Mexico. Error bars represent one standard error of the mean. Different letters above the bars indicate that the means are significantly different ($P \leq 0.05$) according to Tukey's test.

Resistance to disturbance and ability to recover following disturbance (i.e. resilience) may vary considerably among AM fungal taxa, and this may help explain why Glomeromycotan diversity was more sensitive to disturbance than to a loss of plant diversity. This indicates that tillage and other forms of disturbance such as fire and fertilization may be a strong force in structuring AM fungal communities. We are currently exploring effect of the environmental and resources gradients and the history of the land use on the community composition of AM fungal spores. A better understanding of the factors that control the species composition of Glomeromycotan communities is a first step in effectively managing for their benefits in ecosystem restoration and agriculture.

CONCLUSION

Our findings indicate that the factors controlling community diversity differ for plants and mycorrhizal fungi. Disturbance influences AM fungal spore community diversity, and the outcome of these effects varies among ecosystems. Future studies are needed to further elucidate all the factors that structure AM fungal communities. This information will help guide the management of these important symbioses as part of integrated conservation and land management plans.

ACKNOWLEDGMENTS

We acknowledge financial support from Conservation and Sustainable Management of Below-Ground Biodiversity (CSM-BGBD) project coordinated by the Tropical Soil Biology and Fertility Institute of CIAT (TSBF-CIAT), Global Environment Facility (GEF) and United Nations Environment Program (UNEP); SEMARNAT-CONACYT 2002-c01-668; the National Science Foundation, DEB GK-12 and DEB-0842327; DGAPA-PAPIIT IN-200906 and SDEI-PTID-02 Macroproyecto Manejo de Ecosistemas y Desarrollo Humano, Universidad Nacional Autónoma de México, and an ARCS Scholarship to Bala Chauhdary. We thank Peter Reich, Matt Hallderson, Jared Trost, Steve Overby, Amarantha Moreno, Oswaldo Núñez, and Yuriana Martínez for their field and laboratory assistance. Also, we acknowledge to Laura Hernández-Cuevas and Lucía Varela for Mexican species identification, and to Dulce Moreno-Miranda for sites map. We also thank the USDA Forest Service: Rocky Mountain Research and Dolores Forest Service Station, and Cedar Creek field crews.

REFERENCES

Alguacil, MM; Lumini, E; Roldán, A; Salinas-García, JR; Bonfante, P; Bianciotto, V. The impact of tillage practices on arbuscular mycorrhizal fungal diversity in subtropical crops. *Ecological Applications*, 2008, 18, 527-536.

Allen, E; Allen, M; Helm, D; Trappe, J; Molina, R; Rincón, E. Patterns and regulation of mycorrhizal plant and fungal diversity. *Plant and Soil*, 1995, 170, 47-62.

Allsopp, N; Stock, WD. VA mycorrhizal infection in relation to edaphic characteristics and disturbance regime in three lowland plant communities in the south-western Cape, South Africa. *Journal of Ecology*, 1994, 82, 271-279.

Barton, AD; Dutkiewics, S; Flierl, G; Bragg, J; Follows, MJ. Patterns of diversity in marine phytoplankton, *Science*, 2010, 327, 1509-1511.

Bethlenfalvay, GJ. Mycorrhizae and crop productivity. In: Bethlenfalvay G; Linderman R, editors. *Mycorrhizae in Sustainable Agriculture*. Madison, Wisconsin, USA: ASA Special Publication No. 54; 1992; 1-27.

Bever, JD; Morton, J; Antonovics, J; Schultz, P. Host-dependent sporulation and species diversity of arbuscular mycorrhizal fungi in mown grassland. *Journal of Ecology*, 1996, 84, 71-82.

Bever, JD; Schultz, PA; Pringle, A; Morton, JB. Arbuscular mycorrhizal fungi: more diverse than meets the eye, and the ecological tale of why. *BioScience,* 2001, 51, 923-931.

Börstler, B; Renker, C; Kahmen, A; Buscot, F. Species composition of arbuscular mycorrhizal fungi in two mountain meadows with different management types and levels of plant biodiversity. *Biology and Fertility of Soils*, 2006, 42, 286-298.

Brundrett, MC; Bougher, N; Dell, B; Grove, T; Malajczuk, N. 1996. *Working with mycorrhizas in forestry and agriculture*. Canberra, Australia: BPD Graphic Associates; 1996.

Brundrett, MC. Coevolution of roots and mycorrhizas of land plants. *New Phytologist*, 2002, 154, 275-304.

Burrows, RL; Pfleger, FL. Arbuscular mycorrhizal fungi respond to increasing plant diversity. *Canadian Journal of Botany*, 2002, 80, 120-130.

Camargo-Ricalde, SL. Dispersal, distribution and establishment of arbuscular mycorrhizal fungi: a review. *Boletín Sociedad Botánica de México*, 2002, 71, 33-44.

Carey, P; Fitter, A; Watkinson, A. 1992. A field study using the fungicide benomyl to investigate the effect of mycorrhizal fungi on plant fitness. *Oecologia*, 1992, 90, 550-555.

Clapp, JP; Young, JPW; Merryweather, JW; Fitter, AH. Diversity of fungal symbionts in arbuscular mycorrhizas from a natural community. *New Phytologist*, 1995, 130, 259-265.

Cuenca, G; de Andrade, Z; Escalante, G. Arbuscular mycorrhizae in the rehabilitation of fragile degraded tropical lands. *Biology and Fertility of Soils*, 1998, 26, 107-111.

Chen, X; Tang, J; Fang, Z; Shimizu, K. Effects of weed communities with various species numbers on soil features in a subtropical orchard. *Agriculture, Ecosystems and Environment*, 2004, 102, 377-388.

Chaudhary, VB; Lau, M; Johnson, N. Macroecology of microbes-biogeography of Glomeromycota. In: Varma A, editor. *Mycorrhiza*. Berlin: Springer-Verlag; 2008; 529-565.

Daniels, HBA; Skipper HD. Methods for the recovery and quantitative estimation of propagules from soil. In: Schenck NC, editor. *Methods and principles of mycorrhiza research*. St. Paul, Minn, USA: American Society for Phytopathology; 1982, 29-37.

Egerton-Warburton, LM; Allen, EB. Shifts in an arbuscular mycorrhizal communities along an anthropogenic nitrogen deposition gradient. *Ecological Applications*, 2000, 10, 484-496.

Egerton-Warburton, LM; Johnson, NC; Allen, EB. Mycorrhizal community dynamics following nitrogen fertilization: a cross-site test in five grasslands. *Ecological Monographs*, 2007, 77, 527-544.

Entry, JA; Reeves, DW; Mudd, E; Lee, WJ; Guertal, E; Raper, RL. Influence of compaction from wheel traffic and tillage on arbuscular mycorrhizae infection and nutrient uptake by *Zea mays*. *Plant and Soil*, 1996, 180, 139-146.

Gerdemann, JW; Nicolson, TH. Spores of mycorrhizal Endogone species extracted from soil by wet sieving and decanting. *Transactions of the British Mycological Society*, 1963, 46, 235-244.

Hart, MM; Reader, RJ. Do arbuscular mycorrhizal fungi recover from soil disturbance differently? *Tropical Ecology*, 2004, 45, 97-111.

Hartnett, DC; Wilson, GWT. Mycorrhizae influence plant community structure and diversity in tallgrass prairie. *Ecology*, 1999, 80, 1187-1195.

Harrison, MJ. The arbuscular mycorrhizal symbiosis: an underground association. *Trends in Plant Science Reviews*, 1997, 2, 54-60.

Helgason, T; Daniell, TJ; Husband, R; Fitter, AH; Young, JPW. Ploughing up the wood-wide web? *Nature*, 1998, 394,431.

Helgason, T; Merryweather, JW; Denison, J; Wilson, P; Young, JPW; Fitter, AH. Selectivity and functional diversity in arbuscular mycorrhizas of co-occurring fungi and plants from a temperate deciduous woodland. *Journal of Ecology*, 2002, 90, 371-384.

Hempel, S; Renker, C; Buscot, F. Differences in the species composition of arbuscular mycorrhizal fungi in spore, root and soil communities in a grassland ecosystem. *Environmental Microbiology*, 2007, 9, 1930-1938.

Herrera-Peraza, RA; Cuenca, G; Walker, C. *Scutellospora crenulata*, a new species of Glomales from La Gran Sabana, Venezuela. *Canadian Journal of Botany*, 2001, 79, 674-678.

Hillebrand, H. On the generality of the latitudinal diversity gradient. *American Naturalist*, 2004, 163, 192-211.

Hughes, JBM; Bohannan, BJM; Brown, JH; Colwell, RK; Fuhrman, JA; Green, JL; Horner-Devine, MC; Kane, M; Krumins, JA; Kuske, CR; Morin, PJ; Naeem, S; Ovreas, L; Reysenbach, AL; Smith, VH; Staley, JT. Microbial biogeography: putting microorganisms on the map. *Nature Reviews Microbiology*, 2006, 4, 102-112.

Jansa, J; Mozafar, A; Anken, T; Ruh, R; Sanders, IR; Frossard, E. Diversity and structure of AMF communities as affected by tillage in a temperate soil. *Mycorrhiza*, 2002, 12, 225-234.

Jiménez-Esquilina, AE; Strombergera, ME; Massman, WJ; Frank, JM; Shepperd, WD. Microbial community structure and activity in a Colorado Rocky Mountain forest soil scarred by slash pile burning. *Soil Biology and Biochemistry*, 2007, 39, 1111-1120.

Johnson, NC; Zak, DR; Tilman, D; Pfleger, FL. Dynamics of vesicular-arbuscular mycorrhizae during old field succession. *Oecologia*, 1991, 86, 349-358.

Johnson, NC; Tilman, D; Wedin, D. Plant and soil controls on mycorrhizal fungal communities. *Ecology*, 1992, 73, 2034-2042.

Johnson, NC; Wedin, D. Soil carbon, nutrients, and mycorrhizae during conversion of dry tropical forest to grassland. *Ecological Applications*, 1997, 7, 171-182.

Johnson, NC; O'Dell, TE; Bledsoe, CS. Methods for ecological studies of mycorrhizae. In: Robertson GP; Coleman DC; Bledsoe CS; Sollin P, editors. *Standard Soil Methods for Long-Term Ecological Research*. New York, Oxford: University Press; 1999; 378-412.

Johnson, NC. Resource stoichiometry elucidates the structure and function of arbuscular mycorrhizas across scales. *New Phytologist*, 2010, 185, 631-647.

Kardol, P; Bezemer, TM; van der Putten, WH. Temporal variation in plant-soil feedback controls succession. *Ecology Letters*, 2006, 9, 1080-1088.

Kaufman, DM; Willig, MR. Latitudinal patterns of mammalian species richness in the New World: the effects of sampling method and faunal group. *Journal of Biogeography*, 1998, 25, 795-805.

Klironomos, JN; McCune, J; Hart, M; Neville, J. The influence of arbuscular mycorrhizae on the relationship between plant diversity and productivity. *Ecology Letters*, 2000, 3, 137-141.

Klironomos, JN. Feedback with soil biota contributes to plant rarity and invasiveness in communities. *Nature*, 2002, 417, 67-70.

Korb, JE; Johnson, NC; Covington, WW. Slash pile burning effects on soil biotic and chemical properties and plant establishment: recommendations for amelioration. *Restoration Ecology*, 2004, 12, 52-62.

Koske, RE. Distribution of VA mycorrhizal fungi along a latitudinal temperature gradient. *Mycologia*, 1987, 79, 55-68.

Landis, FC; Gargas, A; Givnish, TJ. Relationships among arbuscular mycorrhizal fungi, vascular plants and environmental conditions in oak savannas. *New Phytologist*, 2004, 164, 493-504.

Lovelock, CE; Ewel, JJ. Links between tree species, symbiotic fungal diversity and ecosystem functioning in simplified tropical ecosystems. *New Phytologist*, 2005, 167, 219-228.

McKenney, MC; Lindsey, DL. Improved method for quantifying endomycorrhizal fungi spores from soil. *Mycologia*, 1987, 79, 779-782.

Miller, RM; Jastrow, JD. Mycorrhizal fungi influence soil structure. In: Kapulnik Y; Douds Jr. D, editors. *Arbuscular Mycorrhizas: Physiology and Function*. Netherlands: Kluwer Academic Publishers; 2000; 3-18.

Nadian, H; Smith, SE; Alston, AM; Murray, RS. Effects of soil compaction on plant growth, phosphorus uptake and morphological characteristics of vesicular-arbuscular mycorrhizal colonization of *Trifolium subterraneum*. *New Phytologist*, 1997, 135, 303-311.

Odion, DC; Sarr, DA. Managing disturbance regimes to maintain biological diversity in forested ecosystems of the Pacific Northwest. *Forest Ecology and Management*, 2007, 246, 57-65.

Oehl, F; Sieverding, E; Ineichen, K; Mäder, P; Boller, T; Weimken, A. Impact of land use intensity on the diversity of arbuscular mycorrhizal fungi in agroecosystems of Central Europe. *Applied and Environmental Microbiology*, 2003, 69, 2816-2824.

Öpik, M; Saks, U; Kennedy, J; Daniell, T. Global diversity patterns of arbuscular mycorrhizal fungi-community composition and links with functionality. In: Varma A, editor. *Mycorrhiza: Genetics and Molecular Biology, Eco-Function, Biotechnology, Eco-Physiology, Structure and Systematics*. 3rd edition. Berlin: Springer-Verlag; 2008; 89-111.

Owen, SM; Sieg, CH; Gehring, CA; Bowker, MA. Above- and belowground responses to tree thinning depend on the treatment of tree debris. *Forest Ecology and Management*, 2009, 259, 71-80.

Picone, C. Diversity and abundance of arbuscular-mycorrhizal fungus spores in tropical forest and pasture. *Biotropica*, 2000, 32, 734-750.

Pietikäinen, A; Kytöviita, MM; Husband, R; Young, JPW. Diversity and persistence of arbuscular mycorrhizas in a low-Arctic meadow habitat. *New Phytologist*, 2007, 176, 691-698.

Pringle, A; Bever, JD; Gardes, M; Parrent, JL; Rillig, MC; Klironomos, JN. Mycorrhizal symbioses and plant invasions. *Annual Review of Ecology, Evolution and Systematics*, 2009, 40, 699-715.

Reich, PB; Knops, J; Tilman, D; Craine, J; Ellsworth, D; Tjoelker, M; Lee, T; Wedin, D; Naeem, S; Bahauddin, D; Hendrey, G; Jose, S; Wrage, K; Goth, J; Bengston, W. Plant diversity enhances ecosystem responses to elevated CO_2 and nitrogen deposition. *Nature*, 2001, 410, 809-812.

Rillig, MC; Wright, SF; Shaw, MR; Field, CB. Artificial climate warming positively affects arbuscular mycorrhizae but decreases soil aggregate water stability in an annual grassland. *Oikos*, 2002, 97, 52-58.

Rosenzweig, M. *Species diversity in space and time*. United Kingdom: Cambridge University Press; 1995.

Sanders, IR. Preference, specificity and cheating in the arbuscular mycorrhizal symbiosis. *Trends in Plan Science*, 2003, 8, 143-145.

Smith, SE; Read, DJ. *Mycorrhizal Symbiosis*. 2nd edition. Great Britain: Academic Press; 1997.

Schenck, NC; Perez, Y. *Manual for the identification of VA mycorrhizal fungi*. Gainesville, FL, USA: Synergistic Publications; 1990.

Schüßler, A; Schwarzott, D; Walker, C. A new fungal phylum, the Glomeromycota: phylogeny and evolution. *Mycological Research*, 2001, 105, 1413-1421.

Stutz, JC; Copeman, R; Martin, CA; Morton, JB. Patterns of species composition and distribution of arbuscular mycorrhizal fungi in arid regions of southwestern North America and Namibia, Africa. *Canadian Journal of Botany*, 2000, 78, 237-245.

Trappe, JM. Phylogenetic and ecologic aspects of mycotrophy in the angiosperms from an evolutionary standpoint. In: Safir GR, editor. *Ecophysiology of VA Mycorrhizal Plants*. Boca Raton, Florida: CRC Press; 1987; 5-25.

Treseder, KK. A meta-analysis of mycorrhizal responses to nitrogen, phosphorus and atmosferic CO_2 in field studies. *New Phytologist*, 2004, 164, 347-355.

Vandenkoornhuyse, P; Husband, R; Daniell, TJ; Watson, IJ; Duck, JM; Fitter, AH; Young, JPW. Arbuscular mycorrhizal community composition associated with two plant species in a grassland ecosystem. *Molecular Ecology*, 2002, 11, 1555-1564.

van der Heijden, MGA; Klironomos, JN; Ursic, M; Moutoglis, P; Streitwolf-Engel, R; Boller, T; Wiemken, A; Sanders, IR. Mycorrhizal fungal diversity determines plant biodiversity, ecosystem variability and productivity. *Nature*, 1998, 396, 69-72.

van der Heijden, MGA; Weimken, A; Sanders, I. Different arbuscular mycorrhizal fungi alter coexistence and resource distribution between co-occurring plant. *New Phytologist*, 2003, 157, 569-578.

Walker, C. Systematics and taxonomy of the arbuscular endomycorrhizal fungi (Glomales) - a possible way forward. *Agronomie*, 1992, 12, 887-897.

Wardle, DA; Bardgett, RD; Klironomos, JN; Setala, H; van der Putten, WH; Wall, DH. Ecological linkages between aboveground and belowground biota. *Science*, 2004, 304, 1629-1633.

Warner, NJ; Allen, MF; MacMahon, JA. Dispersal agents of vesicular-arbuscular mycorrhizal fungi in a disturbed arid ecosystem. *Mycologia*, 1987, 79, 721-730.

Willig, MR; Kaufman, DM; Stevens, RD. Latitudinal gradients of biodiversity: pattern, process, scale, and synthesis. *Annual Review of Ecology, Evolution, and Systematics*, 2003, 34, 273-309.

Zar, JH. *Biostatistical Analysis*. London, Great Britain: Prentice Hall; 1999.

In: Mycorrhiza: Occurrence in Natural and Restored...
Editor: Marcela Pagano

ISBN: 978-1-61209-226-3
© 2012 Nova Science Publishers, Inc.

Chapter 3

ARBUSCULAR MYCORRHIZAS AND THEIR IMPORTANCE FOR TROPICAL FOREST FORMATION IN BRAZIL

Waldemar Zangaro[*]
Universidade Estadual de Londrina, Departamento de Biologia Animal e Vegetal
Londrina, 86051-990 Paraná, Brasil

ABSTRACT

The importance of the arbuscular mycorrhizal (AM) fungus for establishment, survival and growth of the tropical native woody species belonging to different ecological groups of succession from Brazilian tropical forests is discussed. The response to AM inoculation and the degree of AM root colonization of the plant species decreased with the advance among successional ecological groups. High response to inoculation and high colonization among early successional woody species (pioneer and early secondary) emphasize the importance of AM association for the initial growth of these woody species, which are evolved in the initial tropical forest structuring. The late successional woody species (late secondary and climax), which dominates in the mature forests, showed absence or low both responsiveness to inoculation and degree of root colonization, indicating limitations to use of the AM fungus as strategies for mineral acquisition in early growth stages. In this chapter, some experimental results of AM inoculation in tropical native woody species from different ecological groups will be presented and discussed, especially in relation to the woody species development, the vegetal succession and the importance of AM association on the reclamation of degraded areas.

---

[*] E-mail: wzangaro@uel.br

1. INTRODUCTION

Arbuscular mycorrhizal (AM) fungus play important role in soil nutrient uptake, especially those of low mobility in the soil like phosphorus. The external hyphae work as an extension of the host's roots, increasing its effectiveness on soil exploration (Gilroy and Jones 2000). The soil P can be reached beyond the depletion zone around the root surface (Miller and Jastrow 1994). The low-diameter AM hyphae are more able to uptake P from small sites that cannot be accessed by the root hairs (Jakobsen 1995). AM fungus make symbiosis with many plant species, including herbaceous and woody (Janos 1983, Sanders et al. 1996, Zangaro et al. 2003). This symbiosis play important role in plant nutrient acquisition, growth and survival (Smith and Read 2008), besides increasing the photosynthetic rate and roots longevity (Linderman 1988, Comas et al. 2002). The AM association is important for rehabilitation of degraded lands and succession process. The association is highly promising for inoculation of native woody species, especially in low-fertility soils (Perry et al. 1987, Zangaro et al. 2000).

2. RESPONSE OF TROPICAL NATIVE WOODY SPECIES TO AM INOCULATION

The influence of the inoculation with AM fungus in the initial growth response of the approximated 150 native woody species belonging to different ecological successional groups from Brazilian tropical forests is discussed. The experimental results with seedlings response to AM inoculation are from studies of Carneiro et al. (1996), Siqueira et al. (1998), Zangaro et al. (2000, 2002, 2003, 2005, 2007), Pouyú-Rojas and Siqueira (2000), Zangaro and Andrade (2002), Siqueira and Saggin Júnior (2001), Matsumoto et al. (2005), Patreze and Cordeiro (2005), Pasqualini et al. (2007) and Vandresen et al 2007. The ecological successional groups classification (pioneer, early secondary, late secondary and climax) of the tropical native woody species was determined according to studies by Chagas e Silva and Soares-Silva (2000), Dias et al. (1998), Ferretti et al. (1995), Gandolfi et al. (1995), Kageyama (1992), Leitão-Filho (1993) and Salis et al. (1994). The response to AM inoculation (Figure 1) and the degree of AM root colonization (Figure 2) of the native woody species decreased with the advance among successional ecological groups. The means of plant responses to AM inoculation (responsiveness) based in shoot dry matter production (Plenchette et al. 1983) was 93.1, 75.2, 30.1 and 19.3 % and the means of AM root colonization was 72.5, 62.8, 25.4 and 15.6 % for pioneers, early secondary, late secondary and climax species, respectively. Plant response and AM root colonization was very contrasting from the early successional species (pioneer and early secondary) to the late successional woody species (late secondary and climax). High response to inoculation and high colonization among early successional woody species emphasize the importance of AM fungus for the initial growth of this woody species, which are the plant species evolved in the initial tropical forest structuring. The late successional woody species, which dominates in the mature forests, showed absence or low both responsiveness to inoculation and degree of root colonization, indicating that woody species of the mature forest display limitations to use of the AM fungus as strategies for mineral acquisition in early growth stages.

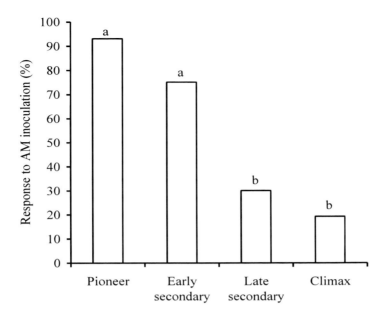

Figure 1. Plant response to AM fungi inoculation of native woody species belonging to different successional stages. Means followed by same letter are not different by Tukey-Kramer HSD test at 0.05 level. Data from of the 93 plant species.

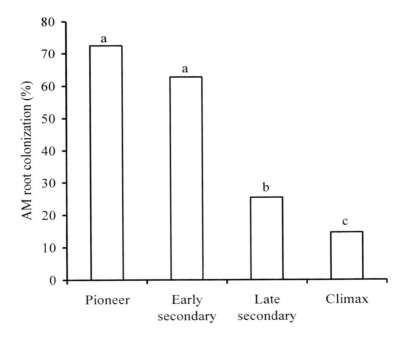

Figure 2. AM root colonization of native woody species belonging to different successional stages. Means followed by same letter are not different by Tukey-Kramer HSD test at 0.05 level. Data from of the 121 plant species.

Different AM fungus inoculums type were used by the above studies: one effective fungi species or mixed of fungus species or inoculation with native populations. The symbiotic effectiveness of the fungus and the host plant response is determined by features of the host plant and mycorrhizal fungus and regulated by soil and environmental factors (Smith and Read 2008). Considering that there is little or no evidence for host-AM fungi specificity, the wide variation in root colonization and plant responsiveness among early and late successional woody species verified for different authors, difference in plant susceptibility to mycorrhiza formation and responsiveness can be attributed mostly to the host genome rather than to that of the fungus (Siqueira and Saggin-Júnior 2001).

3. NUTRIENT ACQUISITION BY PLANTS

In absence of AM fungus, the pioneer and early secondary species showed lower P, Ca and K concentration in leaves than the species from the late secondary and climax species (Table 1). Conversely, when the pioneer and early secondary species were inoculated with AM fungus, the concentration of these nutrients in leaves increased strongly in relation to the uninoculated plants. For the late secondary and climax species there was no increase of nutrient concentration due to mycorrhiza.

The AM fungus generally improves some nutrients uptake that contributes for plant growth (Merryweather and Fitter 1996, Smith and Read 2008). In the uninoculated plants, the concentrations of P, Ca and K in the late secondary and climax were, respectively, 1.8, 1.7 and 1.4 times greater than in pioneers and early secondary species. This suggests that the pioneer and early secondary species are less able to acquire nutrients from a low-fertility soil in the absence of AM fungus, as compared to later successional species. However, in presence of AM fungus the pioneer and early secondary species had an improvement in nutrient uptake capacity. In the case of P and Ca the concentrations in pioneers and early secondary were similar to the late secondary and climax. For K the increase in uptake was 1.28 times greater than the late secondary and climax. This observation towards of the importance of AM association in helping the pioneer and early secondary species in up take nutrients from low-fertility soils.

The mycorrhizal symbiosis increased the uptake and accumulation of nutrients in plant leaf (Table 1) only in the pioneer and early secondary plants. In relation to accumulation of P and Ca, the pioneer, early secondary and late secondary accumulated similar amounts of nutrients, but surpassed the climax plants. The most K accumulation was found in the pioneer plants.

The high nutrient accumulation among the early successional species suggests its high external demand for nutrients. In low-fertility soils the dependence of pioneer and early secondary plants on AM fungus is highlighted. In contrast, most of the late secondary and climax species had no advantage with the AM symbiosis. Their growths were small independently of AM fungus. In this case, the low soil fertility may have limited the plant growth. Nevertheless, these species generally develop well in soils with moderate to high fertility, such as the case of the mature forests conditions.

Arbuscular Mycorrhizas and their Importance for Tropical Forest Formation ... 53

Table 1. Leaf nutrient concentration and content of the native woody species belonging to different ecological groups, inoculated (+AM) and uninoculated (-AM) with arbuscular mycorrhizal fungi. Nutrient concentration ratio between +AM / -AM: NCR[1]. Nutrient content ratio between +AM / -AM: NCR[2]. Data from Zangaro et al. (2003)

Ecological groups	Nutrient concentration (mg 100 mg^{-1})			Nutrient content (mg per leaf)		
	-AM	+AM	NCR[1]	-AM	+AM	NCR[2]
Phosphorus						
Pioneer	0.07 bB	0.14 aA	2.0	0.39 bB	5.16 aA	13.23
Early secondary	0.09 bB	0.14 aA	1.56	0.56 bB	4.29 aA	7.66
Late secondary	0.14 aA	0.15 aA	1.07	2.42 aA	3.29 aA	1.36
Climax	0.15 aA	0.15 aA	1.0	0.99 aB	1.12 aB	1.13
Calcium						
Pioneer	0.32 bB	0.59 aA	1.84	1.79 bB	19.45 aA	10.87
Early secondary	0.37 bB	0.52 aA	1.41	3.86 bB	15.69 aA	4.06
Late secondary	0.54 aA	0.56 aA	1.04	6.86 aA	14.56 aA	2.12
Climax	0.65 aA	0.66 aA	1.02	4.79 aB	5.61 aB	1.17
Potassium						
Pioneer	0.72 bB	1.54 aA	2.14	4.62 bB	53.63 aA	11.61
Early secondary	0.82 bB	1.43 aA	1.74	10.12 bB	39.36 aB	3.89
Late secondary	1.13 aA	1.15 aB	1.02	17.21 aA	20.75 aB	1.21
Climax	1.11 aA	1.17 aB	1.05	8.11 aB	9.32 aC	1.15

Means followed by the same latter (small in row and capital in column) are not different by Tukey-Kramer HSD test at 5%.

3.1. Root Morphology and AM Root Colonization for Nutrient Acquisition

The P uptake by plants is dependent on the surface area of their absorbing structures in the soil (Marschner 1998) and P is the important limiting factor for plant growth in natural ecosystems (Brundrett 2002) specially in tropical soils (Vitousek 1984). The fine root morphology influence nutrient uptake potential (Baylis 1975, Hetrick et al. 1992, White and Westoby 1999, Raghothama and Karthikeyan 2005) and root surface area influence more than root mass (Eissenstat 1992). The fine root diameter influences the P influx rate of root surface, since as root diameter decreased, P influx increased (Itoh and Barber 1983, Eissenstat 1992). The root surface area increases with the increase in root-hair length and incidence (Föehse et al. 1991, Gahoonia et al. 2001) and with AM root colonization (Marschner 1998). The primary function of fine roots is the acquisition of resources for plant (Comas et al. 2002) and the AM association contributes for increase of the root uptake potential (Zangaro et al. 2007). The AM hyphae explore soil volume more extensive and more effective than roots themselves (Brundrett 2002). The plant benefit provided by mycorrhizas increase when the degree of mycorrhizal colonization of roots increases (Smith and Read 2008, Brundrett 2004).

Plant species with high root length, high specific root length, low diameter, more root tips for unit of root length (Comas and Eissenstat 2004) and high incidence of long root hairs (Föehse et al. 1991, Peterson and Farquhar 1996, Raghothama and Karthikeyan 2005) permit the plant root to increase the soil volume exploration by unit of biomass invested in fine roots. High specific root length has been correlated with a quick root proliferation rate (Eissenstat 1992).

AM root colonization varied markedly with different native woody species, the biochemical and physiological mechanisms and the genetic control of these interactions have not been elucidated (Siqueira and Saggin-Júnior 2001). Thus, the AM colonization and plant responsiveness among different successional status of the tropical native woody species can be addressed with the morphological root characteristics and plant metabolic demand. It has been widely accepted from research in temperate regions that the plant root architecture controlled mycorrhizal benefit. Plant species that explore large soil volume display long fine roots, highly branched, low diameter, and cover with numerous root hairs, and are expected to exhibit low levels of AM root colonization (Manjunath and Habte 1991; Schweiger et al. 1995, Brundrett 2002). On the other hand, plant species with coarse root systems and few root hairs tend to have high AM colonization (Baylis1975, Graham and Syvertsen 1985, Hetrick et al. 1992, Manjunath and Habte 1991, Reinhardt and Miller 1990, Schweiger et al. 1995). The results of Siqueira and Saggin-Júnior (2001) and Zangaro et al. (2005, 2007) for plant grown in greenhouse and Zangaro et al. (2008, 2011) for field results from tropical herbaceous and native woody species do not support this hypothesis. Generally, plant species belonging to early successional stages with finer roots and abundant long root hairs displayed high AM root colonization and responsiveness. By contrast, late successional woody species with coarse roots and few root hairs displayed low AM colonization and responsiveness.

In tropical soils, that generally have low P available, the early successional woody species with apparent root morphology for high uptake capacity, do not able to ensure adequate nutrition for maintain its inherent fast growth rate, become the AM root colonization essential for nutrient acquisition. Zangaro et al. (2005, 2007, 2008) suggested that in addition to high carbon allocation to the fine root construction, early successional species maintain more AM fungus because they presented morphological root traits (high root length, high specific root length, low diameter, long and dense root hairs) with high interface that favored the contact with mycorrhizal propagules in the soil. Indeed, the high nutrient accumulation exhibited for early successional species express its high external demand for nutrients. Thus, fast-growing species display inherent intensive plant metabolism that demands higher amounts of nutrients to support the higher growth rates and, therefore, maintain more AM fungus to supply their nutritional needs. The relatively higher investment in leaves by the early succesional species increases the amount of photoassymilates that can be exported to roots (Nielsen et al. 1998, Lynch and Ho 2005) and more carbohydrates can be available to AM fungus in roots.

By contrast, the plants from mature forest present less AM colonization in fine roots in addition to less sporulation in soil. Several features such as shading, slow growth rate of woody species and lower metabolic demand, may result in less available carbohydrates to AM fungus in roots, and consequently less mycorrhizal root colonization and sporulation in the mature forest. Late successional woody species display root morphology for low nutrient acquisition capacities (low root length, low specific root length, short and sparse root hairs) and are able to maintain growth in absence of AM fungus, especially in deficient P soils. Therefore, the fine roots alone may be responsible for the nutrient acquisition among late species. This indicates that slow-growing species may exhibit others strategies for nutrient acquisition instead of AM fungus, as an additional enzymatic nutrient acquisition mechanisms (Chapin 1980), nutrient use efficiency (Manjunath and Habte 1991, Koide 1991), low requirement due to both low growth rate and metabolic demand (Zangaro et al. 2007) and high seed reserves for seedlings growth (Siqueira et al. 1998, Zangaro et al. 2000, Pasqualine et al. 2007).

4. AM FUNGUS AMOUNT DURING PLANT SUCCESSION

Fine root morphological traits, AM root colonization and AM spores density was accessed in sites cover with grass plants, recent secondary forest and mature forest in Londrina county, Paraná state, southern Brazil (Figure 3), accessed in grassland, secondary forest, and mature forest in Atlantic, Araucaria and Pantanal biomes in Brazil (Figure 4) and were also accessed in sites cover with grassland, secondary forest with 15 year, secondary forest with 50 year and mature forest in Telêmaco Borba county, Paraná state, southern Brazil (Figure 5). Generally, the fine root biomass and diameter increased with the succession progress. By contrast, total fine root length, specific root length, root-hair length and root-hair incidence decreased with the succession advance. Similarly, the AM root colonization and the density of AM spores in the soil decreased along the succession, independently of studied biome. AM root colonization and spores density were positively correlated with fine root length, specific root length, root-hair length, root-hair incidence and bulk density, and was negatively correlated with fine root diameter and concentration of some nutrients both in soil and root tissues. The fine root properties and levels of AM colonization and AM spores density changed during successional stages, where grass plants and early successional woody species from secondary forest exhibited maximum effectiveness of carbon use for building root system and more effective soil exploration than woody species of mature forest. Therefore, the efficient investment of carbon in the fine roots (i.e. low diameter and high specific root length) and its appendices (i.e. root hair and AM fungus), in order to improve the nutrient uptake effectiveness, is typical of fast-growing species of the early succession phases, especially in low soil fertility conditions.

The AM root colonization and AM spores density in soil displayed strong reduction during succession progress in all areas studied. These results reflect the greater investment in AM symbiosis by hosts of the early succession phases than native woody species of mature forest. The higher AM fungus variables in early successional stages than mature forests is in accordance with the higher AM spores density found in pasture or natural grasses sites than forest in Australia (Jasper et al. 1991), than humid secondary forest (Fischer et al. 1994) and mature forest (Johnson and Wedin 1997) in Costa Rica, than in dry forest in Mexico (Allen et al. 1998), than in natural sites in Venezuela (Cuenca et al. 1998), than in lowland evergreen forests in Nicaragua and Costa Rica (Picone 2000), in tropical forest in southern Brazil (Zangaro and Andrade 2002), and in low fertility soil at mature forest in Costa Rica (Lovelock et al. 2003). Besides, Zangaro et al. (2000) found low density of AM spores and root colonization in plants from mature forest in southern Brazil and suggested that slow-growing species are less able to keep AM fungus due to growth in relative high soil fertility, shading and low metabolic activity. Aidar et al. (2004) verified that AM root colonization and

AM spores density decreased with increasing soil fertility in a chronosequence of an Atlantic forest in southeast Brazil. Powers et al. (2005) related that the amount of AM fungi hyphae in the soils of four tropical forests in Central and South America was unexpected quite low and suggested that plants of mature forests must rely on itself fine roots instead of AM fungus for nutrient uptake. In an analyse in 15 published papers, Zangaro and Moreira (2010) verified that the amount of AM spores in soils from mature forests and old secondary forests of the Atlantic forest biome was lower than in recent secondary forests and open areas.

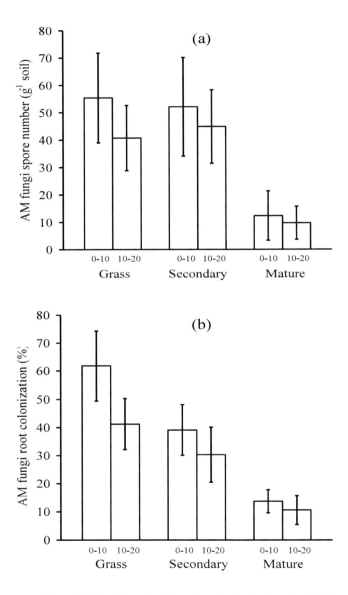

Figure 3. Spores number of the AM fungi in soils (a) and root colonization by AM fungi (b) of native plants from site cover with grass, secondary and mature forest in Londrina county, Paraná State, southern Brazil. Soils samples were taken at 0-10 cm and 10-20 cm depth. Error bars represent ± 1 SD. Data from Zangaro et al. (2008).

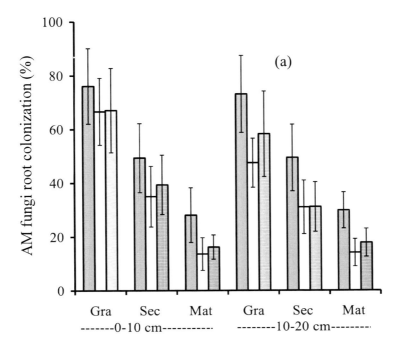

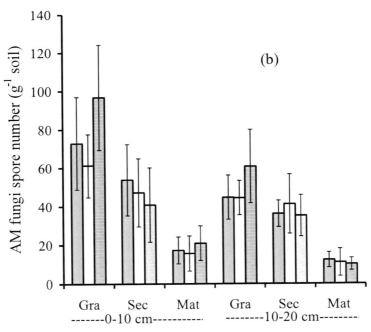

Figure 4. AM root colonization (a) and AM fungi spores density in soil (b) in 0-10 cm and 10-20 cm depth from grasslands (Gra), secondary forests (Sec) and mature forests (Mat) in Brazil. Grey columns, succession in Pantanal biome; open columns, succession in Atlantic biome; dotted columns, succession in Araucaria biome. Error bars are ± SD of the means (n = 15 per site and depth). Data from Zangaro et al. (2011).

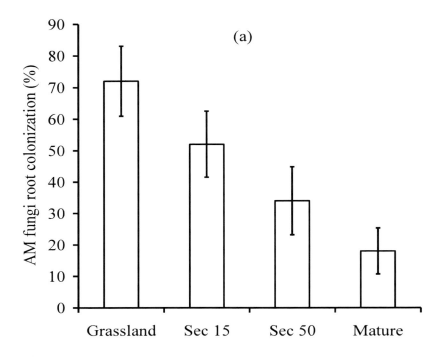

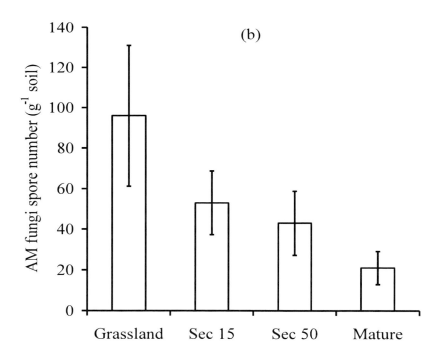

Figure 5. AM root colonization (a) and AM fungi spores number (b) in grassland, secondary forest with 15 year, secondary forest with 50 year and mature forest from Telêmaco Borba county in south Brazil. Roots were taken at 0-5 cm soil depth. Error bars represent ± 1 SD.

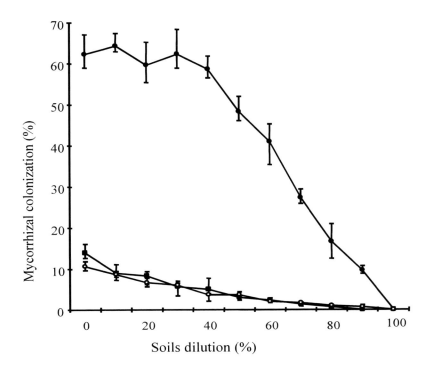

Figure 6. Soils dilution effect from the areas of early succession (●), gap (■) and mature forest (o), on root mycorrhizal colonization of *Cecropia pachystachya*. Vertical bars show the ranges. Data from Zangaro et al. (2000).

5. INOCULUM OF AM FUNGUS AVAILABLE IN SOIL

The inoculum potential of AM fungi in the soils from the mature forest, in a gap in the same mature forest and in the recent secondary forest is showed in Figure 6. The natural soils were diluted with a mix of sand-subsoil and absence of AM fungus. Seedlings of woody species highly micotrophic of the *Cecropia pachystachya* were used as trap plant. The inoculum potential of AM fungus in the soil of the secondary forest was approximately five times greater than in the mature forest and its gap. The secondary forest had large amounts of AM propagules and the mycorrhizal colonization decreased after 50% of soils dilution. At this dilution level, the mycorrhizal colonization fell by 19%, lower than the soil dilution rate. In the gap area, the colonization level was 14% in the original soil and fell by 73% at 50% of soil dilution rate. In the mature forest, the amount of propagules in the original soil was lower than in the gap and the mycorrhizal colonization was 11%. At 50% of soil dilution rate, the root colonization also fell by 73%. In another experiment (Figure 7) the AM root colonization and response to inoculation was accessed in seedlings of woody mycotrophic species *Heliocarpus americanus* grown in 15 soils classes for 40 days and subsequent was planted in infertile soil. Seedlings grown in soils from early stages of succession and secondary forests displayed higher AM root colonization and response than seedlings grown in mature forests soils. These results emphasize the high AM inoculum potential in the soils from early stages of succession and the young secondary forests and the low potential in mature forest conditions. These results suggest that herbaceous plant from open environment and woody pioneer and early secondary plants may be able to multiply the AM fungus in large amounts, allowing high inoculum potential for their offspring's. On the other hand, in the mature forest, the inoculum potential was much lower as compared to the early stages of succession and secondary forest, indicating that the late secondary and climax species that compose the most part of the vegetation in a mature forest have lower levels of micotrophy and as a consequence the potential of AM inoculum is low in mature forest conditions.

6. CHANGE IN PLANT STRATEGY FOR NUTRIENT ACQUISITION DURING SUCCESSION PROGRESS

This part of chapter is based in Zangaro et al. (2011) that focus on the plant strategy for nutrient acquisition and some of the biotic and abiotic factors that could be involved in the gradual decrease the AM root colonization and AM spores density in soil with the advance of succession when plant ecological groups changed. Such behaviour suggests a greater reliance on AM association by plant species belonging to early successional stages than plants from late-successional stages. Zangaro et al. (2008, 2011) verified positive correlations between AM root colonization and AM spores density in soils during tropical succession advance and suggested that this relation reflect: a) the mycelial biomass of AM fungus in the soil; b) indicate the AM inoculum potential in the field and; c) permit to evaluate the potential of plant species from different ecological groups to support the symbiosis. The AM root colonization and AM spores density in soils showed a close relation with fine root morphology (fine root length, specific root length, fine root diameter, density and length of root-hairs), but not with fine root dry mass production, suggesting that the characteristics related to fine root morphology are more important for increasing the AM fungal than the fine root mass found in the soil.

Besides showing higher AM root colonization and AM spores density, early successional plant species from open habitat and young secondary forests also had fine roots with a longer specific root length, smaller fine root diameter, lower root tissue density, and higher root-hair incidence and root-hair length than plant species of late successional stages. These fine root morphological traits permit more effective soil exploration per unit of biomass invested for building the fine root system (Comas et al. 2002, Comas and Eissenstat 2004, Zangaro et al. 2007). In addition to the high mycotrophic status, the fine root morphology suggests that early successional plant species invest in structures that are more effective for greater soil exploration that are at the same time cheaper in terms of C-costs for the plant. These species typically have short lifespan, early flowering times, high photosynthetic rates and high light requirements (Reich et al. 1998, Khurana and Singh 2006). As a consequence of their accelerated metabolism, carbon must be invested efficiently in absorbing structures (fine roots traits and AM fungus), in order to produce a larger absorptive system to ensure a quick and high nutrient uptake rate for these fast-growing species. Moreover, the carbon cost for producing AM hyphae is about two orders of magnitude less than for root production (Hodge 2004), although root hair production and maintenance also represent a low metabolic cost (Lynch and Ho 2005). However, in conditions in which low soil fertility limits the photosynthetic rate and plant growth, the carbon cost for fine root production may be more expensive than for the maintenance of AM hyphae (Smith and Read 2008). In contrast, plant species of mature forests have longer lifespan, slower growth rates, lower light requirements, lower photosynthetic rates, lower metabolic demands, and lower nutrient requirements (Comas et al. 2002, Zangaro et al. 2003). These attributes correspond to lower investment in fine root traits and AM association because nutrient demands are lower in these plants.

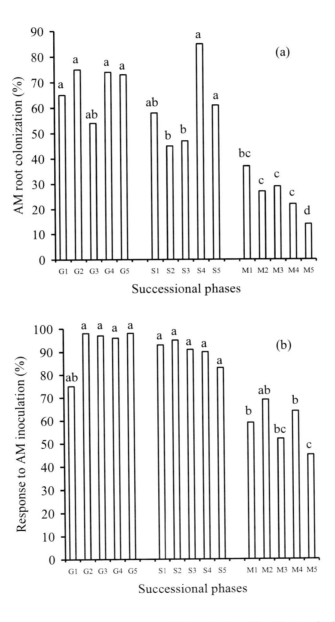

Figure 7. AM root colonization (a) and response to AM inoculation (b) of the tropical native woody *Heliocarpus americanus* grown in infertile soil. AM fungi inoculums are from five early successional areas (G), five secondary forests (S) and five mature forests (M). Means followed by same letter are not different by Tukey test at 0.05 level.

The morphological characteristic of fine roots of fast-growing species belonging to open habitats and secondary forests have specific adaptations for more effective nutrient uptake than slow-growing species in mature forest conditions. This faster fine root spreading potential and higher root area increase the soil exploration for nutrients as well as increasing the likelihood of meeting infective AM fungal propagules in the soil and thus increasing the root colonization (Zangaro et al. 2005, 2007, 2008), as well the great root length and more rooting tips increase the potential for AM infection unit (Allen 2001). Thus, the higher colonization in early successional species is attributable to more abundant targets, given that AM fungus generally does not distinguish between roots of different plant species (Brundrett 2002). In contrast, plant species of mature forests displayed fine root morphological aspects towards to reduce the spread of roots through the soil of mature forests and produce fewer roots branching, decreasing the likelihood of encountering AM propagules. Therefore, the degree of the fine root colonized by AM fungus and their sporulation shows a close relationship with fine root morphological traits and root spreading potential.

The formation of a nutrient depletion zone around the fine roots of fast growing species is expected due to high growth rates and P demand (Smith and Read 2008). As a strategy, plants roots from open habitats and secondary forests may rely on AM hyphae in order to explore the soil beyond the depleted zone around the rhizosphere, especially in low-fertility soils. In this case, AM association become essential for plant survival. On the other hand, due to slow growth and low nutrient demands, the late secondary and climax woody species could develop a nutrient accumulation zone around the fine roots (Gobran et al. 1998, Séguin et al. 2004). In this case, the lower demand for nutrients like P results in a surplus in the rhizosphere and consequently reliance on AM fungus could be reduced, mainly in mature forest soils which have higher rates of organic matter turnover, in addition to higher levels of available P.

The root exudation increase when plants grow in soil with low P available (Grayston et al. 1996) that to stimulate AM hyphal proliferation and increase AM root colonization (Giovannetti et al. 1996). The polysaccharides is exuded by fine roots and root hairs (Peterson and Farquhar 1996, Jones et al. 2004) and the fine root morphological traits of the early successional species such as the higher root length, higher specific root length and higher incidence of longer root hairs are propitious for exudation over longer distances from the root surface than in plant species of mature forests due to their shorter root length and fewer and shorter root hairs (Zangaro et al. 2008). A more active root system due to more exudation would be more attractive to symbiotic fungi, resulting in higher root colonization levels as observed in early successional species. In contrast, the fine root morphology usually found in plant species from mature forests is not propitious for exudation and as a consequence they are less attractive to AM fungi.

Fast growing species have a high photosynthetic capacity (Kitajima 1994, Lusk et al. 2008) and when AM root colonization become high also increases the carbon sink from the host plant and as a consequence the photosynthetic rate is increased (Grayston et al. 1996, Smith and Read 2008). This higher metabolic rate implies in greater production of soluble carbohydrates, which are exported to the roots and become available for AM fungus and early successional plants maintain higher AM root colonization and AM spores production. Conversely, plants from shaded environments as such mature forest conditions generally have lower photosynthetic rates (Lusk et al. 2008) and the low irradiance results in less carbon being translocated to the roots (Grayston et al. 1996), decreasing both exudation (Modjo et al.

1987, Smith and Read 2008) and the concentration of soluble carbohydrates in the roots (Gamage et al. 2004) reflecting in reduction of the AM colonization and AM spores production. The slow growth rate is an adaptation to shaded habitats (Chapin 1980, Reich et al. 1998) and also when trees reach the adult stages, and the AM root colonization can be reduced due to the slow growth rate of the host plant. At the same time, the younger trees and seedlings in the understorey of a mature forest are partially or highly shaded, and as a consequence become carbon-limited and consequently to make difficult the maintaining high AM colonization and production of the spores. Therefore, the low AM root colonization and sporulation in mature tropical forest could be due to the combination of the low occurrence of micotrophic light-demanding species, the low growth rate of adult trees, the shading of young trees, and the intense shading in the forest understorey, limiting seedling growth.

The accumulation of phenolic compounds in roots is a strategy for increasing plant resistance against predators. Fast-growing plant species have lower concentrations of phenolic compounds in roots than slow-growing species (Comas and Eissenstat 2004). A lower concentration of phenolic compounds in roots is expected in the early successional species, allowing higher levels of AM colonization (Giovannetti et al. 1996, Allen 2001). In contrast, in roots of plant species from mature forests, AM root colonization may decrease or be eliminated due to a gradual build-up of phenolics that make the root uninhabitable for AM fungus (Allen 2001).

The shaded environment requires investment in more durable plant structures, and more resistance to herbivore and pathogen attacks in an environment in which the carbon cost for regeneration of damaged structures is costly due to energy limitation (Coley et al. 1985, Coley 1993, Kitajima 1994). Fine roots with a larger diameter, higher root tissue density, lower specific root length and more lignification result in root tissues that are more robust and longer living (Eissenstat 1992, Eissenstat et al. 2000, Comas et al. 2002, Leuschner et al. 2009). These fine root characteristics that are found in the late successional species can reduce root predation, damage, and drying, and thus extend the root lifespan (Kitajima 1994, Comas et al. 2002, Leuschner et al. 2009). The fine root longevity also results in energy economy for the late successional species because their fine, longer living roots remain active in the soil for longer than the fine roots of early successional species. In addition, due to the lower functional capacity for nutrient uptake of the fine roots of slow-growing species, respiration and metabolic maintenance are lower (Comas et al. 2002). The energy economy of plant species in mature forests is driven towards leaf production, as the shade-tolerant species produce leaves with a small symplastic component and larger structural investment than light-demanding species, reducing both construction and maintenance costs (Kitajima 1994, Lusk et al. 2008). Therefore, the cost to maintain AM association by the late successional shade-tolerant species can be high, while limiting AM root colonization levels in mature forests can be important for plant energy conservation and can be considered as an adaptation for plant survival and maintenance in such conditions.

The pattern of plant biomass allocation changes during the succession process. Plants belonging to the early successional phases have relatively more biomass allocated to organs for resource acquisition, like fine roots and leaves, than structural organs (Guariguata and Ostertag 2001). The higher investment in leaves by early successional species increases the amount of photosynthates that can be exported to the roots (Nielsen et al. 1998, Lynch and Ho 2005) and, in addition to the low root tissue density, suggests that less structural carbon is used for cell-wall construction and that more soluble carbon can be available to AM fungus.

In contrast, plants from mature forests allocate more carbon to structural organs, like stems and thicker roots, than organs for resource acquisition (Guariguata and Ostertag 2001). The higher carbon investment in structural organs may result in the lower availability of carbohydrates to AM fungus in roots, decreasing the AM colonization and sporulation in the mature forest. Indeed, the higher root tissue density of plant species of mature forests suggests that more carbon has been used for the construction of fine root tissue and that less soluble carbon is available for AM fungi. In addition, roots with thicker cell walls are less favourable for AM colonization due to mechanical resistance (Giovannetti et al. 1996).

The levels of AM root colonization in seedlings of different tropical species grown in greenhouse correlated positively with plant growth responses, and these AM variables correlated negatively with soil organic C, mineral N, and available P (Zangaro et al. 2008, 2011). Thus, the high nutrient demand of the early successional species that grow in low-fertility tropical soils must be supplemented by mycorrhizal symbiosis as a strategy to improve the acquisition of nutrients. In contrast, late successional species are usually found on more fertile soils in mature forest conditions, and this soil fertility opposes with their low nutrient demand, leading to lower AM association. Indeed, there is more available P in soils of mature forests, and thus even with lower AM colonization rates the roots can take up P itself more efficiently than heavily colonized roots due to the higher carbon cost for maintaining the AM symbiosis under shaded conditions. On the other hand, the increase in soil fertility may either reduce or increase the AM colonization levels in a wide range of plant species (Smith and Read 2008, Gamage et al. 2004, Vandresen et al. 2007). As the levels of AM root colonization are always lower in the mature forest, is suggested that plants in mature environments restrict the establishment of AM fungi in their roots in order to reduce the high carbon cost for maintaining the symbiosis, an effect that is not regulated due to the higher soil fertility levels. Thus, more than soil fertility, light availability seems to be an important abiotic factor determining AM colonization and sporulation in mature forests.

6. IMPORTANCE OF AM FUNGUS FOR INITIAL FOREST STRUCTURING

The high AM root colonization and plant responsiveness among early successional woody species grown in greenhouse when allied with elevated AM root colonization and AM spores in soils from sites cover with grasses and young secondary forest, reflect the host-plant potential for multiplies the AM fungus among herbaceous plants of open habitats and fast-growing woody species. The possible high host AM cropping potential toward large density of AM propagules production, which may be favourable for facilitation both installation and plant recruitment this fast-growing plant species. The early successional woody species are highly responsive for AM colonization, regardless of soil fertility level (Siqueira et al. 1998, Zangaro et al. 2007,) exhibit great aggressiveness during establishment in open and disturbed areas and are hosts with a high degree of AM mycotrophy (Zangaro et al. 2003). Thus, AM fungus may be the main biotic factor for the establishment and growth acceleration of the native woody species that lead for initial tropical forest structuring (Zangaro et al. 2000). During succession progress, the rapid growth and turnover of these fast growing woody species provides a continuous soil organic enrichment and improves the soil structure (Uhl et al. 1982, Zangaro et al. 2003, 2009). The increase on the soil surface fertility along the succession progress is attributable to the biomass accumulation and decomposition as the vegetation develops along the time (Silver et al. 1996, Guariguata and Ostertag 2001, Lugo and Helmer 2004, Boeger et al. 2005). Decomposition of soil organic matter produced by fast-growing woody species is important because of its critical role in the cycling of essential plant nutrients (Degens et al. 2000).

The transformations above and below ground during initial forest structuring permit the posterior establishment and growth of slow-growing woody species, which are dominant in mature forests. Therefore, the adult native woody species that dominate the late succession phases of the tropical forests maintain low amount of AM fungus due to low AM association requirements. Plants in mature forests conditions with low light intensity can be more limited to the carbon than available nutrients in soils, limiting the carbohydrates for AM fungus and decrease the root colonization, suggesting that the low levels of AM fungus in mature forests could be due to high carbon cost for maintain the AM symbiosis than nutrient available in soils. Therefore, the low plant metabolic demand and the light availably appear to be important factor determining AM colonization and sporulation in mature forests. Besides, lipid-rich spores and fungal hyphae are subject to predation and parasitism as they serve as a food source for a wide range of soil animals (Rabatin and Stinner 1988, Stümer et al. 2006). The soil organisms increase during succession (Coleman et al. 2004), and the competition with soil organisms and the hyphae and spores predation may be important aspects that contribute for decrease the AM fungus in soils of mature forests.

7. AM ASSOCIATION ON REVEGETATION PROGRAMS AND RECOVERY OF DEGRADED LANDS

The soil and water degradation are some of the consequences of unsuitable soil use and management. Erosion is one of the main causes of soil degradation in tropical region, where the superficial horizon is generally lost, decreasing the soil fertility due to losses of nutrients and organic matter. Besides the physical and chemical degradation, the biological degradation also restricts the plant reestablishment. Soil fertility also decreases when tropical forests are destroyed and substituted by crops and pastures.

In areas where the natural succession cannot occur due to long distances from the propagules sources, one alternative is the artificial reforestation. The reforestation successful depend on the capacity of the seedling in obtain water and nutrients to establishment, survival and growth in the field (Perry et al. 1987). The success of fast growing species in establishing in degraded areas depends on an adequate nutrient supply. The N and P are the most required plant nutrients, and are generally limiting in degraded soils. N is easily lost from soil due to its dynamics, while P have low-mobility in soil and is fixed in clay and oxidic surfaces. These properties make necessary the use of fertilizers containing N and P for revegetation of degraded areas, which increases the revegetation costs. In addition, the excessive or inadequate use of mineral nutrients may result in environmental pollution, mainly when these nutrients reach the underground water, streams or lakes.

The seedling survival after transplanting to the field is generally increased when they are inoculated with AM fungi (Figure 8). This fact is attributed to an increased capacity for water and nutrients uptake, improved root longevity and resistance against pathogens (Smith and

Read 2008). The establishment of AM association affects the plant physiologically, mainly in relation to C allocation. The extra radical AM mycelium grows throughout the soil and increases the surface area for nutrient acquisition such as P, Ca, Zn and even N (Hawkins and George 2001, Neumann and George 2005). The external AM hyphae can be considered a way that shortens the distance between the nutrients and the plant (Zangaro et al. 2005). Such structures can reach up to 11 cm far from the root surface (Li et al. 1991), and increase the plant potential in obtaining water and nutrients. The external hyphae contribute to soil aggregate stability (Neumann and George 2005), by means of a typical protein termed glomalin (Caravaca et al. 2005). The external hyphae also protect the host plant against excessive heavy metal in the environment and serve as inoculum source for root colonization.

The AM fungus availability in the degraded soils affects the establishment, growth and competition among plant species and may affect the reforestation (Asbjornsen and Montagnini 1994, Zangaro et al. 2000). For the success of the reestablishment of a diverse and self-sustainable ecosystem, an adequate inoculum potential of AM fungus is desirable (Jasper et al. 1991). In areas with poor inoculum potential, the use of appropriate host plants is the key for the reestablishment of a diverse AM fungus population. In the case of areas deprived of AM inoculum, its reintroduction may be done by means of plantlets previously inoculated in nursery (Vandresen et al. 2007, Zangaro et al. 2009). Thus, the knowledge on the degree of mycotrophy among the species that will be employed in the reforestation is very important for the reforestation success.

The results showed here reinforce the importance of AM association for the success of tropical native woody pioneers and early secondary species. These plant species are responsible for the initial forest structuring (Montagnini and Sancho 1994, Guariguata and Ostertag 2001). The establishment of these plants result in improvement of the environment to be colonized by later successional species, such as less water loss, improvement of soil structure and aeration, fewer occurrences of pathogens (Sylvia and Williams 1992, Zangaro et al. 2003). The characteristics of woody early successional species, such as intolerance to shade, high capacity for uptake and accumulation of nutrients and small and abundant seeds, fit as a good partner for AM fungi (Zangaro et al. 2000, 2003). The high mycotrophism of pioneer and early secondary species makes the AM association one of the main tools for installation and start the vegetal succession. This reinforces the importance of AM inoculation in plantlets to be used in reforestation programs (Zangaro et al. 2000).

Spores of AM fungus are generally found in large amounts in pastures, areas that may be useful for reforestation programs (Johnson and Wedin 1997). In the southern region in Brazil, the forest fragments are generally surrounded by pastures or monocrops that are frequently colonized by pioneer and early secondary species originated from these fragments. This suggests that the AM propagules in these soils are enough to promote the establishment, growth and survival of such plants. In case of degraded areas with low AM fungi propagules, the plantlets to be used in the reforestation should be inoculated in the nursery. Nevertheless, the plant response to mycorrhizal inoculation must be known previously. The use of plants that respond to AM inoculation is a tool for the success of the establishment of plants to start the programs designated to rehabilitation of degraded areas.

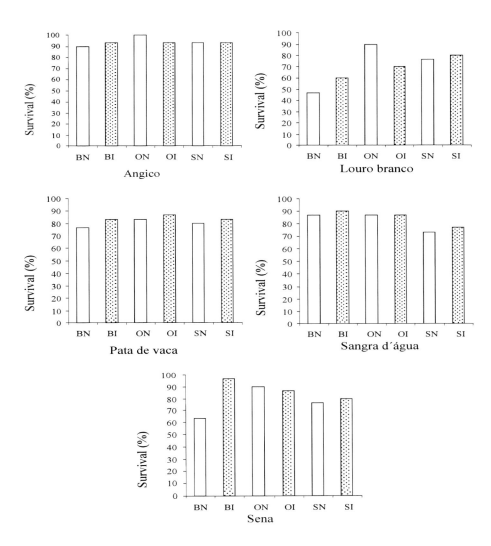

Figure 8. Percent survival for seedlings planted in the field after six months of growth in nursery. Angico (*Anadenanthera colubrina*), Louro branco (*Bastardiopsis densiflora*), Pata de vaca (*Bauhinia forficata*), Sangra d´água (*Croton urucurana*) and Sena (*Senna macranthera*). The seedlings grown in nursery in a base substrate of the organic *Pinus* spp bark compost (B), in base substrate with Osmocote® addition (O) and base substrate with Osmocote® and super phosphate addition (S), inoculated (I) and not inoculated (N) with AM fungi. Data from Vandresen et al. (2007).

REFERENCES

Aidar, MPM; Carrenho, R; Joly, CA. Aspects of arbuscular mycorrhizal fungi in an Atlantic Forest chronosequence. *Biota Neotropica*, 2004, 4,1-15.

Allen, EB; Rincon, E; Allen, MF; Perez-Jimenez, A; Huante, P. Disturbance and seasonal dynamics of mycorrhizae in a Tropical Deciduous Forest in Mexico. *Biotropica*, 1998, 30, 261-274.

Allen, MF. Modeling arbuscular mycorrhizal infection: is % infection an appropriate variable? *Mycorrhiza*, 2001, 10, 255-258.

Asbjornsen, H; Montagnini, F. Vesicular-arbuscular mycorrhizal inoculum potential affects the growth of *Stryphnodendron microstachyum* seedlings in a Costa Rican human tropical lowland. *Mycorrhiza*, 1994, 5,45-51.

Baylis, GTS. The magnolioid mycorrhiza and mycotrophy in root systems derived from it. In: Sanders FE, Mosse B, Tinker PB editors. *Endomycorrhizas*. New York: Academic Press; 1975; 373-389.

Boeger, MRT; Wisniewski, C; Reissmann, CB. Nutrientes foliares de espécies arbóreas de três estádios sucessionais de floresta ombrófila densa no sul do Brasil. *Acta Botanica Brasílica*, 2005, 19,167-181.

Brundrett, MC. Coevolution of roots and mycorrhizas of land plants. *New Phytologist*, 2002, 154,275-304.

Brundrett, MC. Diversity and classification of mycorrhizal associations. *Biological Review*, 2004, 79,473-495.

Caravaca, F; Alguacil, MM.; Barea, JM; Roldán, A. Survival of inocula and native AM fungi species associated with shrubs in a degraded Mediterranean ecosystem. *Soil Biology & Biochemistry*, 2005, 37,227-233.

Carneiro, MAC; Siqueira, JO; Davide, AC; Gomes, LJ; Curi, N; Vale, FR. Fungo micorrízico e superfosfato no crescimento de espécies arbóreas tropicais. *Scientia Florestalis*, 1996, 50,21-36.

Chagas e Silva, F; Soares-Silva, LH. Arboreal flora of the Godoy Forest State Park, Londrina, PR. Brazil. *Edinburgh Journal of Botany*, 2000, 57,107-120.

Chapin, FS. The mineral nutrition of wild plants. *Annual Review Ecology System*, 1980, 11, 233-260.

Coley, PD. Gap size and plant defences. *Trends in Ecology and Evolution*, 1993, 8,1-2.

Coley, PD; Bryant, JP; Chapin, FS. *Resource availability and plant anti-herbivore defence.* Science, 1985, 230,895-899.

Coleman, DC; Crossley Jr, DA; Hendrix, PF. *Fundamentals of Soil Ecology*. San Diego: Elsevier Academic Press; 2004.

Comas, LH; Bouma, TJ; Eissenstat, DM. Linking root traits to potential growth rate in six temperate tree species. *Oecologia*, 2002, 132,34-43.

Comas, LH; Eissenstat, DM. Linking fine root traits to maximum potential growth rate among 11 mature temperate tree species. *Functional Ecology*, 2004, 18,388-397.

Cuenca, G; Andrade, Z; Escalante, G. Diversity of glomalean spores from natural, disturbed and revegetated communities growing on nutrient-poor tropical soils. *Soil Biology and Biochemistry*, 1998, 30,711-719.

Degens, BP; Schipper, LA; Sparling, GP; Vojvodic–vukovic, M. Decreases in organic C reserves in soils can reduce the catabolic diversity of soil microbial communities. *Soil Biology and Biochemistry*, 2000, 32,189-196.

Dias, MC; Vieira, AOS; Nakajima, JM; Pimenta, JA; Lobo, PC. Composição florística e fitossociologia do componente arbóreo das florestas ciliares do rio Iapó, na bacia do rio Tibagi, Tibagi, PR. *Revista Brasileira de Botânica*, 1998, 21,183-195.

Eissenstat, DM; Wells, CE; Yanai, RD; Whitbeck, JL. Building roots in a changing environment: implications for root longevity. *New Phytologist*, 2000, 147,33-42.

Eissenstat, DM. Costs and benefits of constructing roots of small diameter. *Journal of Plant Nutrition*, 1992, 15,763-782.

Ferretti, AR; Kageyama, PY; Arbocz, GF; Santos, JD; Barros, MIA; Lorza, RF; Oliveira, C. Classificação das espécies arbóreas em grupos ecológicos para revegetação com nativas no Estado de São Paulo. *Florestar Estatístico*, 1995, 3,73-77.

Fischer, CR; Janos, DP; Perry, DA; Linderman, RG. Mycorrhiza inoculum potentials in tropical secondary succession. *Biotropica*, 1994, 26,369-377.

Föehse, D; Claassen, N; Jungk, A. Phosphorus efficiency of plants II. Significance of root radius, root hairs and cation-anion balance for phosphorus influx in seven plant species. *Plant and Soil*, 1991, 132,261-271.

Gahoonia, TS; Nielsen, NE; Joshi, PA; Jahoor, A. A root hairless barley mutant for elucidating genetics of root hairs and phosphorus uptake. *Plant and Soil*, 2001, 235,211-219.

Gamage, HK; Singhakumara, BMP; Ashton, MS. Effects of light and fertilization on arbuscular mycorrhizal colonization and growth of tropical rain-forest *Syzygium* tree seedlings. *Journal of Tropical Ecology*, 2004, 20,525-534.

Gandolfi, S; Leitão-Filho, HF; Bezerra, CLF. Levantamento florístico e caráter sucessional das espécies arbustivo-arbóreas de uma floresta mesófila semidecídua do município de Guarulhos, SP. *Revista Brasileira de Biologia*, 1995, 55,753-767.

Gilroy, S; Jones, DL. Through form to function: root hair development and nutrient take up. *Trends in Plant Science*, 2000, 5,56-60.

Giovannetti, M; Sbrana, C; Avio, L; Citwernesi, AS; Logi, C. Differential hyphal morphogenesis in arbuscular mycorrhizal fungi during pre-infection stages. *New Phytologist*, 1996, 133,65-71.

Graham, JH; Syvertsen, JP. Host determinants of mycorrhizal dependency of citrus rootstock seedlings. *New Phytologist*, 1985, 101,667-676.

Grayston, SJ; Vaughan, D; Jones, D. Rhizosphere carbon flow in trees, in comparison with annual plants: the importance of root exudation and its impact on microbial activity and nutrient availability. *Applied Soil Ecology*, 1996, 5,29-56.

Gobran, GR; Clegg, S; Courchesne, F. Rhizospheric processes influencing the biogeochemistry of forest ecosystems. *Biogeochemistry*, 1998, 42,107-120.

Guariguata, MR; Ostertag, R. Neotropical secondary forest sucession: changes in structural and functional characteristics. *Forest Ecology and Management*, 2001, 148,185-206.

Hawkins, HJ; George, E. Reduced N-15-nitrogen transport through arbuscular mycorrhizal hyphae to Triticum aestivum L. supplied with ammonium vs. nitrate nutrition. *Annals of Botany*, 2001, 87,303-311.

Hetrick, BAD; Wilson, GWT; Todd, TC. Relationships of mycorrhizal symbiosis, root strategy, and phenology among tallgrass prairie forbs. *Canadian Journal of Botany*, 1992, 70,1521-1528.

Hodge, A. 2004. The plastic plant: root responses to heterogeneous supplies of nutrients. *New Phytologist*, 2004, 162,9-24.

Itoh, S; Barber, SA. Phosphorus uptake by six plant species as related to root hairs. *Agronomy Journal*, 1983, 75,457-461.

Jakobsen, I. Transport of phosphorus and carbon in VA Mycorrhizas. In: Varma A, Hock B editors. *Mycorrhiza: Structure, Function, Molecular Biology and Biotechnology*. Berlin: Springer-Verlag; 1995; 297-324.

Janos, DP. 1983. Tropical mycorrhizas, nutrient cycles and plant growth. In: Sutton SL, Whitmore TC, Chadwick AC editors. *Tropical Rain Forest: Ecology and Management*. Oxford: Blackwell Scientific Publications; 1983; 327-345.

Jasper, DA; Abbott, LK; Robson, AD. The effect of soil disturbance on vesicular–arbuscular mycorrhizal fungi in soils from different vegetation types. *New Phytologist*, 1991, 118,471-476.

Johnson, NC; Wedin, DA. Soil carbon, nutrients, and mycorrhizae during conversion of dry tropical forest to grassland. *Ecological Applications*, 1997, 7,171-182.

Jones, DL; Hodge, A; Kuzyakov, Y. Plant and mycorrhizal regulation of rhizodeposition. *New Phytologist*, 2004, 163,459-480.

Kageyama, PY. Recomposição da vegetação com espécies arbóreas nativas em reservatórios de usinas hidroelétricas da CESP. *Série Técnica IPEF (Piracicaba)*, 1992, 8,1-43.

Koide, RT. Nutrient supply, nutrient demand and plant response to mycorrhizal infection. *New Phytologist*, 1991, 117,365-386.

Khurana, E; Singh, JS. Impact of life-history traits on response of seedlings of five tree species of tropical dry forest to shade. *Journal of Tropical Ecology*, 2006, 22, 653-661.

Kitajima, K. Relative importance of photosynthetic traits and allocation patterns as correlates of seedling shade tolerance of 13 tropical trees. *Oecologia*, 1994, 98,419-428.

Leitão-Filho, HF. Ecologia da mata atlântica em Cubatão. São Paulo: Editora da UNESP-UNICAMP; 1983.

Leuschner, C; Harteveld, M; Hertel, D. Consequences of increasing forest use intensity for biomass, morphology and growth of fine roots in a tropical moist forest on Sulawesi, Indonesia. *Agriculture, Ecosystems and Environment*, 2009, 129,474-481.

Li, XL; George, E; Marschner, H. Extension of phosphorus depletion zone in VA mycorrhizal white clover in a calcareous soil. *Plant and Soil*, 1991, 136,41-48.

Linderman, RG. VA (vesicular-arbuscular) mycorrhizal simbiosis. *Atlas Sci. Anim. Plant Sci.*, 1988, 1,183-188.

Lovelock, CE; Andersen, K; Morton, JB. Influence of host tree species and environmental variables on arbuscular mycorrhizal communities in tropical forests. *Oecologia*, 2003, 135,268-279.

Lugo, AE; Helmer, E. Emerging forest on abandoned land: Puerto Rico's new forests. *Forest Ecology and Management*, 2004, 190,145-161.

Lusk, CH; Reich, PB; Montgomery, RA; Ackerly, DD; Cavender-Bares, J. Why are evergreen leaves so contrary about shade? *Trends in Ecology and Evolution*, 2008, 23,299-303.

Lynch, JP; Ho, MD. Rhizoeconomics: carbon costs of phosphorus acquisition. *Plant and Soil*, 2005, 269,45-56.

Manjunath, A; Habte, M. Root morphological characteristics of host species having distinct mycorrhizal dependency. *Canadian Journal of Botany*, 1991, 69,671-676.

Marschner, H. Role of root growth, arbuscular mycorrhiza, and root exudates for the efficiency in nutrient acquisition. *Field Crops Research*, 1998, 56,203-207.

Matsumoto, LS; Martines, AM; Avanzi, MA; Albino, UB; Brasil, CB; Saridakis, DP; Rampazo, LGL; Zangaro, W; Andrade, G. Interactions among functional groups in the cycling of, carbon, nitrogen and phosphorus in the rhizosphere of three successional species of tropical woody trees. *Applied Soil Ecology*, 2005, 28,57-65.

Merryweather, J; Fitter, A. Phosphorus nutrition of an obligately mycorrhizal plant treated with the fungicide benomyl in the field. *New Phytologist*, 1996, 132,307-311.

Miller, RM; Jastrow, JD. Vesicular-arbuscular mycorrhizae and biogeochimical cycling. In: Pfleger FL, Linderman RG editors. *Mycorrhizae and Plant Health*. Minnesota: APS Press; 1994;189-212.

Modjo, HS; Hendrix, JW; Nesmith, WC. Mycorrhizal fungi in relation to control of tobacco stunt disease with soil fumigants. *Soil Biology and Biochemistry*, 1987, 19,289-295.

Montagnini, F; Sancho, F. Nutrient budgets of young plantations with native trees: strategies for sustained management. In: Bentley W, Gowen M editors. *Forest Resources and Wood-based Biomass Energy as Rural Development Assets*. New Delhi: Oxford & IBH Publishing; 1994; 213-233.

Neumann, E; George, E. Extraction of extraradicular arbuscular mycorrhizal mycelium from compartments filled with soil and glass beads. *Mycorrhiza*, 2005, 15,533-537.

Nielsen, KL; Bouma, TJ; Lynch, JP; Eissenstat, DM. Effects of phosphorus availability and vesicular-arbuscular mycorrhizas on the carbon budget of common bean (*Phaseolus vulgaris*). *New Phytologist*, 1998, 139,647-656.

Pasqualini, D; Uhlmann, A; Stürmer, SL. Arbuscular mycorrhizal fungal communities influence growth and phosphorus concentration of woody plants species from the Atlantic rain forest in South Brazil. *Forest Ecology and Management*, 2007, 245,148-155.

Patreze, CM; Cordeiro, L. Nodulation, arbuscular mycorrhizal colonization and growth of some legumes native from Brazil. *Acta Botanica Brasilica*, 2005, 19,527-537.

Perry, DL; Molina, R; Amaranthus, MP. Mycorrhizae, mycorrhizospheres, and reforestation: current knowledge and research needs. *Canadian Journal of Forest Research*, 1987, 17,929-940.

Peterson, RL; Farquhar, ML. Root hairs: specialized tubular cells extending root surfaces. *Botanical Review*, 1996, 62,1-40.

Picone, C. Diversity and abundance of arbuscular-mycorrhizal fungus spores in tropical forest and pasture. *Biotropica*, 2000, 32,734-750.

Plenchette, C; Fortin, JA; Furlan, V. Growth response of several plants species to mycorrhiza in a soil of moderate P fertility. I. Mycorrhizal dependency under field conditions. *Plant and Soil*, 1983, 70,191-209.

Pouyú-Rojas, E; Siqueira, JO. Micorriza arbuscular e fertilização do solo no desenvolvimento pós-transplante de mudas de sete espécies florestais. *Pesquisa Agropecuária Brasileira*, 2000, 35,103-114.

Powers, JS; Treseder, KK; Lerdau, MT. Fine roots, arbuscular mycorrhizal hyphae and soil nutrients in four neotropical rain forests: patterns across large geographic distance. *New Phytologist*, 2005, 165,913-921.

Rabatin, SC; Stinner, BR. Indirect effects of interactions between VAM fungi and soil-inhabiting invertebrates on plant processes. *Agriculture, Ecosystem & Environment*, 1988, 24,135-146.

Raghothama, KG; Karthikeyan, AS. Phosphate acquisition. *Plant and Soil*, 2005, 274,37-49.

Reinhardt, DR; Miller, RM. Size classes of root diameter and mycorrhizal fungal colonization in two temperate grassland communities. *New Phytologist*, 1990, 116,129-136.

Reich, PB; Tjoelker, MG; Walters, MB; Vanderklein, DW; Buschena, C. Close association of RGR, leaf and root morphology, seed mass and shade tolerance in seedlings of nine boreal tree species grown in high and low light. *Functional Ecology*, 1998, 12,327-338.

Salis, SM; Tamashiro, JY; Joly, CA. Florística e fitossociologia do estrato arbóreo de um remanescente de mata ciliar do rio Jacaré-Pepira, Brotas, SP. *Revista Brasileira de Botânica*, 1994, 17,93-103.

Sanders, IR; Clapp, JP; Wiemken, A. The genetic diversity of arbuscular mycorrhizal fungi in natural ecosystems - a key to understanding the ecology and functioning of the mycorrhizal symbiosis. *New Phytologist*, 1996, 133,123-134.

Schweiger, PF; Robson, AD; Barrow, N. Root hair length determines beneficial effect of a *Glomus* species on shoot growth of some pasture species. *New Phytologist*, 1995, 131,247-254.

Séguin, V; Gagnon, C; Courchesne, F. Changes in water extractable metals, pH and organic carbon concentrations at the soil-root interface of forest soils. *Plant and Soil*, 2004, 260,1-17.

Silver, WL; Scatena, FN; Johnson, AH; Siccama, TG; Watt, F. At what temporal scales does disturbance affect below-ground nutrient pools? *Biotropica*, 1996, 28,441-457.

Siqueira, JO; Carneiro, MAC; Curi, N; Rosado, SCS; Davide, AC. Mycorrhizal colonization and mycotrophic growth of native woody species as related to successional groups in southeastern Brazil. *Forest Ecology and Management*, 1998, 107,241-252.

Siqueira, JO; Saggin-júnior, OJ. Dependency on arbuscular mycorrhizal fungi and responsiveness of some Brazilian native woody species. *Mycorrhiza*, 2001, 11,245-255.

Smith, SE; Read, DJ. *Mycorrhizal symbiosis*. 3rd ed. London: Academic Press; 2008.

Stürmer, SL; Klauberg Filho, O; Queiroz, MH; Mendonça, MM. Occurrence of arbuscular mycorrhizal fungi in soils of early stages of a secondary succession of Atlantic Forest in South Brazil. *Acta Botanica Brasilica*, 2006, 20,513-521.

Sylvia, DM; Williams, SE. Vesicular-arbuscular mycorrhizae and environmental stress. *Mycorrhizae in Sustainable Agriculture*. In: Bethlenfalvay GJ, Linderman RG editors. ASA Special Publication Number 54; 1992; 101-124.

Uhl, C; Jordan, C; Clark, K; Clark, H; Herrera, R. Ecosystem recovery in Amazon caatinga forest after cutting and burning, and bulldozer clearing treatment. *Oikos*, 1982, 38,313-320.

Vandresen, J; Nishidate, FR; Torezan, JMD; Zangaro, W. Inoculação de fungos micorrízicos arbusculares e adubação na formação e pós-transplante de mudas de cinco espécies arbóreas nativas do sul do Brasil. *Acta Botanica Brasilica*, 2007, 21,753-765.

Vitousek, PM. Litterfall, nutrient cycling, and nutrient limitation in tropical forests. *Ecology*, 1984, 65,285-298.

Wright, I J; Westoby, M. Differences in seedling growth behavior among species: trait correlations across species, and trait shifts along nutrient compared to rainfall gradients. *Journal of Ecology*, 1999, 87,85-97.

Zangaro, W; Andrade, G. Micorrizas arbusculares em espécies arbóreas nativas da bacia do rio Tibagi. In: Medri ME, Bianchini E, Pimenta JA, Shibata O editores. *A bacia do rio Tibagi*. Londrina: Edição dos editores; 2002; 171-210.

Zangaro, W; Assis, RL; Motta, AM; Rostirola, LV; Souza, PB; Gonçalves, MC; Andrade, G; Nogueira, MA. Arbuscular mycorrhizal association and fine root traits changes during succession in southern Brazil. *Mycorrhiza*, 2008, 19,37-45.

Zangaro, W; Bononi, VLR; Trufen, SB. Mycorrhizal dependency, inoculum potential and habitat preference of native woody species in South Brazil. *Journal of Tropical Ecology*, 2000, 16,603-622.

Zangaro, W; Nisizaki, SMA; Domingos, JCB; Nakano, EM. Micorrizas arbusculares em espécies arbóreas da bacia do rio Tibagi, Paraná. *Cerne*, 2002, 8,77-87.

Zangaro, W; Nisizaki, SMA; Domingos, JCB; Nakano, EM. Mycorrhizal response and sucessional status in 80 woody species from south Brazil. *Journal of Tropical Ecology*, 2003, 19,315-324.

Zangaro, W; Nishidate, FR; Camargo, FRS; Romagnoli, GG; Vandresen, J. Relationships among arbuscular mycorrhizas, root morphology and seedling growth of tropical native woody species in southern Brazil. *Journal of Tropical Ecology*, 2005, 21,529-540.

Zangaro, W; Nishidate, FR; Vandresen, J; Andrade, G; Nogueira, MA. Root mycorrhizal colonization and plant responsiveness are related to root plasticity, soil fertility and successional status of native woody species in southern Brazil. *Journal of Tropical Ecology*, 2007, 23,53-62.

Zangaro, W; Nogueira, MA; Andrade, G. Arbuscular mycorrhizal fungi used as biofertilizers in revegetation programmes. In: Rai M editor. *Advances in fungal biotechnology*. New Delhi: I.K. International Publishing House; 2009; 351-378.

Zangaro, W; Moreira, M. Micorrizas arbusculares nos biomas floresta atlântica e floresta de araucária. In: Siqueira, JO; Souza FA; Cardoso, EJBN; Tsai, SM, editors. *Micorrizas: trinta anos de pesquisa no Brasil*. Brasília: Editora UFLA; 2010; 279-310.

Zangaro, W; Alves, RA; Lescano, LEAM; ansanelo AP; Nogueira, MA. Investment in fine roots and arbuscular mycorrhizal fungi decrease during succession in three Brazilian ecosystems , Biotropica 2011; (*in press*).

In: Mycorrhiza: Occurrence in Natural and Restored…
Editor: Marcela Pagano

ISBN: 978-1-61209-226-3
© 2012 Nova Science Publishers, Inc.

Chapter 4

ARBUSCULAR MYCORRHIZAL FUNGI IN THE AMAZON REGION

Clara Patricia Peña-Venegas[*]
[1]Instituto Amazónico de Investigaciones Científicas Sinchi,
Av. Vasquez Cobo Calles 15 y 16 Leticia, Amazonas, Colombia

ABSTRACT

The Amazonia is the highest continuous tropical rain forest ecosystem with an area of approximately 7.413.827 Km^2. Seventy percent of the soils of the Amazon basin are Oxisols and Ultisols. They are characterized by being old, highly weathered and leached, with low pH (3-5.5), poor cation exchange capacity and limited availability of phosphorus content. Due to these particular conditions, plants have developed strategies to supply nutrient limitations such as mycorrhizas. Mycorrhizal community in the Amazon region is dominated by Arbuscular mycorrhizal fungi (AMF). *Glomus* is the dominant genus. AMF occupies the first 30cm of the topsoil in which colonize superficial fine roots of plants. This distribution is related to root distribution and nutrient availability into the soils. According to spore description high AMF diversity exists in the region. Low pH of the soil and soil phosphorus availability do not affect mycorrhizal colonization capacity. Effective AM symbiosis, quantified by the number of arbuscules, might be related to specific soil phosphorus ranges. More arbuscules and therefore a better capacity of phosphorus mobilization to the plant has been observed in soils with available phosphorus concentrations between 15-20ppm. AM symbiosis occurs across the region from natural to anthropic ecosystems. AMF diversity and spore abundance change according to soil intervention. Highest diversity and low spore numbers occur in natural forest ecosystems. No distinguishable patterns exist to anthropic environments. Number of spores, arbuscular mycorrhizal fungi diversity, and % colonization varies depending on soil management, vegetation composition and physical and chemical conditions of the soil.

Keywords: Arbuscular mycorrhiza, Amazonia, AMF community, symbiosis, tropical rain forest

*E-mail: cpena@sinchi.org.co

Amazonia is the largest area of continuous tropical forest. It occupies an extension of 7.413.827 Km^2 (ACTO, 2008). This patch of forest, reservoir of a huge biodiversity, acts as the biggest reserve of carbon and the biggest supplier of water to the Atlantic Ocean (it supplies the15-20% of the total drained water). That is why the Amazonia is considered the main world climate regulator.

Although the Amazon has an exuberant richness of vegetation, 70% of it soils have nutritional limitations due to their formation and prolonged periods of weathering and leaching experimented. Amazonian soils can be classified as Oxisols and Ultisols (Soil Survey Staff 1975) in which iron and aluminium are important elements in the composition of clays. Because of weathering conditions, these soils are acidic (pH 3-5.5), with limited basic cations amounts, low cation exchange capacity and very low phosphorous availability. Only the varzea occupying 7% of the total Amazon basin (527.000 Km^2) has moderate fertile conditions. Its fertility is the result of periodical deposition of rich sediments coming from Andean soils and transported by rivers which cross the Amazon region before reaching the Atlantic Ocean.

Exuberant forest over Amazonian unfertile soils only could be possible by the establishment of effective associations between edaphic micro organisms and plants. The widespread plant-micro organism symbiosis in the Amazonia is arbuscular mycorrhiza. More than 80% of the Amazonian indigenous vegetation establishes this symbiosis (Moyersoen 1993; Alves da Silva *et al.* 2001; Harrier 2001; Peña-Venegas *et al.* 2006). Few non-mycorrhizal plants exist in the Amazon region. Most of them are weeds with short cycle life as well as some species of the Lecythidaceae and Sapotaceae families (St John 1980). Ectomycorrhizas occur in 17-31% of Amazonian indigenous plants (Moyersoen 1993).

The reason why arbuscular mycorrhizal fungi dominate the Amazon forest is not well understood. It had been observed that ectomycorrhizal fungi often dominate low-diverse boreal and temperate forests in which they establish specific relationships with their host plants. In contrast, arbuscular mycorrhiza often dominates high-diversity tropical forest (Janos 1992).

Arbuscular mycorrhizal community composition in natural and disturbed Amazonian soils is characterized by the dominance of *Glomus* (Cáceres 1989; Siqueira *et al.* 1990; Caproni *et al.* 2003; Posada *et al.* 2006; Peña-Venegas *et al.* 2007, Stürmer & Siqueira 2008), or *Acaulospora* (Lopes *et al.* 2009). In any case, the dominant genus represents more than 50% of the total arbuscular mycorrhizal community. Common dominance of *Glomus* and *Acaulospora* might be the consequence of an antique co-evolution between flora and Glomeromycota phylum. There is enough evidence that these two genera of arbuscular mycorrhizal fungi are two of the most antique (De Souza *et al.* 2008), and their dominance could represent non modified antique relationships between plants and fungi (Helgason & Fitter 2005).

The study of arbuscular mycorrhizal fungi (AMF) diversity from natural soils is difficult. Clapp *et al.* (2002) describes the three main processes used with this purpose: 1) The direct extraction of spores from the soil for identification; 2) Isolation of spores from trap cultures and their identification; and 3) Molecular techniques based on extraction, amplification and characterization of nucleic acids from spores or colonized plant roots.

Results showed that any of three methods can not reveal 100% of the AMF diversity: AMF identification from spores obtained directly from the soil might underestimate AMF diversity because spores are not produced constantly (Oliveira & Oliveira 2005), and AMF species might not produce spores or produce spores in low numbers. Better estimation of AMF diversity can be done from trap cultures. This methodology can provide healthy, young spores in enough numbers. Better results can be obtained if the preparation of inoculum for trap cultures uses spores and plant roots (De Souza *et al.* 2008). However, personal experience showed that rare spores are difficult to multiply in trap cultures. Similar difficulties have been expressed by Dr. Cuenca from the Instituto Venezolano de Investigaciones científicas IVIC (pers. com.) with morphotypes from the Gran Sabana-Venezuela. According to INVAM (2009) some indigenous fungi sporulate after 2-3 successive propagation cycles. It is also important to say that not all the AMF species are able to be multiplied in this way (Clapp *et al.* 1995).

Differences on spore multiplication have been observed when different plant hosts are used. Common trap plants used in spore multiplication are mycotrophic C3 plants. The International Culture Collection of VA Mycorrhizal Fungi - INVAM uses Sudangrass (*Sorghum bicolor*), for their AMF multiplications but they affirm that sometimes they have to change the host plant for another non-related with Sudangrass to estimulate multiplication (INVAM 2009). Other host plants commonly used are Bahiagrass (*Paspalum notatum*), Guineagrass (*Panicum maximum*), Cenchrusgrass (*Cenchrus ciliaris*), clover (*Trifolium subterraneum*), strawberry (*Fragaria* sp.), sorghum (*Sorghum vulgare*), corn (*Zea mays*), onion (*Allium cepa*), coprosma (Coprosma robusta), coleus (*Coleus* sp.), Rhodesgrass (*Chloris gayana*). For tropical collections the Instituto Venezolano de Investigaciones Científicas - IVIC from Caracas-Venezuela uses the legume *Vigna luteola*, and the Centro de Agrocultura Tropical - CIAT from Palmira-Colombia uses the legume Kudzu (*Pueraria lobata*) as host plant for trap cultures. Kudzu, Brachiariagrass (*Brachiaria decumbens*) and corn have been used as host plants for the multiplication of native Amazonian AMF morphotypes by the Instituto Amazonico de Investigaciones Científicas Sinchi with moderate successfulness.

Molecular techniques for AMF diversity estimation are useful to known communities of fungi colonizing one specific host plant without taking into account if a species sporulates or not. Although approximation to community composition is more readable by this method, there are difficulties on the clear identification of AMF species. Only from 80´s molecular techniques are popular to study AMF, therefore not all the methods described on literature are accurate to isolate good DNA and obtain unique bands to identify genus and species. Some of the problems are the impossibility to obtain enough and clean DNA from spores or roots, the wrong selection of primers for the PCR which do not permit accurate discrimination between morphotypes, and the poor information available on international genomic databases from AMF species coming from tropical regions including Amazonia.

One good example of this is the study developed by Leon (2006). She studied native AMF communities in cassava (*Manihot esculenta*) roots from the Colombian Amazon region by molecular techniques. In her work she used 14 different methods reported for AMF DNA extraction: DiBonito *et al.* 1995; Sanders *et al.* 1995; Dodd *et al.* 1996; Redecker *et al.* 1997; Antoniolli *et al.* 1998; Tuinen *et al.* 1998; Manian & Sreenivasaprasad 2001; Kowalchuk *et al.* 2002; Öpik 2004; Gadkar *et al.* 2006; the kit ultraclean DNA soil from MO BIO Laboratories, a personal method from AMF DNA extraction from one spore suggested by De Souza (pers. com.); a personal method from AMF DNA extraction from 10-60 spores suggested by De Souza (pers. com.); and a method cited by CTAB. No successful results were obtained using the former methods.

A second alternative was to extract DNA directly from cassava roots. She obtained DNA using the ultraclean DNA soil kit from MO BIO Laboratories. Amplification of DNA was obtained using Sanders *et al.* (1995) method. The amplifications were cloned and an ADRA pattern analysis was done. From 15 sequences cloned only 2 of them were compatible with known sequences reported on the GenBank database.One was related to a Glomus isolated from a *Prunus Africana* in Ethiopia (Wubet *et al.* 2003) and the other was related to a Gigaspora isolated from a *Inga parteno* from a disturbed soil of Costa Rica (Shepherd *et al.* 2007).

Most of the reports of AMF richness from the Amazon region are based on AMF spore description. The principal reason is that the spore of arbuscular mycorrhizal fungi is the only structure with enough morphological characteristics to differentiate species and although the help of a mycologist expert on AMF taxonomy is necessary, the procedure is cheap. It is also important to notice that by the use of spore description not always is it possible to identify 100% of the materials because of the reasons exposed before in this chapter. Peña-Venegas *et al.* (2006), Stürmer & Siqueira (2008), and Lopes et al. (2009) identified the 58%, 68.5% and 73.9% of the total morphotypes isolated respectively. Even though spore description has limitations to estimate AMF diversity, it is the most used method with this purpose. Molecular approaches still are expensive and results limited.

According to references cited in this chapter, 60 species of AMF (Table 1) have been reported in the Amazonia, which represents 31% of the total number of AMF known until today (194 according to INVAM database http://invam.caf.wvu.edu). It confirms that the Amazonia is not only a high diverse ecosystem in flora and fauna but also it is an important reservoir of micro organisms such as arbuscular mycorrhizal fungi. It is important to take into account that these two genera are also the genera that produce more spores into the soil, therefore observation of its dominance might be biased by the methodology used. However, molecular analysis of AMF colonizing roots of plants coming from traditional agroecosystems in the Colombian Amazon region show the dominace of *Glomus* and *Acaulospora*, supporting results obtained by AMF spore analysis.

For Amazonian countries high AMF diversity in their Amazonian territories should be a strategic condition for the conservation and sustainable use of tropical rain forest: From the total AMF existing in Brazil and known until today (99 species according to Stürmer & Siqueira 2008), 34% are present in the Amazon region. From the total AMF known for Colombia (around 44 according to published bibliography) the Amazon region possesses 77%. Moreover, Stürmer & Siqueira (2008) found that if the arbuscular mycrorrhizal community was compared by the Similarity Index of Sørenson to other ecosystems of Brazil, arbuscular mycorrhizal community from Amazon region diverges from other mycorrhizal communities, indicating a particular diversity of the Amazon region.

It is possible to affirm that native arbuscular mycorrhizal species from Amazonian soils do not have host specificity in the way that any AMF might be capable to colonize any plant. Molecular analysis of roots of cassava (*Manihot esculenta*) to determine species of arbuscular mycorrhizal fungi responsible for root colonization shows that more than one arbuscular mycorrhizal species colonized the roots of the plant (León 2006). Similar results were observed in agroforestry systems and traditional indigenous agroecosystems in which it was not possible to conclude that one arbuscular mycorrhizal species selects one plant from others as a better host plant.

Table 1. Diversity of arbuscular mycrorrhizal fungi in the Amazonia based on spore description

Arbuscular Mychorrizal Fungi Species	Country				Reference cited
	Brazil	Colombia	Perú	Venezuela	
Acaulospora bireticulata Rothwell & Trappe	x				Stürmer & Siqueira 2008; Lopes et al. 2009
Acaulospora delicata Walker, Pfeiffer & Bloss	x				Stürmer & Siqueira 2008; Lopes *et al.* 2009
Acaulospora foveata Trappe & Janos	x	x		x	Moyersoen 1993; Ochoa 1997; Peña-Venegas *et al.* 2006; Lopes et al. 2009
Acaulospora laevis Gerdemann & Trappe	x			x	Moyersoen 1993; Stürmer & Siqueira 2008
Acaulospora mellea Spain & Schenck	x	x		x	Moyersoen 1993; Peña-Venegas et al. 2006; Lopes et al. 2009
Acaulospora morrowiae Spain & Schenck	x	x			Salamanca & Silvia 1998; Peña-Venegas et al. 2006; Lopes et al. 2009
Acaulospora rehmii Sieverding & Toro	x	x			Salamanca & Silvia 1998; Peña-Venegas et al. 2006; Lopes et al. 2009
Acaulospora scrobiculata Trappe	x				Stürmer & Siqueira 2008; Lopes et al. 2009
Acaulospora spinosa Walker & Trappe	x				Stürmer & Siqueira 2008; Lopes et al. 2009
Acaulospora tuberculata Janos & Trappe	x	x			Salamanca & Silvia 1998; Peña-Venegas et al. 2006; Stürmer & Siqueira 2008; Lopes et al. 2009
Acaulospora undulata Sieverding	x				Stürmer & Siqueira 2008
Archeospora leptoticha Morton & Redecker	x	x			Peña-Venegas et al. 2006; Lopes et al. 2009
Archeospora trappei Morton & Redecker	x				Stürmer & Siqueira 2008; Lopes et al. 2009
Entrophospora colombiana Spain & Schenck	x	x			Salamanca & Silvia 1998; Peña-Venegas et al. 2006; Lopes et al. 2009
Entrophospora infrequens Ames & Schneider	x				Lopes et al. 2009
Glomus australe Berch	x				Stürmer & Siqueira 2008

Table 1 (Continued).

Arbuscular Mychorrizal Fungi Species	Country				Reference cited
	Brazil	Colombia	Perú	Venezuela	
Glomus aggregatum Schenck & Smith		x	x		Janos, Sahley & Emmons 1995; Ochoa 1997; Salamanca & Silvia 1998
Glomus brohultii Sieverding		x		x	Moyersoen 1993; Ochoa 1997; Peña-Venegas et al. 2006
Glomus chimonobambusae Wu & Liu	x				Stürmer & Siqueira 2008
Glomus clarum Nicholson & Schenck	x				Lopes *et al.* 2009
Glomus claroideum Schenck & Smith	x				Lopes *et al.* 2009
Glomus clavisporum Almeida & Shenck	x				Stürmer & Siqueira 2008
Glomus coremoides Berk & Broome			x		Janos, Sahley & Emmons 1995
Glomus etunicatum Becker & Gerdemann		x			Salamanca & Silvia 1998
Glomus fasciculatum Gerdemann & Trappe	x	x			Pinto 1993; Ochoa 1997; Stürmer & Siqueira 2008
Glomus fuegianum Trappe & Gerdemann	x				Stürmer & Siqueira 2008
Glomus fulvum Trappe & Gerdemann		x			Ochoa 1997
Glomus gendermanii Rose, daniels & Trappe				x	Moyersoen 1993
Glomus geosporum Walker		x	x		Janos, Sahley & Emmons 1995; Ochoa 1997; Salamanca & Silvia 1998
Glomus glomerulatum Sieverding		x			Ochoa 1997; Peña-Venegas et al. 2006
Glomus intraradices Schenck & Smith	x	x			Peña-Venegas et al. 2006; Stürmer & Siqueira 2008
Glomus invermaium Hall		x			Ochoa 1997
Glomus lacteum Rose & Trappe	x				Stürmer & Siqueira 2008
Glomus macrocarpum Tulasne & Tulasne	x	x	x		Pinto 1993; Janos, Sahley & Emmons 1995; Salamanca & Silvia 1998; Stürmer & Siqueira 2008
Glomus magnicaule Hall	x				Stürmer & Siqueira 2008
Glomus manihotis Howeler, Sieverding & Schenck		x			Peña-Venegas *et al.* 2006
Glomus microaggregatum Koske, Gemma & Olexia		x			Peña-Venegas *et al.* 2006
Glomus microcarpum Gendermann & Trappe			x		Janos, Sahley & Emmons 1995
Glomus mosseae Gerdemann & Trappe		x			Pinto 1993
Glomus multicaule Gerdemann & Bakshi		x			Pinto 1993

Species				Reference
Glomus pachycaulis Wu & Chen		x		Pinto 1993
Glomus rubiforme Almeida & Schenck	x	x		Pinto 1993, Peña-Venegas et al. 2006, Stürmer & Siqueira (2008)
Glomus sinuosum Armedia & Schenck		x		Peña-Venegas et al. 2006
Glomus taiwanensis Almeida & Schenck	x			Stürmer & Siqueira (2008)
Glomus tenebrosum Berch		x		Salamanca & Silvia 1998
Glomus tortuosum Schenck & Smith	x	x		Peña-Venegas et al. 2006, Stürmer & Siqueira (2008)
Glomus viscosum Walker, Giovannetti, Avio, Citernesi & Nicolson		x		Peña-Venegas et al. 2006
Glomus versiforme Berch			x	Moyersoen 1993
Gigaspora albida Schenck & Smith		x		Ochoa 1997
Gigaspora decipiens Hall & Abbott	x			Stürmer & Siqueira (2008)
Gigaspora erythropa Koske & Walker		x		Pinto 1993
Gigaspora gigantea Gerdemann & Trappe		x		Pinto 1993, Ochoa 1997
Gigaspora margarita Becker & Hall	x	x		Pinto 1993, Lopes et al. 2009
Paraglomus occultum Morton & Redecker	x			Stürmer & Siqueira (2008), Lopes et al. 2009
Scutellospora calospora Gerdemann & Trappe			x	Moyersoen 1993
Scutellospora heterogama Gerdemann & Trappe			x	Moyersoen 1993
Scutellospora pellucida Walker & Sanders	x	x		Peña-Venegas et al. 2006, Stürmer & Siqueira (2008)
Scutellospora persica Koske & Walker		x		Salamanca & Silvia 1998
Scutellospora scutata Walker & Diederichs	x			Stürmer & Siqueira (2008)
Scutellospora spinosissima Walker, Cuanca & Sanchez		x		Peña-Venegas et al. 2006

Highly leached Oxisols and Ultisols are characterized by deep unfertile mineral horizons. Over them, a thin Horizon A, no more than 30cm deep, encloses all the fertility these soils have. The fertility of these soils depends on the quality and quantity of available organic matter and the transformations micro organisms do to it. Therefore, most of the root systems of Amazonian indigenous plant species are superficial. By its obligate endosymbiotic condition AMF has the distribution of plant root distribution. It exists around 4 times more spores in the first 20cm of the soil than on the subsequent 20-30cm depth (Peña-Venegas *et al.* 2007). Although significant differences existed on the number of spores versus depth, species richness distribution is not affected.

Low pH of the soil does not affect mycorrhizal colonization capacity. AMF root colonization has been found in plants growing in Colombian Amazon soils between pH 3.5 to 6.4. Although a direct relationship between P availability and AMF colonization have been proposed, it has been shown that the use of P-fertilization does not inhibit arbuscular mycorrhizae formation (Cuenca *et al.* 2003). In the same way phosphorus status condition seems do not affect mycorrhization. One example to illustrate this is presented in figure 1. Percentage root colonization in rubber (*Hevea brasiliensis*) and red pepper (*Capsicum* sp.) from the Amazon was compared according to available phosphorus at soil. No significant differences were obtained in any of the two plants related to differences on phosphorus status.

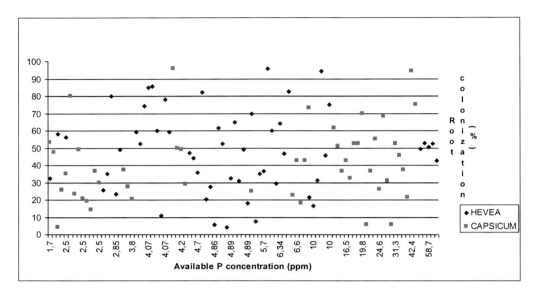

Figure 1. AMF root colonization of two native plants under different available phosphorus concentrations of the soil.

Colonization percentage by itself does not say anything about the efficiency of the symbiosis established. An indirect way to measure it is by the quantification of arbuscules. Arbuscules are the structures in which nutrient exchange between plant and fungi occurs. Therefore, more efficient symbioses are those that present root colonization with a significant number of arbuscules (Collins 1993). Some authors had been shown that presence of arbuscules in roots of Amazonian plants is infrequent (St. John 1980; Restrepo *et al.* 1993; Peña-Venegas *et al.* 2006; Kalinhoff *et al.* 2009) and with low percentage values with respect to total percentage of root colonization.

Particularly low available phosphorus in Amazonian soils and a high frequency of AM symbiosis in Amazonian plants supports that this symbiosis plays an important role in Amazonian ecosystems, although arbuscules are infrequent. Moreover, better root colonization of plants has been observed under non-flooding soils (Tierra firme) than in the varzea when better nutritional soil conditions exist. Peña-Venegas *et al.* (2006) observed for the Colombian Amazon region a higher number of arbuscules under soil available phosphorus concentrations between 15-20ppm, suggesting that specific ranges of phosphorus are related to efficient AMF symbiosis. However, studies currently done are insufficient to support or deny it.

AM symbiosis occurs across the region from natural to anthropic ecosystems. AM symbiosis in natural ecosystems (Primary or secondary forests) is characterized by a low presence of spores into the soil. In undisturbed zones, the main way AMF colonize new host plants is by the expansion of the extra cellular net of mycelia. Although in natural environments the spore number is low, high diversity of AMF exists and rare spores can be found.

AM symbiosis in disturbed ecosystems varies. On indigenous plots of itinerant agriculture Peña-Venegas *et al.* (2007) found that arbuscular mycorrizal symbiosis does not occur in the first months of cultivation because propagules are dramatically reduced; spores are burn and the net of mycelium damaged. After 4 to 6 months adjacent mycorrhized plants and alive pieces of roots of cut trees (Restrepo *et al.* 1993) start to restore extracellular net of mycelium and remaining spores to colonize new host plants. Kalinhoff *et al.* (2009) observed that when the indigenous plots of itinerant agriculture are on production, roots of cultivated species are strongly colonized by AMF and spore numbers in the soil are high.

Oliveira & Oliveira (2005) observed an increased root colonization of copoazu (*Theobroma grandiflorum*) and guarana (*Paullinia cupana*) growing on agroforestry systems in Central Amazonia during the rainy season. At the same time a highest spore number of AMF spores in the soil were observed in the rainy season. Restrepo et al. (1993) observed an increased number of AMF spores in the dry season for Caquetá-northeastern of the Colombian Amazon region, therefore although co-relation between spore number and raining dynamics had been observed other external factors can have more importance in the stimulation of AMF sporulation.

Transformation of natural forest to pastures for cattle ranching in the Amazon region is one of the principal threatens Amazonian countries have to handle. At introduced pastures low AMF diversity exist, suggesting that some species of AMF are selected during the transformation of the forest. Restrepo *et al.* (1993) evaluated native species of AMF obtained from natural forest and proved them in two species of grass: *Brachiaria dyctioneura* and *Centrosema acutifolium*. Results showed a poor adaptation and performance of native AMF when their natural hosts and soil conditions are changed. Posada *et al.* (2006) estimated that only after 10 years of settlement of an introduced pasture, the AMF populations showed some diversification.

There is still little knowledge about AMF in the world, especially in low disturbed ecosystems such as Amazonia. Stressed soil conditions, diverse vegetation and different degrees of ecosystem disturbance can be a great opportunity to understand better how this symbiosis develops and responds under different conditions. It can help people to use this natural resource in an efficient manner. The information presented in this chapter wants to contribute in this way.

REFERENCES

Alves da Silva G., Acioli dos Santos B. A., Alves M.V., Maia L. 2001. Arbuscular mycorrhiza in species of Commelinidae (Liliopsida) in the state of Pernambuco (Brazil). *Acta Botanica Brasilica* 15 (2): 155-165.

Amazonian Cooperation Trate Organization-ACTO. 2008. Geoamazonia project.

Antoniolli, Z.I., Smith, S.E., Ophel-Keller, K., Schachtman, D. P., and Zeze, A. (1998). Assessment of variation in sequences of ribosomal DNA internal transcribed spacers in *Glomus mosseae* and *Gigaspora margarita* from a pasture community in " *Second*

Internacional Conference on Micorriza" (U. AhonenJonnarth, E. Danell, I. Fransson, O. Karen, B. Lindahl, I. Rangel and R. Finlay, eds.)Uppsala.

Cáceres, A. 1989. Las micorrizas vesiculo-arbusculares en un bosque húmedo tropical y su evolución luego de la perturbación (conuco) y la sucesión por 60 años en San Carlos de Río Negro. Tesis de Maestría. *Centro de Estudios Avanzados, Instituto Venezolano de Investigaciones Científicas IVIC,* Caracas, Venezuela, 252pp.

Caproni, A. L.; Franco, A. A.; Berbara, R. L. L.; Truferri, S. B.; De Oliveira, J. R. D.; Monteiro, A. B. 2003. Ocorrência de fungos micorrizicos arbusculares en áreas revegetadas após mineração de bauxita en Porto Trombetas, Pará. *Pesq. Agropec. Bras. Brasilia* 38(12):1409-1418.

Clapp J.P., J.P.W. Young, J. W. Marryweather, A.H. Fitter. 1995. Diversity of fungal symbionts in arbuscular mycorrhizas from natural community. *New Phytologist* 130: 259-265.

Clapp J.P., T. Helgason, T. J. Daniell, J. P. W. Young. 2002. Genetic studies of the structure and diversity of arbuscular mycorrhizal fungal communities. In: *Mycorrhizal ecology. Van Der Heijden and Sanders* (eds.). Springer-Verlag. 201-224.

Collins N. 1993. Can fertilization of soil select less mutualistic mycorrhizae? *Ecological applications* 3(4):749-757

Cuenca, G.; De Andrade, Z.; Lovera, M.; Fajardo, L.; Meneses, E. 2003. Mycorrhizal response of *Clusia pusilla* growing in two different soils in the field. *Trees,* 17: 200-206.

De Souza F. A., Lima da Silva I. C., Berbara L. L. 2008. Fungos micorrízicos arbusculares: muito mais diversos do que se imaginava. Em: Biodiversidade do solo em Ecossistemas brasileiros. Moreira, Siqueira & Brussaard (Eds.). *Universidade Federal de Lavras.* p 483-536.

DiBonito, R., Elliott, M. L., and Des Jardin, E. A. *(*1995*) Detection of arbuscular mycorrhizal fungi in roots of different plant species with PCR.* Applied environmental Microbiology.

Dodd, J., Rosendahl, S. Giovannetti, M., Broome, A., Lanfranco, M. Y Walker, C. 1996. Inter- and intraspecific variation within the morphologically-similar arbuscular mycorrhizal fungi Glomus mosseae and Glomus coronatum. *New Phytol.*133. p 113-122.

Gadkar V., J. D. Driver, M.C. Rillig. 2006. A novel in vitro cultivation system to produce and isolate soluble factors released from hyphae of arbuscular mycorrhizal fungi. *Biotechnology letters* 28: 1071-1076.

Harrier L.A. 2001. The arbuscular mycorrhizal symbiosis: a molecular review of the fungal dimension. *Journal of Experimental Botany* 52: 469-478.

Helgason T., A. Fitter. 2005. The ecology and evolution of the arbuscular mycorrhizal fungi. *Mycologist* 19 (3): 96-101.

International Culture Collection of VA Mycorrhizal Fungi – INVAM. 2009 http://invam.caf.wvu.edu.

Janos, D. P. 1992. Heterogeneity and scale in tropical vesicular–arbuscular mycorrhiza formation. *In* D. H. Read, D. H. Lewis, A. H. Fitter, and I. J. Alexander (Eds.). *Mycorrhizas in ecosystems,* pp. 276–282. CAB International, Wallingford, England.

Janos D.P., Sahley C. T., Emmons L. H. 1995. Rodent dispersal of vesicular-arbuscular mycorrhizal fungi in Amazonian Peru. *Ecology* 76 (6): 1852-1858.

Kalinoff C., Cáceres A., Lugo L. 2009. Cambios en la biomasa de raíces y micorrizas arbusculares en cultivos itinerantes del Amazonas Venezolano. *Interciencia* 34 (8): 571-576

Kowalchuk G. A., F. A. De Souza, J. A. Van Veen. 2002. Community análisis of arbuscular mycorrhizal fungi associated with Ammophila arenaria in Dutch coastal sand dunes. *Molecular ecology* 11: 571-581.

León D. 2006. Evaluación y caracterización de micorrizas arbusculares asociadas a yuca (*Manihot esculenta*) en dos regiones de la Amazonia colombiana. *Tesis de Microbiología Agrícola y Veterinaria*. Facultad de Ciencias. Pontificia Universidad Javeriana. 127pp.

Lopes P., Stürmer S.L., Siqueira J. O. 2009. Occurence and diversity of arbuscular mycorrhizal fungi in trap cultures from soils under different land use systems in the Amazon, Brazil. *Brazilian Journal of Microbiology* 40: 111-121.

Manian, S., Sreenivasaprasad, Mills, P.R. 2001. DNA extraction method for PCR in mycorrhizal fungi. *Letters in Applied Microbiology:* 33.p.307-310.

Moyersoen, B. 1993. Ectomicorrizas y micorrizas vesiculo-arbusculares en Caatinga Amazónica del Sur de Venezuela. *Scientia Guaianae,* 3: 82 pp.

OCHOA, O. C. 1997. Reconocimiento de hongos formadores de micorriza vesiculo-arbuscular (MVA) en cacao (*Theobroma cacao*), Maraco (*T. bicolor*) y copoazu (*T. grandiflorum*), en condiciones de campo en San José del Guaviare - Colombia. Departamento de Biología. *Universidad Nacional de Colombia.* Tesis de Pregrado. 96p.

Oliveira A. N., Oliveira L. A. 2005. Seasonal dynamics of arbuscular mycorrhizal fungi in plants of Theobroma grandiflorum Schum and Paullinia cupana Mart. of an agroforestry system in Central Amazonia, Amazonas State, Brazil. *Brazilian Journal of Microbiology* 36 (3): 262-270

Öpik, M. 2004. Diversity of arbuscular micorrhizal fungi in the roots of perennial plants and their effect on plant performance. *Dissertationes Biologicae Universittatis Tartuensis.* Tartu. pp 80.

Peña-Venegas C. P., Cardona G. I., Mazorra, A. 2006. *Micorrizas Arbusculares de la Amazonia colombiana. Catálogo Ilustrado.* Instituto Amazónico de Investigaciones Científicas SINCHI. 90pp.

Peña-Venegas C. P., G. I. Cardona, J. H. Arguelles, A. L. Arcos. 2007. Micorrizas arbusculares del sur de la Amazonia colombiana y su relación con algunos factores fisicoquímicos y biológicos del suelo. *Acta Amazónica* 37 (3): 327-336.

Pinto J. B. 1993. Evaluación de las poblaciones micorrizales en suelos degradados y de bosque maduro en Araracuara - Amazonas. Departamento de Biología. *Universidad Nacional de Colombia. T*esis de grado. 151p.

Posada R. H., Franco L. A. 2006. El tiempo de establecimiento de pasturas y su relación con la micorriza arbuscular en paisajes de loma y vega. *Acta Biológica Colombiana* 11: 55-64

Redecker, D., Thierfelder, H., Walker, C Y Werner, D. 1997. Restriction Analysis of PCR-Amplified Internal Transcribed Spacers of Ribosomal DNA as a Tool for Species Identification in Different Genera of the Order Glomales. *Applied and Environmental Microbiology.* Vol 63.No 5.p 1756-1761.

Restrepo M. C., J. E.Martinez, J. E. Montenegro, A. Caicedo, E. Torres. 1993. Análisis sobre la actividad de hongos formadores de micorrizas vesículo arbusculares. *En:* Aspectos ambientales para el ordenamiento territorial del occidente del departamento del Caquetá.

Estudios en la Amazonia colombiana. Instituto Geográfico Agustín Codazzi - IGAC-. p 698-736.

SalamancaC. R., M. R Silva. 1998. *Las micorrizas como alternativa para el manejo sostenible de los agroecosistemas tropicales.* Boletín Técnico No. 12. CORPOICA. 26p.

Sanders I.R. , Alt M., Groppe K., Boller T., and Wiemken A. 1995. Identification of ribosomal DNA polymorphisms among and within spores of the Glomales: application to studies on the genetic diversity of arbuscular mycorrhizal fungal communities. *New Phytologist* 130 (3): 419-427.

Shepherd M., Nguyen L., Jones M. E.,Nochols J. D., Carpenter F.L. 2007. A method for assessing arbuscular mycorrhizal fungi group distribution in tree roots by intergenic transcribed sequence variation. Plant and Soil 290: 259-268.

Siquiera J.O., Rocha W.F., Oliveira E., Colozzi-Filho A. 1990. The relationship between vesicular-arbuscular mycorrhiza and lime: Associated effects on the growth and nutrition of Brachiaria grass. *Biol. Fertil. Soils* 10 (1): 65-71.

St. John T. V. 1980. A survey of mycorrizal infection in an amazonian rain forest. *Acta Amazonica* 10(3): 527-533.

Stürmer S. L., Siqueira J. O. 2008. Diversidade de fungos micorrízicos arbusculares em Ecossistemas brasileiros. Em: *Biodiversidade do solo em Ecossistemas brasileiros. Moreira, Siqueira & Brussaard* (Eds.). Universidade Federal de Lavras. p 537-583.

Tuinen, D., Jacquot, E., Zhao, B., Gollotte, A., Gianinazzi-Pearson, V. 1998. Characterization of root colonization profiles by a microcosm community of arbuscular mycorrhizal fungi using 25S rDNA-targeted nested PCR. *Molecular Ecology* 7(7), 879-887.

Wubet T., Weiss M., Kottke I., Teketay D., Oberwinkler F. 2003. Molecular diversity of arbuscular mycorrhizal fungi in *Prunus africana*, an endangered medicinal tree species in dry Afromontane forest of Ethiopia. *New Phytologist* 161: 517-528.

In: Mycorrhiza: Occurrence in Natural and Restored… ISBN: 978-1-61209-226-3
Editor: Marcela Pagano © 2012 Nova Science Publishers, Inc.

Chapter 5

OCURRENCE OF MYCORRHIZAS
IN HIGHLAND FIELDS

Marcela C. Pagano and Marta Noemi Cabello[2]*
[1]Federal University of Ceará, Fortaleza, Ceará, Brazil,
[2]Instituto Spegazzini, Researcher from CIC, National University of La Plata, Argentina

ABSTRACT

The southeastern Brazilian highlands are centers of endemism and diversity; however, outcrop plant communities still lack systematic studies. Research in these areas, which are subject to mining, and where metal-tolerant plant species may exist, is essential to successful restoration. The aim of this review is to explore current information on the occurrence of mycorrhizas in outcrops, and to speculate about their mutualistic interactions. Root colonization in the studied Brazilian plant species as well as the arbuscular mycorrhizal (AM) spore diversity found in their rhizospheres are illustrated. The relevant findings are emphasized, such as the presence of vesicles and auxilliary cells in most roots. Occurrence of different types of AM fungi from six outcrop plants belonging to 6 families and 6 genera are presented. The associations of the fungi with different types of host plants as well as the soil properties are discussed. The mycorrhizal status of two angiosperms is reported for the first time. Highland field's plant species contains natural AM fungal species richness that can be affected by mining, and this supplies important information for restoration programs. Moreover, AM dependent plants in natural and restored highland field ecosystem, like *Eremanthus incanus*, can present higher potential for regeneration in habitats subject to disturbance. Research directions needed to increase understanding of mycorrhizal associations in these environments are indicated. The chapter discusses benefits and problems encountered, in order to highlight the need for a continual and integrated study of the highland ecosystems.

Keywords: Arbuscular mycorrhizal fungi; native species; Highlands; outcrops; root colonization

---

* marpagano@gmail.com pagano@netuno.lcc.ufmg.br

INTRODUCTION

Plant communities on rock outcrops remain some of the best preserved terrestrial habitats. However, they are rarely included in floristic inventories due to the difficulties of access or to low economic interest. Moreover, their potential for solving still unanswered questions is still to be explored as ecophysiology, phytosociology, ecology, evolution, floristic, biogeography and conservation could be better understood in these habitats (reviewed by Scarano 2007).

In general, endemism may be related to the degree of isolation of the rock outcrop in relation to similar habitats and also to how intensely the ecological conditions on the rock surface represent a barrier for establishment of plants from the surrounding vegetation (Burke 2002).

Rock outcrops frequently support a very specialized vegetation (Burke 2002) due to a combination of factors such as low water retention, scarcity of nutrients, difficulty for seed retention and germination, and increased exposure to winds and insolation (Larson et al. 2000). In rupestrian landscapes, soils are shallow, acid, nutrient-poor, and have excessively drained sands that are highly erodible (reviewed by Scarano 2007).

A detailed study by Porembski (2007) describing the plant cover of inselbergs (isolated rock outcrops that rise abruptly above the surrounding plains) which bear a flora rich in endemics showed three hot spots of global inselberg plant diversity: a) southeastern Brazil, b) Madagascar and c) southwestern Australia. He stressed the high beta diversity (degree of floristic differentiation over small distances) of the inselberg saxicolous vegetation rich in dry resistant perennial species, relatively rare annuals, xerophytic and succulent bromeliads, cacti and orchids.

The southeastern Brazilian Highlands have shrubby, tortuous and sclerophyllous vegetation or open grasslands, which replace the savanna vegetation at 1,000 m altitude. Plants grow in stones, in sandy soils and present varied adaptations (Rizzini 1997). Mori (1989) termed "campo rupestre" the savanna-like vegetation over 600 m elevation in Bahia, Goiás and Minas Gerais States, occurring in rockier soils, and presenting endemic species of Asteraceae, Euphorbiaceae, Melastomataceae and Velloziceae. The "campo rupestre" vegetation is protected in the national parks and reserves (Mori, 1989).

Safford (1999) named "campos de altitude" (Highland Fields) a series of humid, subalpine grasslands restricted to the highest peaks and plateaux of southeastern Brazilian Highlands.

Ironstone outcrops, commonly known as "Canga" crusts, are rich in dicots, and present monocot aggregations, but distinct plant communities are found associated to different microhabitats (Jacobi et al. 2007).

This chapter focuses on a review of the vegetation growing on rock outcrops: 1) at the Serra do Cipó (43°30' W, 19°10' S) and 2) at an ironstone outcrop at (43°30'W, 19°10'S), in Minas Gerais State, in southeastern Brazil.

Serra do Cipó is at the southernmost portion of Espinhaço Mountains, a predominantly quartzitic range extending for 1,100 km in central Brazil. This region, characterized by quartzitic mountains with altitudes varying between 1,000-1,400 m, has shrubby, tortuous and sclerophyllous vegetation or open grasslandsand following the cerrado (savanna) vegetation from around 900-1,000 m altitude (Rizzini 1997). Climate is characterized by dry winters (3-

5 months) and rainy summers with an average annual rainfall of 1,500 mm and mean temperature of 17.4-19.8 °C. According to Köppen, the climate of the region is Aw type (tropical).

The second highland region, were an ironstone outcrop, in the extreme southern part of Espinhaço Range, called "Quadrilátero Ferrífero", due to exposed iron oxide deposits, provides habitat for many saxicolous species (Rizzini 1997). Studies from an undisturbed site in highland fields in São Gonçalo do Rio Abaixo, Minas Gerais State, and from a nearly highland field under restoration (43°26'W, 19°53'S) 1100 m a.s.l., are also discussed.

The role of AM fungi in improving the mineral nutrition of their host plants using the small diameter hyphae accessing microsites that roots cannot reach becomes relevant in outcrops, where the complex network of fungal mycelia and plant roots extends vertically into the soil and the rock substrate (Egerton-Waburton et al. 2003).

The mycorrhizal associations of native species growing at highland fields have been scarcely studied. Nogueira et al. (2005) showed the mycorrhizal status of some orchids; Pagano and Scotti (2009) showed AM colonization of two other species (*Paepalanthus bromelioides* and *Bulbostylis* sp.); and Matias et al. (2009) reported AM colonization for *Centrosema coriaceum* and estimated the spore numbers in the rhizosphere of the native species *C. coriaceum* and *Tibouchina multiflora*.

This fragile region has suffered human impact on a large scale, and conservation action needs to be developed to protect their fauna and flora (Costa et al. 1998). Outcrops, which are subject to mining, and harbour metal-tolerant and hyperaccumulator plant species (Jacobi et al. 2007) can be monitoring sites for the effect of climate change (reviewed by Scarano 2007).

In this chapter I report on the mycorrhizal associations of some plant species of the rock outcrops at the Serra do Cipó, and at an ironstone outcrop at southeastern Brazil, and present data about their life-form distribution and soil characterization.

SOIL CONDITIONS IN THE ROCK OUTCROPS IN MINAS GERAIS, BRAZIL

In the rock outcrop at the Serra do Cipó (rhizospheric soil of two co-occurring plant species *Paepalanthus bromelioides* and *Bulbostylis* sp.) the basic properties of the rhizospheric soil were as follows: the pH 5.3 and organic matter content 2.72% (Table 1). Sand >78% and clay >7% were other properties. Base saturation was low, P content was low, and acidity was moderated. The texture of the fine soil showed high content of sand (68%) and low content of clay (7%) , belonging to the sand textural class (Pagano and Scotti 2009). Soil samples were analyzed at IMA (Instituto Mineiro de Agronomia) Agropecuary Chemical Laboratory (Brazil).

In the ironstone outcrop, the soil was sandy loam (0-30 cm depth). Some basic properties of the soil were as follows: the soil was strongly acid (4.2), base saturation was low, and P content, very low. The texture of the fine soil showed low content of clay and higher percent of silt (Table 1).

Table 1. Soil characteristics and plants species studied in southeast of Brazil

Rock outcrop type	Soil pH	SOM	C	P	Sand[b] (%)	Plant species	Reference
Quartzite-sandstone, "Serra do Cipó"	5.3	2.72	1.58	1.3	78.74	*Paepalanthus bromelioides*; *Bulbostylis* sp.	Pagano and Scotti (2009)
Ferruginous rocks	4.2	3.09	1.79	1.7	63.26	*Eremanthus incanus*; *Centrosema coriaceum*; *Pavonia viscosa*; *Tibouchina multiflora*	This study; Pagano et al. (2010)

[a] Mean of two measures from one composite sample. pH (H_2O) 1:1; SOM = Soil organic matter (%); C = carbon (%); Available P (mg dm^{-3}); [b]Particle size distribution = sand 2-0.02 mm.

Table 2. Mycorrhizal status of the plants studied in southeast of Brazil

Family	Plant species	Growth form	NS	PC	AM type	AMF %	Previous report
Asteraceae	*Eremanthus incanus* Less.	Arboreal	+	ac, iv, rh,	ND	V	AM[1]
Cyperaceae	*Bulbostylis* sp.	Perennial herbaceous	+	ar, iv	*Arum*	III	AM[2]
Leguminosae	*Centrosema coriaceum* Benth	Herbaceous (climbing)	+	ar, c, v, ac, eh	*Arum*	IV	AM[3]
Malvaceae	*Pavonia viscosa* A. St.-Hil.[a]	Perennial herbaceous	+	ar, rh, iv, ov	*Arum*	IV	NR
Melastomataceae	*Tibouchina multiflora* Cogn. [a]	Arboreal	+	ac, ov	ND	III	NR
Eriocaulaceae	*Paepalanthus bromelioides* Silveira	Perennial herbaceous	-	ac, c, ov	ND	IV	AM[2]

[a]New records of AM type. Note: Indicated are the families. NS non-septate hyphae, PC: Patterns of AM colonization; ar: arbuscules, ac: auxiliary cells, c: hyphal coils, eh: extraradical hyphae, h: intra- or intercellular aseptate hyphae, iv: irregular vesicles, ov: oval vesicles, rh: root hairs. Structures shown as: + always present, – not detected. AM type: colonization type: *Arum* type, *Paris* type, I= intermediate. Arbuscular mycorhizal (AM) colonization, class: I, 1-5%; II, 6-25%; III, 26-50%; IV, 51-75% and V, 76-100%. Rh = root hairs. Ac = auxiliary cells. (+) = presence, (-) = absence. [1]Pagano et al. (2010), [2]Pagano and Scotti (2009), [3]Matias et al. (2009). NR= No report.

AM SPORE DISTRIBUTION AND DIVERSITY IN THE ROCK OUTCROPS

Occurrence of different types of arbuscular mycorrhizal fungi from six native plants were recorded in different seasons during 2003 and 2006 (Pagano and Scotti, 2009, Matias et al. 2009, Pagano et al. in press). Field samplings consisting in soil and root fragments were taken in several field trips to the sites. AMF spores were analyzed for species identification and roots samples for mycorrhizal colonization.

A least eight AMF taxa were present in the rhizospheres of the surveyed plants in the two studied outcrops (Table 2). AM fungal species belonging to five genera were recorded.

Reported AM fungi (AMF) taxa found in the studied rooting zones were: *Acaulospora spinosa*, *A. elegans*, *A. foveata*, *Acaulospora* sp., *Gigaspora margarita*, *Glomus* sp., *Dentiscutata biornata*, *D. cerradensis*, *D. heterogama*, *Dentiscutata* sp. and *Racocetra verrucosa*. AMF spore richness was higher in the rhizospheric soils of *Eremanthus incanus* and the legume *C. coriaceum;* however these species were also isolated from restored sites.

All plant species showed AMF mycotrophy. Three AMF's genera found in the rooting-zone soils of *Paepalanthus bromelioides* and *Bulbostylis* sp. were: *Glomus* (two species), *Acaulospora* and *Scutellospora* (one species each). *Glomus* was the dominant genus and *Glomus brohultii* was the most common species. The average AMF spore number was 77-139 per 100 g dry soil and the species richness was 3 to 4 AMF species per sample. Both plant species showed high spore numbers and dominant hyphae and vesicle colonization. AMF diversity was found to be low. In the roots of *P. bromelioides* only hyphae and vesicles were observed.

Table 3. AM spore diversity on rooting-soils of some studied species isolated from natural or restored highland fields soils in Brazil

AMF Species	Pb	B	Cc	Ei
Gigasporaceae				
Gigaspora margarita W.N. Becker & I.R. Hall			X	X
Dentiscutataceae				
Dentiscutata sp.1				X
Dentiscutata biornata (Spain, Sieverd. & Toro) Sieverd., F.A. Souza & Oehl	X	X	X	
Racocetraceae				
Racocetra verrucosa (Koake & C. Walker) Oehl, F.A.Souza & Sieverd.				X
Cetraspora pellucida (T.H. Nicolson & N.C. Schenck) Oehl, F.A. Souza & Sieverd.			X	
Acaulosporaceae				
Acaulospora sp. 1	X	X		X
A. spinosa Walker & Trappe			X	X
A. foveata Trappe & Janos				X
Glomeraceae				
Glomus sp. 1	X	X		
Species richness[*]	3	3	4	6

Pb = *Paepalanthus bromelioides*, B = *Bulbostylis* sp., Cc = *C. coriaceum*, Ei = *E. incanus*.

Eremanthus incanus Less. (Asteraceae), a common species of highlands regions of the extreme southern part of Espinhaço Range presented high AM colonization (*Arum*-type) (Pagano et al. 2010).

It has been showed that some vegetal species, like *E. incanus*, form a persistent soil seed bank, contributing to a higher potential for regeneration in habitats subjected to disturbance (Velten and Garcia 2007). The additional fact that *E. incanus* are AM dependent in natural and restored highland field ecosystems (Pagano et al. 2010) supplies important information for restoration programs. Highland fields contain natural AM fungal species associated with the rhizosphere of plants that can be affected by land use (mining). Since the bare soil did not present any AM fungal propagule, restoration using selected native plants can be facilitated by AM fungal inoculation (spores) (Pagano et al. 2010).

In the ironstone outcrop, as expected, different AM species were found in the rhizosphere of plants under experimental (restored sites) and natural conditions. Under natural conditions five AM species were also observed (Table 3); however, only two species were in common.

In the ironstone outcrop, ten taxa of AM fungi (Figure 1) were distinguished in the rooting zone soil samples, of which 7 were identified at the species level and 3 at the genus level (Table 3). Of the ten taxa, one belonged to the genus *Glomus*, four to *Acaulospora*, three to *Scutellospora* one to *Racocetra* and one to *Gigaspora*. *A. spinosa* and *S. cerradensis* were the most common species in the ironstone outcrop.

It has been frequently observed that plants growing under stress conditions show higher AM dependence particularly in semiarid and arid environments (Varma 1995, Nouaim and Chaussod 1996).

Figure 1. Partial view of the restored Ferruginous rock outcrop. It is showed the vegetal species planted (a). *E. incanus* (b), *T. multiflora* (a dominant species) (c), *C. coriaceum* (d). Quartzite-sandstone outcrop: *Bulbostylis* sp. (e) *Paepalanthus bromelioides* (f). Spores of species of AMF found in natural highland fields soils in Brazil: *Scutellospora biornata* spore with germination shield (g-h), *Scutellospora* sp.1 (i), *Glomus* sp. (j), *Acaulospora spinosa* (k). Bars for g, i, k = 50 µm; h, k: = 10 µm; I = 25 µm.

Acaulospora, *Glomus* and *Scutellospora* were reported by Gai et al. (2006) for Tibetan grasslands (arid or semi-arid type of high-altitude frigid zone). The altitude is from 3,500 to 4,800 m, the mean annual temperature is 0-8°C and the annual precipitation 304-542 mm. These authors found the following AMF species (corresponding to farmland, montane scrub grassland, alpine steppe and alpine meadow): *Glomus aggregatum* Schenck & Smith, *G. etunicatum* Becker & Gerdemann, *G. geosporum* (Nicol. & Gerd.) Walker, *G. intraradices* Schenck & Smith, *G. luteum* Kennedy, Stutz, & Morton, *G. mosseae* (Nicol. & Gerd.) Gerd. & Trappe, *G. rubiformis* (Gerd. & Trappe) Almeida & Schenck, *G. versiforme* (Karsten) Berch, *Glomus* sp.1, *Acaulospora appendicula* Spain, Sieverding & Schenck, *A. delicata* Morton, *A. elegans* Trappe & Gerdemann, *A. lacunosa* Morton, *A. spinosa* Walker & Trappe, *A. mellea* Spain & Schenck, *A. scrobiculata* Trappe, *Acaulospora* sp.1, *A.* sp.2, *A.* sp.3, *Entrophospora infrequens* (Hall) Ames & Schneider, *Scutellospora aurigloba* (Hall) Walker & Sanders, *S. erythropa* (Koske & Walker) Walker & Sanders, *S. calospora* (Nicol. & Gerd.) Walker, *S. spherica* Koske & Walker, *S. pellucida* (Nicol. & Schenck) Walker & Sanders.

Lugo et al. (2008) found ten AMF species in Puna highlands arid sites varying in altitude from 2,000 to 4,400 m above sea level (masl) in Argentina, and two species (*A. spinosa* and *S. biornata*) were in common with the Brazilian highlands ecosystems. Furthermore, *A. spinosa* was also reported for arid sites in China (Tao and Zhiwei 2005), for Tibetan grasslands (Gai et al. 2006) and for Puna highlands arid sites in Argentina (Lugo et al. 2008).

In the ironstone outcrop, *Gigaspora margarita* was found in the restored site, and also in the undisturbed area. Presence of *Gigaspora*-like auxiliary cells, with narrow projections, observed in the plant roots (Figure 2b,d) suggests a possible effectiveness of *G. margarita* spore inoculation. This AM species has a worldwide distribution and is commonly used as inoculum.

In other reported studies from natural highland fields (Pagano and Scotti 2009, Matias et al. 2009), higher spore number of *Glomus* was observed. In the ironstone outcrop, a higher spore number of *Glomus* was reported in the rhizosphere of *Tibouchina multiflora*.

In the ironstone outcrop, a higher spore number of *Acaulospora* than *Glomus* was found in the rhizosphere of *E. incanus*, suggesting an abiotic or biotic effect on the AMF composition.

The predominance of Acaulosporaceae could be associated with the presence of pioneer plant species such as found by Córdoba et al. (2001) in foredunes. Some authors showed that *Acaulospora* tended to be more frequent in the worse sites (Carpenter et al. 2001).

On the other hand, it is known that most *Scutellospora* species have been described from warmer climates characterized by pronounced rainfall and a dry season (Tchabi et al. 2008). In the ironstone outcrop, in the rhizosphere of *E. incanus*, we recovered some *Scutellospora* species, one in the pristine site and the rest in the restored site. *Dentiscutata biornata* (previously named *Scutellospora biornata* seems to be a common AM species in highland fields. A different morphotype of *Scutellospora* was present in the more pristine site (Pagano et al. 2010).

It has been showed that *E. incanus* form a persistent soil seed bank, having higher potential for regeneration in habitats subjected to disturbance (Velten and Garcia 2007), and the additional fact that *E. incanus* are AM dependent in natural and restored highland field ecosystem supplies important information for restoration programs.

Notably, in the reported studies, all the species presented a >26-50% colonization class, which may be related to the very low P content.

Studies of AM colonization are important for seedling production and preparation of technologies for successful restoration, because of the fact that vegetal species exhibit different AM dependency (Siqueira and Saggin-Junior 2001).

In Brazil, the occurrence of AMF in rupestrian landscapes is not yet well documented. The studies discussed here provides the first detailed report ever published on the mycorrhizal status of some of the species examined, which belong to the Malvaceae and Melastomataceae families, confirming the mycotrophic nature of these species at Brazilian highlands.

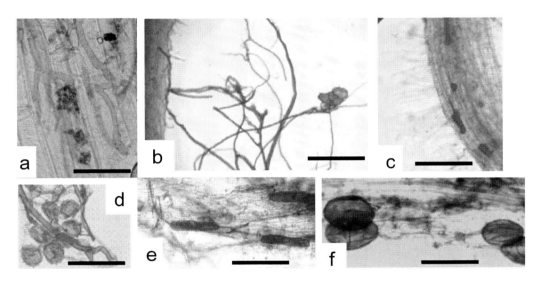

Figure 2. AM colonization in *C. coriaceum* fine roots. Arbuscules in root cells (a) and auxiliary cells (b); *P. viscosa* intra radical hyphae bearing vesicles (c); auxiliary cells in roots of *T. multiflora* (d), intra radical hyphae bearing vesicles in roots of *Bulbostylis* sp. (e) and in *E. incanus* (f). Bars for figures b, c = 50 µm; e, f = 50 µm, a, d = 10 µm.

ECONOMIC IMPORTANCE OF THE VEGETAL SPECIES IN THE ROCK OUTCROPS

Eremanthus incanus is very important in the State of Minas Gerais, Brazil, as source of wood for rafters, posts, firewood and charcoal, and for land reclamation. This species produces essential oil, whose main component is alpha-bisabolol, used in cosmetics and pharmaceuticals manufacturing (Galdino et al. 2006). *Eremanthus incanus* occurs in cerrado savanna, in secondary forests and caatinga, and is dominant at 650 a 1.200 m altitude. Other species, such as *Eremanthus erythropappus,* occurs also in highland fields (900 to 1.700 m altitude) in unfertile soils. The explotation of these species of *Eremanthus*, commonly known as "candeia" could have an efficient forest management in the future.

A complete understanding of plant life histories, including traits related to AM formation, the potential uses of microbiological associations, and the microbial inoculant production may be more studied in these environments in order to attain sustainable practices of plant management.

The environmental uniqueness, high diversity, lack of studies and rapid destruction of these ecosystems pose an immediate challenge for their conservation (Jacobi et al. 2007)

Moreover, no reference to global change-related studies being carried out in Brazilian mountains and inselbergs, has been reported (Scarano 2007), and only Benites et al. (2007) mentioned that marshy peats found at "Serra da Mantiqueira" in Brazil, have a high potential for carbon sequestration.

BENEFITS AND PROBLEMS ENCOUNTERED IN THE STUDY

In order to maximize the study of outcrops more research should be directed towards: (i) understanding the occurrence of beneficial soil microorganisms and survival in soil; (ii) the impact of fertilizers on soil microbiota in the case for restoration of the flora (unnecessary nutrient supply given to less demanding species results in waste of inputs); (iii) the role of legumes in transmitting nutrients to non-legumes through mycorrhizal pathway; and (iv) the presence and effects of allopathic substances on soil biota.

The presence of the commonly *Glomus* spores in their rhizospheres suggest that this AM genus could be a potential inoculum for the *T. multiflora*, *P. bromelioides* and *Bulbostylis* sp. On the other hand, *Gigaspora* could be a potential inoculum for *C. coriaceum* and for *E. incanus*, as well as *Acaulospora* for *P. viscosa*.

Mycorrhizal management is often a better option than mycorrhizal inoculation, considering the problems and costs of large-scale inoculum production. However, *C. coriaceum*, *T. multiflora* and *E. incanus* show greater height growth or cover, when inoculated, suggesting an advantageous response of these species.

Other problems are the taxonomic identification of AMF, which is based on spore morphology, proving difficult. Species such as *S. spherica* Koske & Walker, mentioned in the list of Gai et al. (2006) are not reported in the actual literature. Silva et al. (2005) mentioned 32 described *Scutellospora* species, whereas, Oehl et al. (2008) mentioned 36 *Scutellospora* species, which were organised in three new families. The identification of AMF species involved in various associations is essential for the functional diversity study of AMF populations (Smith and Read 2008).

Some problems facing the study of mycorrhizas in outcrops in Brazil are low financial support, and the detrimental state of highways for field sampling. Complications due to the little time of the field surveys and little technical support provide extra problems. More nursery experiments need to be developed to prevent seedlings growth under different conditions, and there is a lack of greenhouse or nursery availability for these purposes. More detailed studies involving responses to nutrient addition in the field are needed for most native species used in restoration.

Also, studies including molecular identification directly in host plant roots are important to identify efficient AM isolates. Moreover, sampling in the rainy and dry periods would be required to conclude more information on mycorrhizal characterization in these environments.

CONCLUSION

In this chapter, we briefly described the importance to study rock outcrops in Brazil. Research in these areas, which are subject to mining, and where metal-tolerant plant species may exist, is essential to successful restoration programs. Moreover, the wise management of ecosystem goods and services, and the management of species of economic importance, like species of *Eremanthus*, which have a great potential for regeneration in habitats subjected to disturbance, can prevent a deepening of poverty. Additionally, alternatives to preserve this environment by using native species are restoration programs, should take mycorrhizae into account. Moreover, *T. multiflora* (a dominant species) and *C. coriaceum* (a noduliferous legume), are also favourable species for use in restoration of the ironstone outcrop.

Throughout the chapter, I have presented detailed information on plants occurring in the rock outcrops in Minas Gerais State, Brazil, summarizing that these plants maintain higher root colonization and spore numbers, suggesting a beneficial role of AMF, and highlighting the importance of mycorrrhizae as an essential component for establishment and sustainability of plant communities. Nevertheless, further studies are required to achieve maximum benefits from these microorganisms and their associations. All these reports show that most plants in these environments seem to be AM-dependent species.

Highland field plant species contain natural AM fungal species richness that can be affected by land use (mining); restoration using selected plant species can be facilitated with AM fungal inoculation (spores). Several reports revealed that AMF are a common and important component in highland vegetation in Brazil, and should be included in future restoration programs.

Finally, the evidence presented here emphasizes the need to consider the symbiotic fungi in management practices in these environments. The choice of vegetal species would therefore have great implication in the manipulation and conservation of AMF species. The loss of AMF with disturbance and the ability of native fungi to colonize plants in natural conditions require more studies. In general, highly dependent hosts should be selected over mycorrhizal- independent hosts.

ACKNOWLEDGMENTS

Marcela Pagano is grateful to Minas Gerais State Agency for Research and Development (FAPEMIG) and to Council for the Development of Higher Education at Graduate Level (CAPES), Brazil. Marta Noemí Cabello is researcher from Comisión de Investigaciones Científicas de la Provincia de Buenos Aires (CIC), Argentina.

REFERENCES

Benites,VM; Simas, FNB; Schaefer, CEGR; Santos, HG; Mendonça, BAF. Soils associated to rock outcrops in the highlands of Serras da Mantiqueira and Espinhaço, southeastern Brazil. *Revista Brasileira de Botânica*, 2007, 30, 569-577 (In Portuguese).

Burke, A. Island–matrix relationships in Nama Karoo inselberg landscapes. Part I: Do inselbergs provide a refuge for matrix species? *Plant Ecology*, 2002, 160, 79-90.

Carpenter, FL; Mayorga, SP; Quintero, EG; Schroeder M. Land-use and erosion of a Costa Rican Ultisol affect soil chemistry, mycorrhizal fungi and early regeneration. *For. Ecol. Man.*, 2001, 144, 1-17.

Córdoba, AS; Mendonça, MM; Stürmer, SL; Rygiewicz, PT. Diversity of arbuscular mycorrhizal fungi along a sand dune stabilization gradient: A case study at Praia da Joaquina, Ilha de Santa Catarina, South Brazil. *Mycoscience,* 2001, 42, 379-387.

Costa, CMR; Herrmann, G; Martins, CS; Lins, LV; Lamas, IR. *Biodiversidade em Minas Gerais: um atlas para sua conservação.* Belo Horizonte: Fundação Biodiversitas, 1998.

Egerton-Waburton, LM; Graham, RC; Hubbert, KR. Spatial variability in mycorrhizal hyphae and nutrient and water availability in a soil weathered bedrock profile. *Plant Soil*, 2003, 249, 331-342.

Gai, JP; Feng G; Cai, XB; Christie, P; Li, XL. A preliminary survey of the arbuscular mycorrhizal status of grassland plants in southern Tibet. *Mycorrhiza*, 2006, 16, 191-196.

Galdino, APP; Brito, JO; Garcia, RF; Scolforo, JR. Studies on the yield and quality of the "candeia" (*Eremanthus* ssp.) oil and the influence of the distinct commercial origins of the wood. *Rev. Bras. Pl. Med.*, 2006, 8, 44-46.

Jacobi, CM; Carmo, FF; Vincent, RC; Stehmann, JR. Plant communities on ironstone outcrops – a diverse and endangered Brazilian ecosystem. *Biodiversity and Conservation*, 2007, 16, 2185-2200.

Larson, DW; Matthes, U; Kelly, PE. Cliff ecology: pattern and process in cliff ecosystems. Cambridge studies in Ecology, Cambridge University Press, Cambridge, 2000.

Lugo, MA; Ferrero, M; Menoyo, E; Estévez, MC; Siñeriz, F; Anton A. Arbuscular mycorrhizal fungi and rhizospheric bacteria diversity along a altitudinal gradient in South American Puna grassland. *Microb. Ecol.*, 2008, 55, 705-713.

Matias, SR; Pagano, MC; Muzzi, FC; Oliveira, CA; Carneiro, AA; Horta, SH; Scotti, MR. Effect of rhizobia, mycorrhizal fungi and phosphate-solubilizing microorganisms in the rhizosphere of native plants used to recover an iron ore area in Brazil. *Eur. J. Soil Biol.*, 2009, 45, 259–266.

Mori, SA. Eastern, extra-amazonian Brazil. In: Campbell DG and Hammond D, editors. *Floristic inventory of tropical countries.* New York: NYBG/WWF; 1989; 432-434.

Nogueira, RE; Pereira, OL; Kasuya, MC; Lanna, MCS. Mendonça, MP Fungos micorrízicos associados a orquídeas em campos rupestres na região do Quadrilátero ferrífero, MG, Brasil. *Acta Bot. Bras.*, 2005, 19, 417-424.

Nouaim, R; Chaussod, R. Rôle des mycorhizes dans l'alimentation hydrique et minérale dês plantes, notamment des ligneux de zones arides. CIHEAM - Options Mediterraneennes, 1996.

Oehl, F; de Souza FA; Sieverding, E. Revision of *Scutellospora* and description of five new genera and three new families in the arbuscular mycorrhiza-forming *Glomeromycetes*. *Mycotaxon*, 2008, 106, 311-360.

Pagano, MC; Scotti, MR. A survey of the arbuscular mycorrhiza occurrence in *Paepalanthus bromelioides* and *Bulbostylis* sp. in rupestrian fields, Brazil. *Micologia Aplicada International*, 2009, 21, 1-10.

Pagano, MC; Scotti, MR. A survey of the arbuscular mycorrhizaoccurrence in Pa*epalanthus bromel*ioides and *Bulbostylis* sp. in rupestrian fields, Brazil. *Micologia Aplicada International*, 2009, 21, 1-10.

Pagano, MC; Cabello, MN; Scotti, MR. Arbuscular mycorrhizal colonization and growth of Eremanthus incanus Less. in a highland field. Plant Soil Environ., 2010, 56, 9, 412–418.

Porembski, S. Tropical inselbergs: habitat types, adaptive strategies and diversity patterns. Revista Brasil. Bot., 2007, 30, 579-586.

Rizzini, CT. Tratado de Fitogeografia do Brasil: aspectos ecológicos, sociológicos e florísticos, 2nd ed. São Paulo: Âmbito Cultural Edições Ltda.; 1997, (In Portuguese).

Safford, HD. Brazilian Páramos II. Macro- and mesoclimate of the campos de altitude and affinities with high mountain climates of the tropical Andes and Costa Rica. Journal of Biogeography, 1999, 26, 713-737.

Scarano, FR. Rock outcrop vegetation in Brazil: a brief overview. Rev. Bras. Bot., 2007, 30, 561-568.

Siqueira, JO; Saggin-Júnior, OJ. Dependency on arbuscular mycorrhizal fungi and responsiveness of Brazilian native woody species. Mycorrhiza, 2001, 5, 245-255.

Silva, GA; Maia LC; Sturmer, SL. A dichotomous key to Scutellospora species (Gigasporaceae, Glomeromycota) using morphological characters. Mycotaxon, 2005, 94, 293-301.

Tchabi, A; Coyne, D; Hountondji, F; Lawouin, L; Wiemken, A; Oehl F. Arbuscular mycorrhizal fungal communities in sub-Saharan Savannas of Benin, West Africa, as affected by agricultural land use intensity and ecological zone. *Mycorrhiza*, 2008, 18,181-195.

Smith, SE; Read, DJ. *Mycorrhizal symbiosis*. 3rd ed. London: Academic Press; 2008.

Varma, A. Arbuscular mycorrhiza fungi: the state of the art. *Crit. Rev. Biotechnol.*, 1995, 15, 179-199.

Velten, S.B., Garcia, Q.S. Variation between three *Eremanthus* (Asteraceae) species in their ability to form a seed bank. *Revista Brasil. Bot.*, 2007, 30,713-719.

In: Mycorrhiza: Occurrence in Natural and Restored…
Editor: Marcela Pagano

ISBN: 978-1-61209-226-3
© 2012 Nova Science Publishers, Inc.

Chapter 6

CO-OCURRENCE OF ARBUSCULAR MYCORRHIZAS AND DARK SEPTATE ENDOPHYTES IN PTERIDOPHYTES FROM A VALDIVIAN TEMPERATE RAINFOREST IN PATAGONIA, ARGENTINA

Natalia Fernández[a,c], Sonia Fontenla[a,c] and María Inés Messuti[b,c]

Laboratorio de Microbiología Aplicada y Biotecnología[a]; Depto. de Botánica[b];
Universidad Nacional del Comahue[c] –
Instituto de Investigaciones en Biodiversidad y Medioambiente (INIBIOMA),
Quintral 1250, S.C. de Bariloche (R8400 FRF), Río Negro, Argentina.

ABSTRACT

Plant roots are colonized by several fungi species, including those that form arbuscular mycorrhizas (AM). The contribution of AM to plant productivity, plant nutrition and soil composition is crucial, and in natural ecosystems plant diversity and community structure are strongly influenced by these organisms. Despite the importance of pteridophytes (lycophytes and ferns) in the origin and evolution of vascular plants, little is known about AM in this group of plants. The aim of this chapter is to describe the occurrence of AM in different species of pteridophytes present in the Valdivian temperate rainforest of Puerto Blest (Patagonia, Argentina), and to discuss different factors that might determine not only the mycorrhizal status but also the development of different AM morphologies in these plants (*Arum*, *Paris* or *Intermediated*-type). Sporophytes were sampled by random walk from terrestrial, epipetric and epiphytic habitats. The samples were stained and the percentage of root length colonized by these symbiotic fungi was quantified in all of them. A total of 21 species belonging to 10 families and 12 genera were found. Arbuscular mycorrhizas were present in 47.6 % of these species. Fifty percent of the families included only consistently mycorrhizal species (Blechnaceae, Dryopteridaceae, Gleicheniaceae, Lophosoriaceae and Pteridaceae), 20 % facultatively mycorrhizal species (Equisetaceae and Lycopodiaceae) and 30 % non-mycorrhizal species (Aspleniaceae, Grammitidaceae and Hymenophyllaceae). It was observed that while facultative species have an *Intermetiate*-type AM, consistently mycorrhizal species were only colonized by *Paris*-type associations. In addition to AM, another group of fungi were present within the roots of all the species considered in this study. They corresponded to dark septate endophytes (DSE) and were characterized by regularly septate hyphae and microsclerotia. The colonization pattern of DSE is described and its importance as possible "mycorrhizal associations" is discussed.

Keywords: lycophytes and ferns, arbuscular mycorrhizas (AM), morphological type, dark septate fungi, Valdivian temperate rainforest, Patagonia, Argentina

INTRODUCTION

In recent years, the effort to understand belowground biological interactions in forest ecosystems has been considerably increased, since they substantially influence ecosystem processes. Soil fungi are diverse and play wide-ranging roles in forest dynamics, so they are an important part of the belowground biota (Copely 2000). Among these fungi are those that form arbuscular mycorrhizas (AM). Arbuscular mycorrhizal fungi represent some of the most abundant organisms on this planet, and the interaction between them and plant roots is one of the most ancient and important symbiotic relationships in the living world (Brundrett 2002, Rosendahl 2008, Smith and Read 2008). The contribution of AM to soil composition, plant nutrition and productivity is crucial, so plant diversity and community structure are strongly influenced by these organisms (van der Heijden et al. 1998, Read 1999, O'Connor et al. 2002).

Arbuscular mycorrhiza morphology can be divided into two main types, *Arum* and *Paris*. The *Arum*-type association is characterized by extensive intercellular hyphae between cortical cells and short lateral branches that form arbuscules within cortical cells. In contrast, in *Paris*-type colonization the fungi spread directly from cell to cell forming intracellular hyphal coils that frequently have intercalary arbuscules ("arbusculate coils") (Smith and Smith 1997, Brundrett 2004). In addition to these main morphological types, an *Intermediate*-type colonization also has been described, meaning either the occurrence of *Arum*- and *Paris*-type morphologies within the same species (Smith and Smith 1997), or the presence of fungal structures that are in between *Arum* and *Paris* morphologies (Dickson 2004).

There are different factors determining the development of AM morphologies, such as anatomical characters of the host roots (e.g. presence of intercellular spaces within the root cortex) (Brundrett and Kendrick 1990a,b) and plant genotype (Ahulu et al. 2006). A recent study carried our by Cavagnaro et al. (2001a) demonstrated that colonization type is not only under plant control, they found that fungal identity also plays an important role in determining AM morphology. It has also been indicated that AM colonization type is a consequence of the interaction between both, plant and fungus (Dickson 2004, Kubato et al. 2005). Some authors have suggested that *Paris*-type associations might be more advantageous for plants growing slowly and/or in low-nutrient and high-stress environmental conditions, because their spreading rate within roots is slower than that of the *Arum*-type. This phenomenon would be a way of keeping the energy supply to the fungi at a manageable level (Brundrett and Kendrick 1990a,b, Cavagnaro et al. 2001b, Yamato and Iwasaki 2002). This idea is supported by *Arum*-type mycorrhizas being most abundant in agricultural crops (which are fast growing plants), whereas the *Paris*-type has been found to be most frequent in plants from natural ecosystems (Smith and Smith 1997, Yamato and Iwasaki 2002, Ahulu et al. 2005, Muthukumar et al. 2006).

Endophytic fungi are ubiquitous in plants and are referred to fungi that live for all, or at least a significant part, of their life cycle internally and asymptomically within plant tissues. It has been documented in trees, shrubs, herbs and ferns that individual species and even individual plants typically harbor large numbers of endophytic fungal species (Saikkonen et al. 1998). Even more, Brundrett (2002) put forward that these fungi are the most likely source of new plant–fungus associations. Despite the great diversity and abundance of endophytes and their occurrence in roots of several plant species, they have been much less studied than AM. However, increased attention has recently been given to a group of miscellaneous root-colonizing fungi designated as dark septate endophytes (DSE). They do not seem to exhibit any host specificity and have been reported from 600 plant species including mycorrhizal as well as non-mycorrhizal plants. These conidial fungi are characterized for developing melanized, regularly-septate hyphae and microsclerotia in the host roots. The ecological role of DSE is currently unresolved, but their widespread occurrence in different ecosystems, their potential to function as plant growth promoters and to provide defenses against herbivores and pathogens, suggest that these endophytes are a significant component of plant communities. A limited number of studies have so far attempted to evaluate and quantify root colonization by both mycorrhizal and DSE fungi, and based on the scarce information available it appears that DSE are as abundant as mycorrhizal fungi, so it is important to consider them when studying mycorrhizas in natural ecosystems (Jumpponen and Trappe 1998, Jumpponen 2001, Barrow 2003, Mandyam and Jumpponen 2005).

Land colonization by plants may not have been possible without fungal symbionts. From this evolutionary point of view, the mycorrhization of early diverging land plant lineages, such as bryophytes and pteridophytes, is of particular interest (Pirozynski and Malloch 1975, Simon et al. 1993, Remy et al. 1994, Heckman et al. 2001, Read et al. 2000). However, the vast majority of studies on mycorrhizae have focused on seed plants (gymnosperms and angiosperms) because of their ecological and economic dominance. Pteridophytes (lycophytes and ferns or monilophytes)[1] are of ancient origin and occupy an important position in the origin and evolution of vascular plants. They dominated the world from the Silurian to the Paleozoic and remain a major component of many ecosystems to this day (Rothwell 1996). Arbuscular mycorrhizas have been described in some fossil pteridophytes, which means that this symbiosis has persisted for more than 400 million years (Pirozynski and Malloch 1975, Remy et al. 1994). However, mainly because of its low commercial value (Tryon and Lugardon 1990, Lee 2001), this group of plants has been largely ignored and their AM status remains poorly understood, as well as the role that these fungi may play in their ecological and biogeographical distribution (Dhillion 1993, West et al. 2009). Some authors have recently registered the occurrence of AM in sporophytes and gametophytes of pteridophytes, but most of them did not describe the morphological type present in the analyzed species (Gemma et al. 1992, Muthukumar and Udaiyan 2000, Zhao 2000, Lee et al. 2001, Khade and Rodrigues 2002, Zhang et al. 2004, Lehnert et al. 2009, West et al. 2009). According to some

---

[1] Recent phylogenetic studies have revealed a basal dichotomy within vascular plants, separating the lycophytes from the euphyllophytes. The last comprise two major clades: the spermatophytes (seed plants) and the monilophytes (ferns). Plants included in the lycophyte and fern clades are all spore-bearing or "seed-free", and because of this common feature their members have been lumped together in a paraphyletic group named "seedless vascular plants" or "seed-free plants" (Pryer et al. 2004, Smith et al. 2006).

few studies, both *Arum* and *Paris*-types are present in pteridophytes, being the last the most frequent morphological type among these plants (Smith and Smith 1997, Zhang et al. 2004, Kessler et al. 2010). Dark septate endophytes have also been previously cited for some few lycophytes and ferns (Cooper 1976, Berch and Kendrick 1982, Dhillion 1993).

Abundant pteridophytes are present among the exuberant vegetation of the Valdivian temperate rainforests of South America, which are characterized by their high level of endemisms (90 % at the species level and 34 % at the genus level for woody species) and the antiquity of their ecological interactions. Puerto Blest is situated in the Valdivian temperate rainforest region of Argentina (Cabrera 1976, Armesto et al. 1995). Previous studies have demonstrated the occurrence of AM in different plant species within this forest (Chaia et al. 2006). However, the mycorrhizal status of lycophytes and ferns in this ecoregion is completely unknown, as it is for most of the pteridophytic flora of Argentina. The aim of this chapter is to describe and discuss the mycorrhizal status and the AM morphologies of pteridophytes growing in different substrates in Puerto Blest. The occurrence of DSE is also depicted.

VALDIVIAN TEMPERATE RAINFORESTS

Temperate Rainforests: General Description

The term "temperate rainforest" is not new to ecology, but has been applied to a broad range of forest types throughout the world. Alaback (1991) established that these environments are defined by some common distinctive features, such as: (1) annual precipitation greater than 1400 mm; (2) cool summers and wet weather all the year around; (3) infrequent fire events which in consequence do not represent an important factor in forest dynamics; (4) dormant season caused by low temperatures and transient snow. Floristically, they are characterized by large number of species composing the upper canopy layer but only some of them are dominant, a dense and continuous understory layer with absence of annual plants, presence of lianas, and dense mats of epiphytes in the forest floor and in the upper canopy (including mosses, liverworts and vascular plants). They include some of the longest-lived tree species as well as the largest remaining virgin landscapes outside the tropics.

The principal rainforest regions are along the northern Pacific cost of North America and along a similar physiographic pattern in southern Chile and Argentina, the rest of them are in isolated patches in mid-montane regions or in small islands (Figure 1). Of specific interest are the temperate rainforests in southern South America (Chile and Argentina) and southern Australasia (Victoria, Tasmania and New Zealand). These rainforests are notable for the shared presence of the genus *Nothofagus* and other floristic elements, explained by vicariance of a presumed Mesozoic cool temperate Gondwana flora and by terrane displacement. They also exhibit strong climatic and physiognomic similarities (Arroyo et al. 1996, Lücking et al. 2003).

The extremely wet and cool climatic conditions provide unique ecological and physiological conditions to these stable ecosystems and make them very sensitive to climatic change. That is why it is important to consider these pristine ecosystems for monitoring global climatic change. However, it is crucial to describe and understand them before being capable of analyzing the effects of this global phenomenon on rainforests, and then taking decisions on conservation and management options (Alaback 1991).

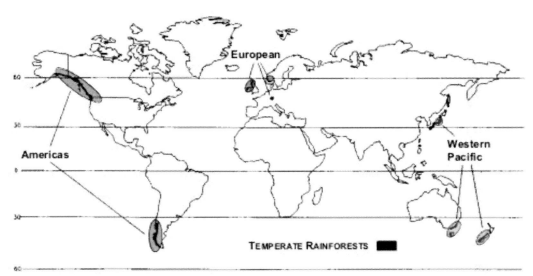

Figure 1. World distribution of temperate rainforests (modified from Alaback 1991).

As it has been previously mentioned, mycorrhizal fungi play key roles in numerous processes and phenomena that determine ecosystem functioning, community structure and dynamics. As environmental conditions change globally, the behavior of those fungi will play a substantial part in the response of ecosystems. There is therefore a need to understand how mycorrhizal fungi themselves respond to global change. In order to make this possible, it is first necessary to study in detail this symbiosis in different plant species, so as to establish the basic information needed to enable this type of studies (Fitter et al. 2004). This chapter adds to the general knowledge of mycorrhizal associations in temperate rainforests. It is focused on pteridophytes, which have been described as good bioindicators of changes occurring in ecosystems, making it possible to use these plants in addition to mycorrhizal behavior as a reference in time to monitor the ecological changes that may occur in the future (Arana et al. 2004).

The Valdivian Temperate Rainforest Region

In South America there are distinct rainforest biomes which are partially or completely isolated from each other (Figure 2a): (1) Amazonian tropical rainforest, reaching its southernmost border in Peru and Bolivia; (2) coastal Atlantic rainforest of south-eastern Brazil; (3) subtropical moist forest in northern Argentina and adjacent areas of Paraguay, Uruguay and Brazil; and (4) Valdivian temperate rainforest in southern Chile and Argentina (Lücking et al. 2003). The last is highly interesting from a biogeographical point of view, since it is a true biogeographic island isolated by 1500–2000 km from other rainforests and separated from climatically similar areas by extensive ocean barriers (Arroyo et al. 1996, Lücking et al. 2003).

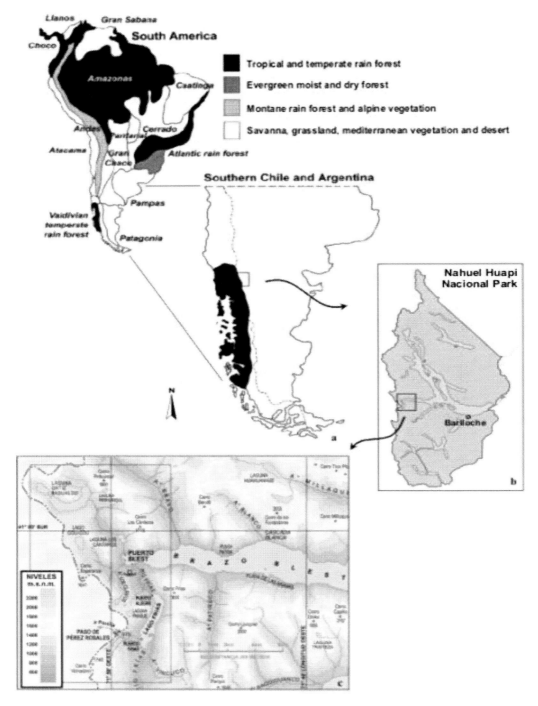

Figure 2. Geographic location of the study area. (a) Valdivian temperate rainforest region in relation to other main vegetation types in South America (modified from Lücking et al. 2003); (b) Nahuel Huapi Nacional Park; (c) Puerto Blest region.

The Valdivian temperate rainforest region is located in the southern cone of South America. It occupies a narrow strip of ~75000 km^2 which runs from 37° 20′ to 47° 00′ S latitude between the Eastern slope of the Andes and the Pacific Ocean, lying mostly in Chile and extending into adjacent areas of Neuquén, Río Negro and Chubut Provinces in Argentina (Figure 2a,b). Maximum annual average temperatures vary between 21 °C and 13 °C and the minimum range from 7 °C to 4 °C in the northern and southern ends of the ecoregion respectively, being the average annual temperature 8–12 °C. Annual precipitation varies between 1400 mm in the north and more than 4900 mm per year in the southern part of the ecoregion (Cabrera 1976, Alaback 1991, Armesto et al. 1995, Arroyo et al. 1996).

There is a high diversity of plant families represented in the Valdivian temperate forests. This ecoregion is particularly noteworthy for its high level of endemisms at species (~90 % of the seed plants) and even at genus level, which suggests geological antiquity, long isolation, and high rates of extinction during the cooler Pleistocene. The flora includes 450 species of vascular plants comprised in 200 genera and 100 families, many of them being of Gondwanan origin, with their nearest relatives occurring in Australia, New Zealand, New Caledonia and Tasmania (Arroyo et al. 1996, Lücking et al. 2003).

Puerto Blest

Puerto Blest (41° 02' S, 71° 49' W; 750 m.a.s.l.) is part of the reduced and fragmented Valdivian temperate region of Argentina. It comprises an area of 3.6 Km^2 and is located within the Nahuel Huapi National Park, Río Negro Province, Patagonia (Figure 2b,c). It is one of the rainiest places in the country, with an average annual rainfall of 3000 mm and an annual average temperature of 9 °C (Dimitri 1972, Brion et al. 1988). In this hydrophilic forest, evergreen species are the most abundant and the dominant tree species is the southern beech *Nothofagus dombeyi* (Fagaceae) (Figure 3). However, in some areas it forms mixed forests with other species (such as *Fitzroya cupressoides*, *Saxegothaea conspicua*, *Dasyphyllum diacanthoides*, *Laureliopsis philippiana*). Within Puerto Blest forest there is a waterlogged peat bog (Figure 3c), which corresponds to a water retaining *Sphagnum magellanicum* prairie (Roig 1998).

In this Valdivian temperate rainforest, trunks and branches are usually covered by different epiphytes such as lichens, mosses and ferns (Figure 3d). The understory is very shaded due to the dense forest canopy throughout the year and it is constituted by a great diversity of plants, including several pteridophytes (Cabrera 1976, Brion et al. 1988, Roig 1998).

ANALYZING ARBUSCULAR MYCORRHIZAS IN PTERIDOPHYTES

In this section we briefly described the methodology we used for analyzing the occurrence of AM and DSE in pteridophytes. More details can be obtained from Fernández et al. (2008, 2010). Sporophytes were collected by random walk from different substrates and habitats of Puerto Blest. They were removed from the substrate with a shovel in order to preserve most of the root system. Epiphytic specimens were collected from 10 to 50 cm height above the ground. Roots were rinsed with tap water, cleaned under a stereoscopic microscope so as to remove all root pieces belonging to other plants, and conserved in 70 % formalin-acetic acid-alcohol (FAA). Samples were stained using a modified Phillips and Hayman (1970) method and then stored in acidic glycerol.

Figure 3. Puerto Blest images representing some of its distinctive characteristics. (a) Upper canopy dominated by *N. dombeyi*; (b) Mountains and mixed forests around lake Frías; (c) Waterlogged peat bog dominated by *F. cupressoides*; (d) Dense understory and epiphytes growing on a trunk and on the forest floor.

The criteria used for the determination of AM was the presence of arbuscules at least in one individual of each species and the occurrence of the rest of typical AM structures (intra or intercellular hyphae, vesicles, coils) in the samples. These structures were used for their classification into *Arum*, *Paris* or *Intermediated*-type (Smith and Smith 1997, Dickson 2004).

Root length colonized was estimated according to the intersect method described by McGonigle et al. (1990) and then it was analyzed using a One-way-ANOVA and Duncan Post-hoc Test (data was first normalize using the square transformation). If mycorrhizal as well as non-mycorrhizal sporophytes were recorded for the same species, it was scored as facultatively mycorrhizal. Fungal structures corresponding to AM and DSE, were documented as brightfield images, which were captured with a digital camera and Image-Pro Plus 4.1.0.0 analysis software for Windows.

MYCORRHIZAL COLONIZATION IN PTERIDOPHYTES OF PUERTO BLEST

Pteridophytes

In Argentina, pteridophytes are represented by 346 native species included in 86 genera and 27 families. There are three pteridophytic diversity hotspots: two subtropical humid centers in the northwest and northeast of Argentina, and a temperate humid center in southern part of Argentina (Figure 4). These three regions concentrate 93 % of the species and 95 % of the endemisms. The southern temperate center, where 79 species corresponding to 33 genera have been described (de la Sota et al. 1998), has the lowest pteridophytic diversity, but the highest number of endemisms (77 %) (Ponce et al. 2002, Arana et al. 2004). As it can be observed in Figures 2 and 4, this zone overlaps with the Valdivian temperate forest region.

The survey described in this chapter includes a total of 21 species, 12 genera and 10 families. They correspond to 27 % of the species and 36 % of the genera cited for the southern temperate region. Two species (*Lycopodium paniculatum* and *Equisetum bogotense*) are eusporangiated pteridophytes (9.5 %), while the rest of them (90.5 %) are leptosporangiated ferns comprised in the Filicopsida Class (Zuloaga and Morrone 1996). All the species have been previously reported for Puerto Blest, and represent most of the pteridophytic flora described for this forest (*Gleichenia cryptocarpa* was not found) (Brion et al. 1988, de la Sota et al. 1998).

A total of 161 sporophytes were collected. In Table 1 it can be noticed that the number of samples encountered per family varied widely. The highest percentage of sporophytes corresponded to the Hymenophyllaceae (30.4 %) and Blechnaceae (28.6 %), followed by the Lycopodiaceae (8.7 %), Equiseataceae (7.5 %) and Aspleniaceae (6.2 %) respectively. This is not surprising since the Hymenophyllaceae and Blechnaceae are the only families in Puerto Blest that include more than 2 species, some of them widely distributed in this hydrophilic forest (*Hymenophyllum pectinatum*, *H. seselifolium*, *Blechnum chilense*, *B. penna-marina*). These findings are in agreement with Ponce et al. (2002), who established that the Blechnaceae and Hymenophyllaceae are the most common families in the southern temperate center, and that *Blechnum* and *Hymenophyllum* are among the richest genera of Argentina, with 19 and 16 species respectively.

Most of the samples (62.1 %) corresponded to terrestrial specimens, whereas 31.7 % and 6.2 % were collected from epiphytic and saxicolous habitats respectively. It was observed that roots of sporophytes collected from the last two habitats had abundant soil particles, probably as a consequence of their proximity to the ground (10 to 50 cm aboveground).

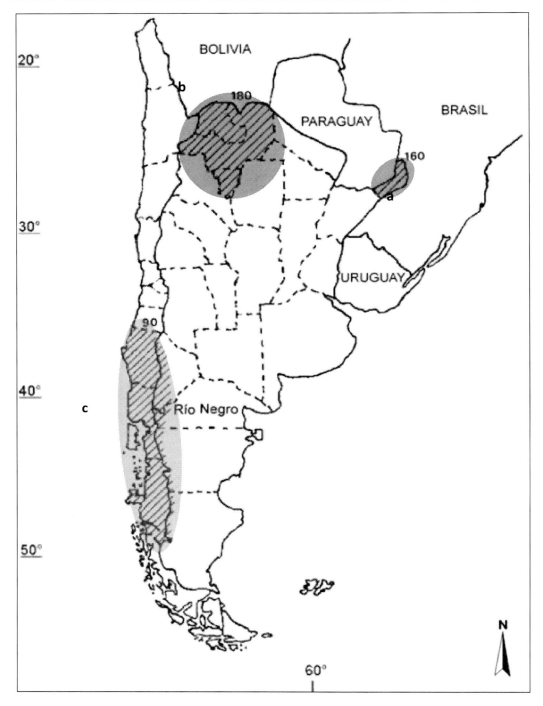

Figure 4. Centers of pteridophytic diversity in Argentina. (a) Subtropical humid center of northeastern Argentina (green); (b) Subtropical humid center of northwestern Argentina (blue); (c) Temperate center of southern Chile and Argentina (orange) (modified from Ponce et al. 2002).

Mycorrhizal Status of Pteridophytes Present in Puerto Blest

Mycorrhizas were registered in 55.3 % of the analyzed sporophytes (47.6 % of the species and 70 % of the families). The presence of intracellular hyphae, coils, vesicles and arbuscules confirmed that in all the cases they corresponded to AM. An interesting observation is that AM fungi entered into the cortical cells by two different ways: directly through the root epidermis or they first penetrate into root hairs and then they spread into the root cortex (Figure 5,6).

Despite the occurrence of mycorrhizas and the percentage of root length colonized varied widely among species and families, all the species included in the same family had the same mycorrhizal behavior (Table 1). Thus, families of pteridophytes present in the Valdivian temperate rainforest of Puerto Blest were divided into three groups: a) non-mycorrhizal, b) facultatively mycorrhizal and c) consistently mycorrhizal.

a) Non-mycorrhizal

The 11 species included in the Aspleniaceae, Hymenophyllaceae and Polypodiaceae presented extensive, thin (less than 2 mm wide) and fibrous root systems with abundant root hairs. Baylis (1975) and St. John (1980) described this root system morphology as "graminoid". A total of 67 sporophytes belonging to these 3 families were sampled, 73 % were collected from epiphytic habitats and the rest of them from the forest soil. All the specimens analyzed lacked AM associations (Table 1).

The absence of mycorrhizae in these mostly epiphytic fern species is concordant with previous studies that have analyzed this symbiosis in other species belonging to the same families (Lesica and Antibus 1990, Gemma et al. 1992, Michelsen 1993, Zhao 2000, Wang and Qiu 2006). Epiphytic habitats are usually considered extreme environments for plants because they are subjected to great variation in temperature and in water and nutrient availability (Benzing 1990). Therefore, it is generally expected that mycorrhizae may be an important adaptation for epiphytes (Benzing 1990, Lesica and Antibus 1990, Janos 1993, Rains et al. 2003). Nevertheless, it has been observed that epiphytic species are usually non-mycorrhizal (Allen et al. 1993) or facultatively mycorrhizal (Maffia et al. 1993, Michelsen 1993, Nadarajah and Nawawi 1993), and in case they have AM the colonization is usually low (Bermudes and Benzing 1989, Lesica and Antibus 1990, Rowe and Pringle 2005). Different explanations have been suggested for the paucity of AM in epiphytic habitats. For instance, it might be that these environments are too dry and exposed to support the fungus or that the rate of photosynthesis, which is relatively low in epiphytes (Benzing 1986), may be insufficient to maintain the symbiosis (Lesica and Antibus 1990). Nutritional insufficiency, and consequently the inability to sustain the energy cost of mycorrhizae, has also been suggested as a major factor determining mycorrhizal success in epiphytes (Benzing 1990, Janos 1993). Another widespread explanation is that epiphytic habitats are deficient in AM inocula, so epiphytes usually do not have the possibility of being colonized (Lesica and Antibus 1990, Janos 1993, Michelsen 1993). However, this suggestion contrasts with several authors who have reported that AM spores and mycelia may be dispersed to epiphytic environments by several vectors (mainly birds, ants and small mammals) (McIlveen and Cole 1976, Janos et al. 1995, Mangan and Adler 2000). It has also been demonstrated that many epiphytes lack AM despite the presence of inocula in the substrate (Lesica and Antibus 1990, Maffia et al. 1993, Rabatin et al. 1993). In Puerto Blest, it has been registered that the AM soil infective capacity ranges from 0.6 to 1.7 Infective Units g/soil (Chaia et al. 2006) and that several other plant species are capable or forming mycorrhizas when they are growing in the soil (Chaia et al. 2006, Fernández et al. 2008). Hence, deficiencies in nutrient acquisition or AM inocula would not explain the non-mycorrhizal behaviour observed in specimens of *Asplenium dareoides* and in different *Hymenophyllum* species rooted in the forest soil. It is therefore necessary to seek other explanations for this phenomenon. Another factor that might explain the absence of AM in epiphytic ferns of Puerto Blest is the root system morphology, since plants with long root length (Koide and Li 1991, Ryser and Lambers 1995), small root diameter (Reinhardt and Miller 1990, Manjunath and Habte 1991, Hetrick et al. 1992) and abundant long root hairs ("graminoid roots") (Baylis 1975, St. John 1980, Schweiger et al. 1995), like those observed in the epiphytic species presented in this chapter, tend to be non-mycorrhizal (Baylis 1975, St. John 1980, Hetrick 1991, Michelsen 1993, Zangaro et al. 2005).

b) Facultatively Mycorrhizal

Equisetum bogotense and *L. paniculatum* are the only representatives of the Equisetaceae and Lycopodiaceae families in Puerto Blest. There are some common features between these species that distinguish them from the rest of the pteridophytes of Puerto Blest. They are the only species that present microphylls, "fleshy roots" (roots that are larger in diameter and softer to the touch than fibrous or wiry roots) and eusporangia (Brion et al. 1988, Gifford and Foster 1989, Sitte et al. 2004). In addition, it was found that both species are facultatively mycorrhizal, a phenomenon that has been observed in other *Equisetum* and *Lycopodium* species in different regions of the world (Berch and Kendrick 1982, Dhillion 1993, Zhao 2000, Winther and Friedman 2007). Typical AM structures were observed in eight specimens of *E. bogotense* (67 %) and six of *L. paniculatum* (43 %). The percentage of root length colonized by AM varied widely within each species. In *E. bogotense* it ranged between 0 and 22.5 %, being the mean colonization value 8.6 %. The AM colonization level ranged from 0 to 50 % in *L. paniculatum*, with a mean value of 9.5 % (Fernández et al. 2008). Mean colonization values did not differed significantly between these species (p = 0.622).

In *L. paniculatum*, no clear relationship between AM colonization and substrate was observed, but in the case of *E. bogotense* it was noticed that it had a different symbiotic behavior depending on the type of soil where the samples were found, so that specimens growing in stony soil had AM but they were absent in those plants collected from moist sandy soil. A similar trend was previously described by Koske et al. (1985) and Dhillion (1993) in other species of *Equisetum*. These authors have suggested that the absence of mycorrhizas in moist sites may be more indicative of the effect of high soil moisture than of the mycotrophic potential of this genus, which constitutes a possible explanation for the facultative *AM* association described for this species in Puerto Blest. Another fact that might explain the facultative behavior observed in these two species is the presence of "fleshy roots". This type of root, considered primitive, is uncommon among modern ferns (Boullard 1979, Nelson 2000) and usually act as nutrients reservoirs (Nelson 2000), making plants less dependent on the symbiosis.

Co-occurrence of Arbuscular Mycorrhizas and Dark Septate Endophytes ... 111

Table 1. Comparison of the mycorrhizal status of lycophytes and ferns present in Puerto Blest according to their evolutionary stage, taxonomy and habitat.

EVOLUTIONARY STAGE and TAXON	HABITAT (N° sporophytes)				AM		
	E	S	T	Total	Status	% (X±SD)	MT
LYCOPHYTES							
Eusporangiate							
Lycopodiaceae[1]							
Lycopodium paniculatum	2	-	12	14	F (43)	9.5 ± 15.3	I
MONYLOPHYTES							
Eusporangiate							
Equisetaceae[1]							
Equisetum bogotense	-	-	12	12	F (67)	8.6 ± 7.8	I
Leptosporangiate							
Aspleniaceae[2]							
Asplenium dareoides	8	-	2	10	NM	-	-
Blechnaceae							
Blechnum chilense	-	2	16	18	C	84 ± 8.2	P
Blechnum magellanicum	1	2	1	4	C	63.1± 7.8	P
Blechnum penna-marina	-	1	23	24	C	72.3 ± 15.2	P
Dryopteridaceae							
Megalastrum spectabile	-	2	1	3	C	86 ± 2.3	P
Polystichum multifidum	-	1	3	4	C	85.4 ± 21	P
Gleicheniaceae							
Gleichenia quadripartita	-	-	1	1	C	66.3 ± 7.9	P
Hymenophyllaceae[2]							
Hymenophyllum cuneatum	-	-	1	1	NM	-	-
Hymenophyllum dentatum	-	-	1	1	NM	-	-
Hymenophyllum falklandicum	1	-	-	1	NM	-	-
Hymenophyllum ferrugineum	2	-	-	2	NM	-	-
Hymenophyllum pectinatum	10	-	5	15	NM	-	-
Hymenophyllum seselifolium	4	-	7	11	NM	-	-
Hymenophyllum tortuosum	5	-	2	7	NM	-	-
Hymenophyllum umbratile	4	-	1	5	NM	-	-
Serpyllopsis caespitosa	6	-	-	6	NM	-	-
Dicksoniaceae							
Lophosoria quadripinnata	-	-	8	8	C	71.8 ± 13.6	P
Polypodiaceae[2]							
Grammitis magellanica	8	-	-	8	NM	-	-
Pteridaceae							
Adiantum chilense	-	2	4	6	C	66.4 ± 5	P
TOTAL	**51**	**10**	**100**	**161**			

E: Epiphytic; S: Saxicolous; T: Terrestrial; F: Facultatively mycorrhizal (% of sporophytes colonized by AM); NM: Non-mycorrhizal; C: Consistently mycorrhizal; % (X ± SD): mean percentage of root length colonized and standard deviation; MT: Morphological type; I: *Intermediate*; P: *Paris* (for more details on the root length colonized by AM and the habitats where each sporophyte was collected see [1] Fernández et al. 2008 and [2] Fernández et al. 2010).

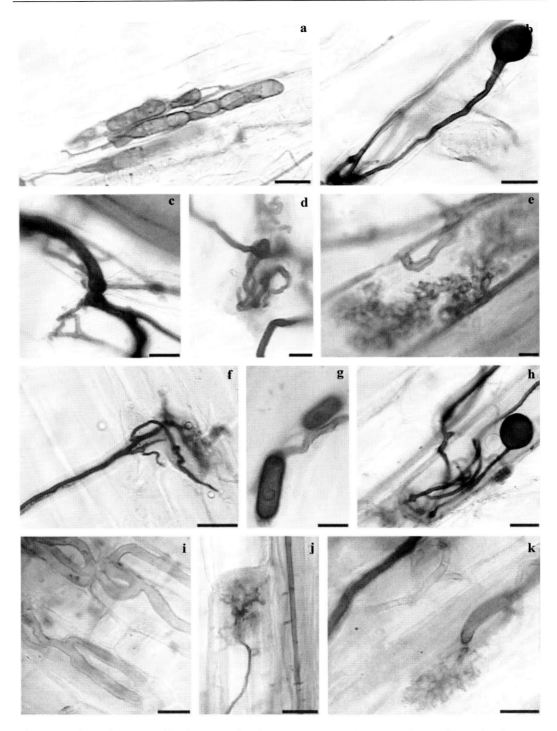

Figure 5. Arbuscular mycorrhiza in roots of *E. bogotense* (a-e) and *L. paniculatum* (f-k). (a,b,g,h) Intracellular hyphae and vesicles where lipidic bodies can be observed; (c,i) Cell-to-cell AM hypha; (d, i) Intracellular coil; (e,j,k) Arbuscule; (f) AM hypha entering the root thought a root hair (*Scale bars*: a = 50 µm; b,c,f,h.i = 20 µm; d,g,j,k = 10 µm; e = 5 µm).

The AM morphology observed in *E. bogotense* and *L. paniculatum* could not be clearly classified as *Arum* or *Paris*-type. Both species had intracellular coils, and *L. paniculatum* also presented some well defined cell-to-cell hyphae. When arbuscules were present, only one within each cell was observed (typical *Arum*-type arbuscules) and "arbusculate-coils" were not detected, as would be expected in a *Paris*-type pattern. Additionally, both species lacked typical *Arum*-type intercellular hyphae. Therefore, the AM morphology observed in these species was considered as an *Intermediate*-type (Figure 5, Table 1). This morphological type, which has been described for other plants (Dickson 2004) including some other pteridophytes (Duckett and Ligrone 1992, Zhang et al. 2004), has not been described before for any other plant species in this Valdivian temperate rainforest.

c) Consistently Mycorrhizal

A total of 68 sporophytes belonging to 8 species included in the families Blechnaceae (67.6 %), Dicksoniaceae (11.8 %), Dryopteridaceae (10.3 %), Pteridaceae (8.8 %), Gleicheniaceae (1.5 %) were sampled. These families are typically terrestrial, but some specimens were also collected from saxicolous (14.7 %) and epiphytic (1.5 %) habitats. The totality of the sporophytes analyzed had typical AM structures, independently of the substrate where they were growing. Due to the presence of extensive "arbusculate coils" and the absence of intercellular hyphae, the AM observed in these plants could be clearly classified as *Paris*-type (Figure 6). The percentage of root length colonized was wide-ranging among species and families, being greater than 63 % in all the species.

Only seven among the eight species colonized by AM were considered for the ANOVA analysis (*G. quadripartita* could not be included because only one specimen was found). The colonization values corresponding to *B. chilense*, *Megalastrum spectabile* and *Polystichum multifidum* were similar but significantly greater than those of *Adiantum chilense* (p = 0.021; p = 0.013; p = 0.008 respectively) and *B. magellanicum* (p = 0.009; p = 0.005; p = 0.003), which did not differ significantly between them. *Lophosoria quadripinnata* and *B. pennamarina* had intermediate colonization percentages and they did not show significant differences with any of the species in the other two groups. The colonization level of the only collected *G. quadripartita* sporophyte was similar to the value obtained for the species in the group with the lower percentages (this species was included in this mycorrhizal behavior group despite only one specimen was found because according to previous studies the Gleicheniaceae family is characterized for a strong presence of mycorrhizae -Boullard 1979, Gemma et al. 1992, Lee et al. 2001, Khade and Rodrigues 2002, Zhang et al. 2004-). At the family level, the highest mean colonization value corresponded to the Dryopteridaceae, which differed significantly from the Aspleniaceae (p = 0,002) and the Dicksoniaceae (p = 0.021), the families with the lowest colonization percentages (Table 1). In her study about mycorrhizae in Hawaiian pteridophytes, Gemma et al. (1992) also recorded the highest colonization level in this family. Even though we have insufficient evidence to make definitive conclusions about the influence of the substrate on AM colonization, some tendencies were observed. For example, the epiphytic *B. magellanicum* specimen and the *P. multifidum* sporophyte growing on a concrete block showed less colonization than the rest of the samples of the same species. The saxicolous specimens of *M. spectabile* and *A. chilense* also had lower values than the conspecific terrestrial sporophytes (Table 1).

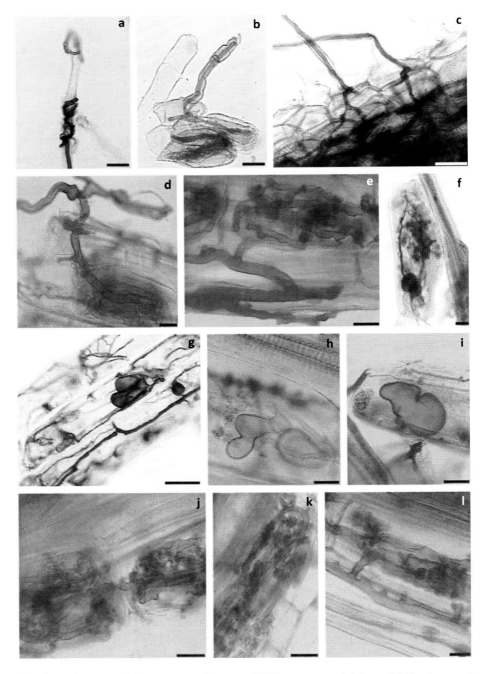

Figure 6. Arbuscular mycorrhiza structures in roots of different terrestrial ferns. (a) Hypha growing around a root hair of *P. multifidum*; (b,c) AM entering the root through a root hair of *M. spectabile* and *G. quadripartita* respectively; (d) Hypha entering through the root epidermins and coiling within the cortex of *A. chilense*; (e) Entry point in the root epidermis and coils in *B. chilense*, the hypha can be observed passing from one cell to the other; (f,g) Coiling hypha, vesicles and arbuscules in roots of *M. spectabile* and *L. quadripinnata* respectively; (h,i) Irregular vesicles observed in *A. chilense* and *L. quadripinata*; (j) Arbusculate coils in *A. chilense*; (k) Arbusculate coil that occupied the major part of the cell volume in *B. penna-marina*; (l) Detail of the arbuscules present in *B. chilense* (*Scale bars*: a = 10 μm; b-e,h-c = 20 μm; f = 50 μm; g = 100 μm).

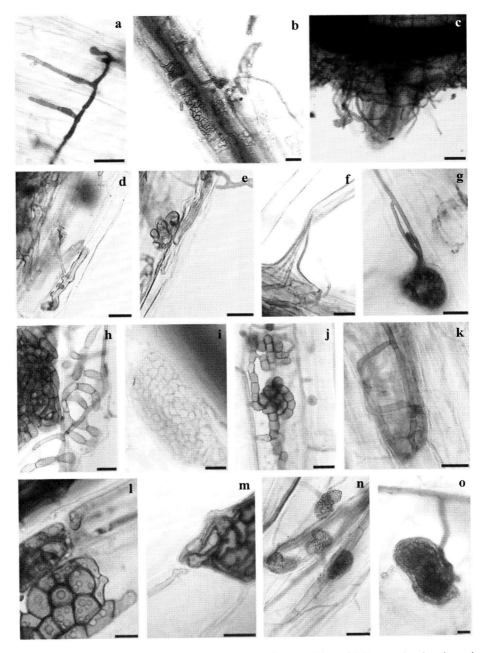

Figure 7. Colonization pattern of DSF in pteridophytes of Puerto Blest. (a) Runner hyphae branched at a 90° angle to the main hypha in *L. paniculatum*; (b) Extraradical hyphae in *S. caespitosa*; (c) Superficial hyphae net in *H. dentatum*; (d,e) Appressoria in roots of *M. spectabile* and *B. magellanicum* respectively; (f,g) DSE entering through the root hairs of *A. chilense* and *G. quadripartita*; (h,i) Intracelular moniliform hyphae in *E. bogotense* and *A. dareoides* respectively; (j) Highly melanized moniliform hyphae in roots of *H. tortuosum*; (k) DSE coiling within the cortical cells of *L. quadripinata*; (l,m) Hypha passing through adjoining epidermal cell walls by forming narrow penetration tubes which arise from appressorium-like structures and microsclerotia in *E. bogotense* and *B. magellanicum*; (o,p) Cerebriform microsclerotia in *H. seselifolium* and *E. bogotense* (*Scale bars*: a = 100 μm; b,c = 50 μm; d-n = 20 μm; o = 5 μm).

These results agree with previous studies that have reported the occurrence of AM in other species belonging to the same families (Berch and Kendrick 1982, Gemma et al. 1992, Muthukumar and Udaiyan 2000, Zhao 2000, Lee et al. 2001, Khade and Rodrigues 2002, Zhang et al. 2004), and indicate that these species are consistently mycorrhizal in Puerto Blest, suggesting a high dependence of these plants on this symbiosis. The occurrence of *Paris*-type morphology in this group of pteridophytes is concordant with different authors (Yamato and Iwasaki 2002, Zhang et al. 2004, Ahulu et al. 2005), including Smith and Smith (1997), who established that representatives of the leptosporangiate Filicales mostly have *Paris*-type AM. It has been suggested that *Paris*-type morphology keeps the energy supply to the fungi at a suitable level under stressful environmental conditions, and therefore it is most beneficial for plants growing slowly (Brundrett and Kendrick 1990a,b, Cavagnaro et al. 2001b, Yamato and Iwasaki 2002, Muthukumar et al. 2006). The fact that pteridophytes are mostly perennial plants with long life cycles and slow growth rates (Page 2002), and that the species considered in this chapter grow in high stress habitats (mainly as a consequence of the lack of light in the forest understory) is consistent with this hypothesis. According to Alaback (1991), the unique cool and wet climate of the temperate rainforests creates distinct ecological stresses to which plants must adapt. Therefore, the occurrence of *Paris*-type morphology can be considered as a good adapting strategy, since it might be contributing to make these plants shade tolerant.

DARK SEPTATE ENDOPHYTES

Root systems of all the sporophytes analyzed in the study presented in this chapter were also colonized by the DSE. These fungi are easily visible even as stained or pigmented structures under light microscopy. They were observed growing inter- and intracellularly within the cortex and on root surfaces (Barrow 2003). The pattern of DSE colonization was similar among the samples, and the standardized vocabulary proposed by Jumpponen and Trappe (1998) will be used in this chapter to describe it.

The initial colonization consisted in superficial narrow, septate and more frequently melanized runner hyphae or loose hyphal networks. Occasionally, runner hyphae were branched at a 90° angle to the main hypha. Superficial hyphae were sometimes extending from the root surface and associated with asexual fungal structures. At the point where DSE penetrated into the cortical cells, they formed different types of appressorium (a swollen structure preceding penetration). As it was observed for AM, these fungi were also capable of entering cortical cells through the root hairs. Once into the epidermal layer the fungi tended to coil, in occasions similar to AM. Then, the fungus pass through adjoining epidermal cell walls by usually forming narrow penetration tubes which arise from appressorium-like structures. While the fungi are growing intercellularly, they developed microsclerotia (intracellular groups of rounded, closely packed, thick-walled and usually darkly pigmented fungal cells), which could occupied the entire volume of the cell. A distinguishing feature of the DSE present in pteridophytes of Puerto Blest was the occurrence of "cerebriform microsclerotia". Dark septate fungi forming all these structures had straight, moniliform or irregularly lobed hyphae which sometimes got stained with Trypan-blue. The root stele was not colonized in any of the examined species and any distortion in the shape or size of the host root cells was observed (Figure 7).

Our results demonstrate that DSE are prevalent in pteridophytes of Puerto Blest, and their presence in non-mycorrhizal epiphytic species suggests that they are better adapted to grow within the roots of these plants than AM fungi. The production of melanized structures may indicate a function of altering environmental tolerances, or deterring insect and mammalian herbivores (Saikonnen et al. 1998, Mandyam and Jumpponen 2005).

This group of fungi cannot be overlooked while assessing the fungal communities of any ecosystem, as their abundance may equal or even exceed that of the AM. To understand evolutionary relationships in plant-endophyte interactions and how these associations are modified by environmental changes, effects of environment, genetic background, and genotype/environment interactions, DSE should be considered (Saikonnen et al. 1998).

CONCLUSION

According to Kessler et al. (2010), up to the moment only ca. 10 % of the known global fern and lycophyte flora have been studied and there are only some few works in which the degree of mycorrhization has been analyzed in pteridophyte assemblages. Therefore, this research yields valuable information about the occurrence of this symbiosis in this evolutionary important group of plants. This is the first time that the mycorrhizal status of a complete assemblage of pteridophytes in a Valdivian temperate rainforest is described, and our findings are outstanding. Something that has to be highlighted is that all the species of the same family tended to have the same mycorrhizal behavior regardless of the substrate on which they were found (terrestrial, epiphytic, saxicoulous). It was also demonstrated, as it was expected, that *Paris*-type is the most common AM morphology among pteridophytes of Puerto Blest. This AM morphology was registered in the eight consistently mycorrhizal species, while the *Intermediate*-type was found in the only two facultatively mycorrhizal species.

In respect to the influence of the substrate on the AM colonization, it can be resumed that AM were recorded in 2 %, 78 % and 100 % of the epiphytic, terrestrial and saxicolous samples respectively. This is in accordance with previous studies among ferns (Boullard 1979, Janos 1993, Nadarajah and Nawawi 1993, Muthukumar and Udaiyan 2000, Kessler et al. 2010) and angiosperms (Maffia et al. 1993, Michelsen 1993), which established that epiphytic plants are non-mycorrhizal or have significantly lower colonization rates than saxicolous and terrestrial plants, suggesting that the substrate plays an important role in determining the AM colonization intensity (Gemma et al. 1992, Janos 1993, Muthukumar and Udaiyan 2000, Kessler et al. 2010). However, it is important to notice that mycorrhizae are not completely absent in epiphytic sporophytes, terrestrial species occasionally growing in epiphytic habitats are mycorrhizal (the root length colonized was 54.2 % in the sporophytes of *B. magellanicum* growing on a trunk). In addition, typical epiphytic species collected from the forest ground completely lack AM. Consequently, these findings support the idea that the distinction between non-mycorrhizal and mycorrhizal states often is species specific (Wang and Qiu 2006, Kesser et al. 2010). The study carried out by Khade (2002) also supports this hypothesis. He did not find any correlation between root colonization percentages of different terrestrial fern species and AM spore density in the soil, and in his work it can be observed that some species are slightly colonized despite the presence of abundant inocula in the substrate. This situation raises the question of whether there are ecological conditions favoring mycorrhization among lycophytes and ferns, if some phylogenetic groups are especially likely to be colonized (Kessler et al. 2010), or if both factors have to be taken in account when analyzing the mycorrhizal status of this group of plants.

Regarding the possible relationship between evolutionary stages (eusporangite, leptosporangiate) and the mycorrhizal status of pteridophytes, Boullard (1979) postulated that lycophytes and primitive eusporangiate ferns, which mostly have fleshy roots, are more highly colonized by AM than more advanced leptosporangiate ferns, that usually have "advanced" fine roots. This means that the mycorrhizal colonization of pteridophytes is paralleled to their evolution, so that phylogenetically primitive species often possessed mycorrhizae, and advanced taxa tended to lack them (Boullard's hypothesis) (Boullard 1979, Berch and Kendrick 1982, Gemma et al. 1992, Zhao 2000). Trappe (1987) found a similar trend toward nonmycotrophy in angiosperms. Nevertheless, some studies (Cooper 1976, Berch and Kendrick 1982, Lee et al. 2001) and the information presented in this chapter do not support Boullard's theory, since it was showed that leptosporangiate ferns can be extensively colonized by AM. Furthermore, the existence of these fungi in all the sporophytes included in the terrestrial families suggests that in Puerto Blest terrestrial leptosporangiate ferns tend to be constantly mycorrhizal, while eusporangiate species are facultatively mycorrhizal.

Summarizing what has been discussed so far, in this hydrophilic forest there are 21 species of pteridophytes, 2 eusporangiate and 19 leptosporangiate. The former are facultatively mycorrhizal species that have an *Intermediate*-type AM association. Among the leptosporangiate ferns, epiphytic species (57.9 %) are non-mycorrhizal whereas terrestrial species (42.1 %) tend to be consistently mycorrhizal with *Paris*-type AM (Table 1). Taking all the species into account, AM were found in ten of them (47.6 %). This percentage is lower than in other parts of the world, where more than 56 % of the lycophytes and ferns have usually been regarded as mycorrhizal (Cooper 1976, Berch and Kendrick 1982, Gemma et al. 1992, Muthukumar and Udaiyan 2000, Lee et al. 2001, Zhang et al. 2004, Kessler et al. 2010, Lehnert et al. 2009). This value is also lower than the 82 % reported for angiosperms (Trappe 1987) or the 80 % estimated for land plants in general (Wang and Qiu 2006). That nine species (42.9 %) are in the Hymenophyllaceae might explain the high percentage of non-mycorrhizal pteridophytes, since members of this family lack AM associations in this forest.

The evolutionary trend that could explain the lower incidence of AM in pteridophytes than in seed plants is that the first originated earlier than gymnosperms and angiosperms, so they have evolved further in the direction of non-mycorrhizal (Zhao 2000). This explanation might be also plausible for the facultative behavior described for *E. bogotense* and *L. paniculatum*, which are generally considered as basal groups of pteridophytes (Pryer et al. 2004). It must be noticed that this evolutionary theory is opposite to the Boullard's hypothesis discussed before, but it definitely fits better to the results presented in this chapter.

Dark septate endophytes have been reported in several plant species from different habitats, including some pteridophytes (Cooper 1976, Berch and Kendrick 1982; Jumpponen and Trappe 1998, Kessler et al. 2010, Jumpponen 2001). The presence of these fungi in every species analyzed in this work accords with such studies and demonstrates that DSE are ubiquitous among pteridophytes of Puerto Blest. Some basic structures seem to be constant for DSE colonization regardless of the host species, such as superficial networks, inter- and intracellular hyphae and typical microsclerotia (Figure 7). The occurrence of cerebriform microsclerotia is especially remarkable, because as far as we are aware they have not been previously cited in the literature. The presence of DSE within the roots of lycophytes and ferns present in Puerto Blest demonstrates that these plants can be co-colonized by a variety of fungi (Winther and Friedman 2007). According to Allen et al. (1993) and Rains et al. (2003), the lack of AM in most of epiphytes in addition to the occurrence of DSE in many of them is notable and may suggest that DSE may function as "mycorrhizal fungi". Even more, some authors that have studied the occurrence of intraradical fungi in several species of lycophytes and ferns have already considered these fungi as mycorrhizal (Kessler et al. 2010, Lehnert et al. 2009). They also discussed that if DSE are excluded as potential mycorrhizal partners, the recorded mycorrhization in epiphytic or saxicolous habitats is low or absent, but if they are regarded as mycorrhizas, this level increase considerably (Schmidt et al. 1995, Kottke 2002, Kessler et al. 2010). However, in our opinion, further study is needed to completely test this hypothesis because it has been demonstrated that these plant-fungal interactions may also be neutral or antagonistic. Besides, they do not develop a specialized interface between the fungus and the host cells. The information presented in this chapter about DSE would be very useful to elucidate the nature and ecological importance of these poorly known root colonizing fungi in temperate rainforests, to understand how they influence the host and their relationship with AM fungi (Jumpponen and Trappe 1998, Saikkonen et al. 1998, Jumpponen 2001, Peterson et al. 2004).

To conclude, it is evident that there are diverse and complex factors contributing to the mycorrhizal status of lycophytes and ferns, such as habitat, environmental conditions, root morphology, genetics and taxonomy. However, one of the most important causes conditioning the mycorrhizal status (facultatively, consistently or non-mycorrhizal) and morphological type (*Intermediate* or *Paris*) of pteridophytes appears to be related to evolutionary stages (eusporangiate or leptosporangiate) (Table 1), as discussed by Zhao 2000. Since it has been demonstrated that pteridophyte–fungus patterns seem to be related to the evolution of plants, it is worthwhile to notice that this feature might be used as a phylogenetic criterion in further phylogenetic studies for pteridophytes (Gemma et al. 1992, Zhao 2000, Lee et al. 2001) and seed plants (Trappe 1987). This chapter contributes to the general knowledge of the mycorrhizal status of pteridophytes and discusses not only the classic literature regarding the occurrence of AM in these plants, but also the latest references on this subject. Altogether, it provides some of the necessary tools needed not only for future research in this field, but also for analyzing how global changes will affect mycorrhizas and pteridophytes in the Valdivian temperate rainforest region.

ACKNOWLEDGMENTS

We are grateful to Parques Nacionales for giving us the opportunity of working in a protected area. We thank Dr. Javier Puntieri and Dr. Cecilia Brion for helping us with the pteridophytes identification. Funds for this research were provided by grants PICT 22200 (FONCyT, ANPCyT), B121 and B143 (Universidad Nacional del Comahue).

REFERENCES

Ahulu, EM; Gollotte, A; Gianinazzi-Pearson, V; Nonaka, M. Cooccurring plants forming distinct arbuscular mycorrhizal morphologies harbor similar AM fungal species. *Mycorrhiza*, 2006, 17, 37–49.

Ahulu, EM; Nakata, M; Nonaka, M. *Arum-* and *Paris*-type arbuscular mycorrhizas in a mixed pine forest on sand dune soil in Niigata Prefecture, central Honshu, Japan. *Mycorrhiza*, 2005, 15, 129–136.

Alaback, PB. Comparative ecology of temperate rainforests of the Americas along analogous climatic gradients. *Rev. Chil. Hist. Nat.*, 1991, 64, 399-412.

Allen, MF; Rincon, E; Allen, EB; Huante, P; Dunn, JJ. Observations of canopy bromeliad roots compared with plants rooted in soils of a seasonal tropical forest, Chamela, Jalisco, Mexico. *Mycorrhiza*, 1993, 4, 27-428.

Arana, MD; Ponce, M; Vischi, NB. Sinopsis de los helechos y grupos relacionados (Pteridophyta) de la provincia de Córdoba, Argentina. *Bol. Soc. Arg. Bot.*, 2004, 39, 89-114.

Armesto, JJ; León-Lobos, P; Arroyo MK. Los bosques templados del sur de Chile y Argentina: Una isla biogeográfica. In: Armesto, JJ; Villagrán, C; Arroyo, MK (Eds.) *Ecología de los bosques nativos de Chile*. Editorial Universitaria Santiago de Chile, 1995, 23-28.

Arroyo, MTK; Riveros, M; Peñaloza, A; Cavieres, L; Faggi, AM. Phytogeographic relationships and regional richness patterns of the cool temperate rain forest of southern South America. In: Lawford, RG; Alaback, PB; Fuentes, ER (Eds.) *High Latitude rain forests and associated ecosystems of the west coast of the Americas: climate, hydrology, ecology and conservation*. Springer, Heidelberg, 1996, 134–172.

Barrow, JR. Atypical morphology of dark septate fungal root endophytes of *Bouteloua* in arid southwestern USA rangelands. *Mycorrhiza*, 2003, 13, 239–247.

Baylis, GTS. The magnolioid mycorrhiza and mycotrophy in root systems derived from it. In: Sanders, FE; Mosse, B; Tinker, PB (Eds.) *Endomycorrhizae*. Academic Press, New York, 1975, 373-379.

Benzing, DH. The vegetative basis of vascular epiphytism. *Selbyana*, 1986, 9, 23-43.

Benzing, DH. *Vascular Epiphytes: General biology and related biota*. Cambridge University Press, Cambridge, 1990.

Berch, SM; Kendrik, B. Vesicular arbuscular mycorrhizae of southern Ontario ferns and fern allies. *Mycologia*, 1982, 74, 769-776.

Bermudes, D; Benzing, DH. Fungi in neotropical epiphyte roots. *Biosystem*, 1989, 23, 65-73.

Boullard, B. Considerations sur les symbioses fongiques chez les Pteridophytes. *Syllogeus*, 1979, 19, 1-58.

Brion, C; Puntieri, J; Grigera, D; Calvelo, S. *Flora de Puerto Blest y sus alrededores*. Universidad Nacional del Comahue, Centro Regional Universitario Bariloche: General Roca, 1988.

Brundrett, M; Kendrick, B. The roots and mycorrhizas of herbaceous woodland plants I. Quantitative aspects of morphology. *New Phytol.*, 1990a, 114, 457–468.

Brundrett, M; Kendrick, B. The roots and mycorrhizas of herbaceous woodland plants II. Structural aspects of morphology. *New Phytol.*, 1990b, 114, 469–479.

Brundrett, MC. Coevolution of roots and mycorrhizas of land plants. *New Phytol.*, 2002, 154, 275–304.

Brundrett, MC. Diversity and classification of mycorrhizal associations. *Biol. Rev.*, 2004, 79, 473–495.

Cabrera, AL. *Territorios fitogeográficos de la República Argentina*. Enciclopedia Argentina de Agricultura y Jardinería. Editorial Acme SACI, Buenos Aires, 1976.

Cavagnaro, TR; Gao, LL; Smith, FA; Smith SE. Morphology of arbuscular mycorrhizas is influenced by fungal identity. *New Phytol.*, 2001a, 151, 469–475.

Cavagnaro, TR; Smith, FA; Lorimer, MF; Haskard, KA; Ayling, SM; Smith SE. Quantiative development of *Paris*-type arbuscular mycorrhizas formed between *Asphodelus fistulosus* and *Glomus coronatum*. *New Phytol.*, 2001b, 149:105–113.

Chaia, EE; Fontenla, SB; Vobis, G; Wall, LG. Infectivity of soilborne *Frankia* and mycorrhizae in *Discaria trinervis* along a vegetation gradient in Patagonian soil. *J. Basic Microbiol.*, 2006, 6, 263-274.

Cooper, KM. A field survey of mycorrhizae in New Zealand ferns. *N. Z. J. Bot.*, 1976, 14, 169-181.

Copely, J. Ecology goes underground. *Nature*, 2000, 406, 452-454.

de La Sota, ER; Ponce, MM; Cassá De Pazos, L. Pteridophyta. In: Correa, MN (Ed.) *Flora patagónica*. Parte I, Tomo I. Colección Científica del INTA, Buenos Aires, 1998, 282-371.

Dhillion, SS. Vesicular-arbuscular mycorrhizae of *Equisetum* species in Norway and USA: occurrence and mycotrophy. *Mycol. Res.*, 1993, 97, 656-660.

Dickson, S. 2004. The *Arum-Paris* continuum of mycorrhizal symbioses. *New Phytol.*, 2004, 63, 187-200.

Dimitri, MJ. *La región de los Bosques Andino-Patagónicos. Sinopsis General*. Tomo 10. INTA, Buenos Aires, 1972.

Duckett, JG; Ligrone, L. A light and electron microscope study of the fungal endophytes in the sporophyte and gametophyte of *Lycopodium cernuum* with observations on the gamethophyte-sporophyte junction. *Can. J. Bot.*, 1992, 70, 58–72.

Fernández, N; Fontenla, S; Messuti, MI. Mycorrhizal Status of Obligate and Facultative Epiphytic Ferns in a Valdivian Temperate Forest of Patagonia, Argentina. *Am. Fern J.*, 2010, 100, 16-26.

Fernández, N; Messuti, MI; Fontenla, S. Arbuscular mycorrhizas and dark septate fungi in *Lycopodium paniculatum* (Lycopodiaceae) and *Equisetum bogotense* (Equisetaceae) in a Valdivian temperate forest of Patagonia, Argentina. *Am. Fern J.*, 2008, 98:117-127.

Fitter, AH; Heinemeyer, A; Husband, R; Olsen, E; Ridgway, KP; Staddon, PL. Global environmental change and the biology of arbuscular mycorrhizas: gaps and challenges. *Can. J. Bot.*, 2004, 82, 1133-1139.

Gemma, JN; Koske, RE; Flynn, T. Mycorrhizae in Hawaiian pteridophytes: occurrence and evolutionary significance. *Am. J. Bot.*, 1992, 79, 843-852.

Gifford, EM; Foster, AS. *Morphology and Evolution of Vascular Plants*. 3rd Ed. WH Freeman and Company, New York, 1989.

Heckman, DS; Geiser, DM; Eidell, BR; Stauffer, RL; Kardos, NL; Hedges, B. Molecular evidence for the early colonization of land by fungi and plants. *Science*, 2001, 293, 1129–1133.

Hetrick, BAD. Mycorrhizae and root architecture. *Experientia*, 1991, 47, 355-362.

Hetrick, BAD; Wilson, GWT; Todd, TC. Relationships of mycorrhizal symbiosis, root strategy and phenology among tallgrass prairie forbs. *Can. J. Bot.*, 1992, 70, 1521-1528.

Janos, DP. Vesicular-arbuscular mycorrhizae of epiphytes. *Mycorrhiza*, 1993, 4, 1-4.

Janos, DP; Sahley, CT; Emmons, LH. Rodent dispersal of vesicular-arbuscular mycorrhizal fungi in amazonian Peru. *Ecology*, 1995, 76, 1852-1858.

Jumpponen, A. Dark septate endophytes – are they mycorrhizal? *Mycorrhiza*, 2001, 11, 207–211.

Jumpponen, A; Trappe, JM. Dark septate endophytes: a review of facultative biotrophic root-colonizing fungi. *New Phytol.*, 1998, 140, 295-310.

Kessler, M; Jonas, R; Cicuzza, D; Kluge, J; Piatek, K; Naks, P; Lehnert, M. A survey of the mycorrhization of Southeast Asian ferns and lycophytes. *Plant Biol.*, 2010, 12: 788-793.

Khade, SW; Rodrigues, BF. Arbuscular mycorrhizal fungi associated woth some pteridophytes from western ghat region of Goa. *Trop. Ecol.*, 2002, 43, 251-256.

Koide, RT; Li, M. Mycorrhizal fungi and the nutrient ecology of three old field annual plant species. *Oecologia*, 1991, 85, 403-412.

Koske, RE; Friese, CF; Olexia, PD; Hauke, RL. Vesicular arbuscular mycorrhizas in *Equisetum*. *Trans. Br. Mycol. Soc.*, 1985, 85, 350–353.

Kottke, I. Mycorrhizae – Rhizosphere determinants of plant communities. In: Waisel, Y; Eshel, A; Kafkafi, U (Eds.) *Plant roots – the hidden half*. Marcel Dekker Inc., New York, USA, 2002, 919–932.

Kubato, M; McGonigle, TP; Hyakumachi, M. Co-occurrence of *Arum*- and *Paris*-type morphologies of arbuscular mycorrhizae in cucumber and tomato. *Mycorrhiza*, 2005, 15, 73–77.

Lee, AE; Lee, SS; Lee, CH. Mycorrhizal symbioses found in roots of fem and its relatives in Korea. *J. Plant Biol.*, 2001, 44, 81-86.

Lehnert, M; Kottke, I; Linda, S; Pazmiño, S; Suárez, JP; Kessler, M. Mycorrhizal Associations in Ferns from Southern Ecuador. *Am. Fern J.*, 2009, 99, 292–306.

Lesica, P; Antibus, RK. The occurrence of mycorrhizae in vascular epiphytes of two Costa Rican rain forests. *Biotropica*, 1990, 22, 250-258.

Lücking, R; Wirth, V; Ferraro, LI; Cáceres, MES. Foliicolous lichens from Valdivian temperate rain forest of Chile and Argentina: evidence of an austral element, with the description of seven new taxa. *Glob. Ecol. Biogeogr.*, 2003, 12, 21–36.

Maffia, B; Nadkarni, NM; Janos, DP. Vesicular-arbuscular mycorrhizae of epiphytic and terrestrial piperaceae under field and greenhouse conditions. *Mycorrhiza*, 1993, 4, 5-9.

Mandyam, K; Jumpponen, A. Seeking the elusive function of the root-colonising dark septate endophytic fungi. *Stud. Mycol.*, 2005, 53, 173–189.

Mangan, SA; Adler, GH. Consumption of arbuscular mycorrhizal fungi by terrestrial and arboreal small mammals in a Panamanian cloud forest. *J. Mammal.*, 2000, 81, 563-570.

Manjunath, A; Habte, M. Root morphological characteristics of host species having distinct mycorrhizal dependency. *Can. J. Bot.*, 1991, 69, 671-676.

McGonicgle, TP; Miller, MH; Evans, DG; Fairchild, GL; Swan, JA. A new method which gives and objective measure of colonization of roots by vesicular-arbuscular mycorrhizal fungi. *New Phytol.*, 1990, 115, 495-501.

McIlveen, WD; Cole, H. Spore dispersal of Endogonaceae by worms, ants, wasps and birds. *Can. J. Bot.*, 1976, 54, 1486-1489.

Michelsen, A. The mycorrhizal status of vascular epiphytes in Bale Mountains National Park, Ethiopia. *Mycorrhiza*, 1993, 4, 11-15.

Muthukumar, T; Udaiyan, K. Vesicular arbuscular mycorrhizae in pteridophytes of Western Ghats, Southern India. *Phytomorphology*, 2000, 50, 132-142.

Muthukumar, T; Senthilkumar, M; Rajangam, M; Udaiyan, K. Arbuscular mycorrhizal morphology and dark septate fungal associations in medicinal and aromatic plants of Western Ghats, Southern India. *Mycorrhiza*, 2006, 17, 11–24.

Nadarajah, P; Nawawi, A. Mycorrhizal Status of epiphytes in Malaysian oil palm plantations. *Mycorrhiza*, 1993, 4, 21-25.

Nelson, G. *The Ferns of Florida: A Reference and Field Guide*. Pineapple Pr, United States, 2000.

O'Connor, PJ; Smith, SE; Smith, FA. Arbuscular mycorrhizas influence plant diversity and community structure in a semiarid herbland. *New Phytol.*, 2002, 154, 209–218.

Page, CN. Ecological strategies in fern evolution: a neopteridological overview. *Rev. Palaeobot. Palynol.*, 2002, 119, 1-33.

Peterson, RL; Massicote, HB; Melville, LH. *Mycorrhizas: Anatomy and Cell Biology*. NRC Research Press, Ottawa, 2004.

Phillips, JM; Hayman, DS. Improved procedures for clearing roots and staining parasitic and vesicular arbuscular mycorrhizal fungi for rapid assessment of infection. *Trans. Brit. Myc. Soc.*, 1970, 55, 158-160.

Pirozynski, KA; Malloch, DW. The origin of land plants: a matter of mycotrophism. *Biosystems*, 1975, 6, 153-164.

Ponce, M; Mehltreter, K; de la Sota, E. Análisis Biogeográfico de la diversidad pteridofítica en Argentina y Chile continental. *Rev. Chil. Hist. Nat.*, 2002, 75, 703-717.

Pryer, MK, Schuettpel, E; Wolf, PG; Schneider, H; Smith, AR; Cranfill, R. Phylogeny and evolution of ferns (monilophytes) with a focus on the early leptosporangiate divergences. *Am. J. Bot.*, 2004, 91, 1582-1598.

Rabatin, SC; Stinner, BR; Paoletti, MG. Vesicular-arbuscular mycorrhizal fungi, particularly *Glomus tenue*, in Venezuelan bromeliad epiphytes. *Mycorrhiza*, 1993, 4, 17-20.

Rains, KC; Nadkarni, NM; Bledsoe, CS. Epiphytic and terrestrial mycorrhizae in a lower montane Costa Rican cloud forest. *Mycorrhiza*, 2003, 13, 257-264.

Read, DJ. The ecophysiology of mycorrhizal symbiosis with special reference to impacts upon plant fitness. In: Scholes, JD; Barker, MG (Eds.) *Physiological Plant Ecology*. Press MC, Blackwell Science, London, 1999, 133-152.

Read, DJ; Duckett, JG; Francis, R; Ligrone, R; Russell, A. Symbiotic fungal associations in 'lower' land plants. *Phil. Trans. R. Soc. Lond. B*, 2000, 355, 815-831.

Reinhardt, DR; Miller, RM. Size classes of root diameter and mycorrhizal fungal colonization in two temperate grassland communities. *New Phytol.*, 1990, 116, 129-136.

Remy, W; Taylor, TN; Hass, H; Kerp, H. Four hundred-million-year-old vesicular arbuscular mycorrhiza. *Proc. Nat. Acad. Sci.*, 1994, 9, 11841-11843.

Roig, FA. La vegetación de la Patagonia. In: Correa, MN (Dir.). *Flora Patagónica*. Parte 1. Colección Científica INTA, Buenos Aires, 1998, 48-166.

Rosendahl, S. Communities, populations andindividuals of arbuscular mycorrhizal fungi. *New Phytol.*, 2008, 178, 253–266.

Rothwell, GW. Pteridophyte evolution: an often under-appreciated phytological success story. *Rev. Palaeobot. Palynol.*, 1996, 209–222.

Rowe, AR; Pringle, A. Morphological and molecular evidence of arbuscular mycorrhizal fungal associations in costa rican epiphytic bromeliads. *Biotropica*, 2005, 37, 245-250

Ryser, P; Lambers, H. Root and leaf attributes accounting for the performance of fast-and slow-growing grasses at different nutrient supply. *Plant Soil*, 1995, 170, 251-265.

Saikkonen, K; Faeth, SH; Helander, M; Sullivan, TJ. Fungal endophytes: a continuum of interactions with host plants. *Annu. Rev. Ecol. Syst.*, 1998, 29, 319–343

Schmid E., Oberwinkler F., Gómez L.D. Light and electron microscopy of a host–fungus interaction in the roots of some epiphytic ferns from Costa Rica. *Can. J. Bot.*, 1995, 73, 991–996.

Schweiger, PF; Robson, AD; Barrow, NJ. Root hair length determines beneficial effect of a *Glomus* species on shoot growth of some pasture species. *New Phytol.*, 1995, 131, 247–254.

Simon, L; Bousquet, J; Levesque, RC; Lalonde, M. Origin and diversification of endomycorrhizal fungi and coincidence with vascular land plants. *Nature*, 1993, 363, 67–69.

Sitte, P; Weiler, EW; Kadereit, JW; Bresinsky, A; Korner, C. Strasburger. *Tratado de Botánica*. 35th Ed. Ediciones Omega, Barcelona, 2004.

Smith, FA; Smith, SE. Tansley Review N° 96: Structural diversity in (vesicular)-arbuscular mycorrhizal symbioses. *New Phytol.*, 1997, 137, 373-388.

Smith, SE; Read, DJ. Mycorrhizal symbiosis. 3rd Ed. Academic Press, London, 2008.

Smith, AR; Pryer, KM; Schuettpelz, E; Korall, P; Schneider, HG. The classification for extant ferns. *Taxon*, 2006, 55, 705-731.

St. John, TV. Root Size, Root Hairs and Mycorrhizal Infection: A Re-Examination of Baylis's Hypothesis with Tropical Trees. *New Phytol.*, 1980, 84, 483-487.

Trappe, JM. Phylogenetic and ecologic aspects of mycotrophy in the angiosperms from an evolutionary standpoint. In: Safir, GR (Ed.) *Ecophysiology of va mycorrhizal plants*. CRC Press, USA, 1987, 5–25

Tryon, AF; Lugardon, B. *Spores of the Pteridophyta*. Springer-Verlag, New York, 1990.

van der Heijden, MGA; Klironomos, JN; Ursic, M; Moutoglis, P; Streitwolf-Engel, R; Boller, T; Wiemken, A; Sanders, IR. Mycorrhizal fungal diversity determines plant biodiversity, ecosystem variability and productivity. *Nature*, 1998, 396, 69–72.

Wang, B; Qiu, YL. Pyhlogenetic distribution and evolution of mycorrhizas in land plants. *Mycorrhiza*, 2006, 16, 299–363.

West, B; Brandt, J; Holstien, K; Hill, A; Hill, M. Fern-associated arbuscular mycorrhizal fungi are represented by multiple *Glomus* spp.: do environmental factors influence partner identity? *Mycorrhiza*, 2009, 19, 295-304.

Winther, JL; Friedman, WE. Arbuscular mycorrhizal associations in Lycopodiaceae. *New Phytol.*, 2007, 177, 790-801.

Yamato, M; Iwasaki, M. Morphological types of arbuscular mycorrhizal fungi in roots of understory plants in Japanese deciduous broadleaved forests. *Mycorrhiza*, 2002, 12, 291–296.

Zangaro W; Nishidate, RF; Spago Camargo, FR; Gorete Romagnoli, G; Vandressen, J. Relationships among arbuscular mycorrhizae, root morphology and seedling growth of tropical native woody species in southern Brazil. *J. Trop. Ecol.*, 2005, 21, 529-540.

Zhang, Y; Guo, L; Liu, R. Arbuscular mycorrhizal fungi associated with common pteridophytes in Dujiangyan, southwest China. *Mycorrhiza*, 2004, 14, 25-30.

Zhao, Z. The arbuscular mycorrhizae of pteridophytes in Yunnan, Southwest China: Evolutionary Interpretations. *Mycorrhiza*, 2000, 10, 145-149.

Zuloaga, FO; Morrone, O. *Catálogo de las Plantas Vasculares de la República Argentina: Pteridophyta, Gymnospermae y Angiospermae (Monocotyledoneae)*. Tomo I. Missouri Botanical Garden, Missouri, 1996.

In: Mycorrhiza: Occurrence in Natural and Restored… ISBN: 978-1-61209-226-3
Editor: Marcela Pagano © 2012 Nova Science Publishers, Inc.

Chapter 7

MYCORRHIZAS, AN IMPORTANT COMPONENT IN SEMIARID SITES

Marcela C. Pagano[1], Maria Rita Scotti[2] and Marta Noemi Cabello[3]

[1]Federal University of Ceará, Fortaleza, Ceará, Brazil
[2]Federal University of Minas Gerais, Belo Horizonte, Minas Gerais, Brazil
[3]Instituto Spegazzini, Researcher from CIC, National University of La Plata, Argentina

ABSTRACT

Tropical dry deciduous forests are under extreme climatic and **edaphic** environmental **conditions**, generally presenting shallow soils with low water availability. Fabaceae, Myrtaceae, and Meliaceae are the most common families with highest species richness in tropical dry forests, where plant communities are supposed to be dominated by mycorrhizal plants; however, the ecological traits of few plant species has been studied and verified as mycotrophic. The aim of this review is to explore the current information on the occurrence of mycorrhizas in the semiarid of Brazil, and to speculate about the benefit of symbiosis in semiarid sites. The diversity of arbuscular mycorrhizal fungi (AMF), analyzed based on spore morphology, revealed higher species richness and at least eighteen AM fungal known species present in the natural dry environments. The AMF spore diversity is illustrated, including those AMF which cannot be described at species level. The study of these areas, which commonly present sandy soils with low fertility, is essential to a successful restoration. For some regions, mixed forest used for land revegetation presents an AMF spore community composition more similar to that of the preserved sites. The spore community of the scrub vegetation usually present in disturbed sites is somewhat different. The semiarid's plant species contains natural AMF fungal species richness that can be affected by human activities, and AM dependent plants in natural ecosystems, may present higher potential for regeneration in habitats subjected to disturbance. We end with research directions that are needed to increase understanding of mycorrhizal associations in these environments. The problems encountered are discussed in this chapter, in order to highlight the need for a continual and integrated study of the semiarid ecosystems.

Keywords: Arbuscular mycorrhizal spores, Dry forest, Species inventory

INTRODUCTION

The plant communities on dry forests have been poorly investigated and are rarely included in research due to the difficulties of access or to low economic interest. Moreover, irrigation programs in the Semiarid of Brazil in Minas Gerais State as well as in other states (Yano-Melo et al. 2003) were created to increment the agricultural production, economic and social growth of the region. Within these programs, projects of wood provision for the local populations in Minas Gerais were established using agroforestry systems of mixed native species and *Eucalyptus* in a disturbed area (Pagano et al. 2009), where the abundance of scrub vegetation after forest clearing or burning, retards natural succession. Preserved areas as the Biological Reserve contains remnant dry deciduous forest (Rizzini 1997). Common tree species include *Anadenanthera colubrina* (Vell) Brenan, *Anadenanthera peregrina* (L.) Speg., *Balfourodendron molle* (Miq.) Pirani., *Myracrodruon urundeuva* Fr. Allem, *Plathymenia reticulata* Benth., *Schinopsis brasiliensis* Engl., *Cavannilesia arborea* K. Schum. (Rizzini 1997). Some of these species have been reported as mycotrophic. Reports for *A. colubrina* and *M. urundeuva* (Silva et al. 2001, Wang and Qiu 2006), *A. peregrina* (Pagano et al. 2008), *P. reticulata* (Pagano et al. 2009) and *S. brasiliensis* (Pagano et al. 2010) suggest a higher AM colonization and probably AMF dependence in these plants. According to Gentry (1995), the number of species in tropical dry deciduous forests is smaller than in semideciduous forests and rainforests, and this is usually attributed to extreme edaphic conditions such as shallow soils with low water availability and high availability of nutrients (Prado and Gibbs 1993). Leguminosae (Caesalpiniaceae, Mimosaceae, and Fabaceae), Myrtaceae, and Meliaceae are the most common families with highest species richness in tropical dry forests (Gentry 1995).

The mixed forests proposed as agroforestry systems for this region (Pagano et al. 2009), containing the legume tree *Plathymenia reticulata* Benth (Fabaceae), *Tabebuia heptaphylla* (Vell.) Tol. (Bignoniaceae) and *Eucalyptus camaldulensis* Dehnh (Mirtaceae) and others plant species (Pagano et al. 2008) were studied in more detail. For other mixed-species plantations with *Eucalyptus* that have also been tested in the Jaíba region, see Pagano (2007).

The dry forest has been seriously and continuously deforested due to adjacent land use, irrigated agriculture, burning and the extraction of wood, resulting in changes in vegetation type after disturbance. Disturbed sites presented vegetation with a floristic composition including herbs, shrubs and trees (Pagano 2007), which common plant species include, Fabaceae and Euphorbiaceae; however, this vegetation is still very poorly known, and most of the occurring species have not been reported on as mycotrophic according to the list of plant species compiled by Wang and Qiu (2006).

Arbuscular mycorrhizas (AM) are the most widespread members of the vast population of rhizospheric fungi that are able to symbiotically associate with roots of the majority of terrestrial plant families (Smith and Read 2008). AM are a critical component in ecosystems because these organisms can increase the access of the plant roots to the nutrients, especially phosphate resulting in an increase of plant growth (Smith and Read 2008), plant water stress tolerance (Gupta and Kumar 2000), and plant health (Gange and West 1994). They have decisive consequences for the survival and functioning of plant communities and ecosystems (van der Heijden 2003). However, plant growth responses depend on the particular host/fungus combination (van der Heijden et al. 1998, Klironomos et al. 2000).

The identification of AM species involved in various associations is essential for the functional diversity study of their populations (Smith and Read 2008). AM fungi have been separated from the polyphyletic phylum Zygomycota and placed in the new phylum Glomeromycota (Schüßler et al. 2001). Identification of AM fungi has relied extensively on the morphology of spores and related structures (Smith and Read 2008).

Only a few works have investigated AM diversity in natural areas of the Caatinga biome and mostly in the northeast region of Brazil (Pernambuco and Bahia States). Few detailed studies have been conducted on the mycorrhizal diversity in the semiarid of southeastern Brazil (Minas Gerais State) (Pagano et al. 2008, 2009).

The aim of this review is to explore the current information on the occurrence and diversity of mycorrhizal communities from semiarid Brazil. We also compare the diversity of AM of an undisturbed dry forest from Semiarid Brazil (Minas Gerais State) with an adjacent disturbed site as well as with mixed-forest plantations in order to analyze the implications for restoration purposes. An exhaustive inventory of AM spores was done taking samples along the year (at the dry and rainy seasons during visits at the sites in 2003-2006) to explore which AM morphospecies are involved in a particular ecological context having similar abiotic conditions but different biotic ones in a preliminary assessment.

REPORTS FOR BRAZIL

Plants growing under stress conditions show higher AM dependence particularly in semiarid and arid environments (Varma 1995, Nouaim and Chaussod 1996); however, few works have dealt with AM occurrence in the semiarid region in Brazil (e.g. Maia and Trufem 1990, Yano-Melo et al. 1997, Yano-Melo et al. 2003, Souza et al. 2003, Maia et al. 2010), most of them showing the diversity in cultivated areas in Pernambuco State, and few in natural areas in this biome.

Maia and Trufem (1990) studied the AM in cultivated areas of Pernambuco, while Yano-Melo et al. (1997) reported the AM in plantations of banana in the same State. In following reports Yano-Melo et al. (2003) reporting 21 taxa of AM in salinized and surrounded areas at the São Francisco submedium valley, in Pernambuco and Bahia states in Brazil (Table 1). In 2006, Maia et al. summarized twenty AMF species in natural caatinga from Brazil.

Also in Pernambuco, Albuquerque (2008) reported 29 AMF species in the caatinga, located in the municipalities of Caruaru, Serra Talhada and Araripina. The AMF community composition and structure differed between those areas, being *Acaulospora* the predominant genus.

Souza et al. (2003) found in areas with caatinga vegetation from Alagoas State that more than 95% of plants formed AMF, presenting from 5 to 80% of colonization, and 24 taxa of AMF.

For semiarid region of Minas Gerais, Brazil, Pagano et al. (2008, 2009) reported on the mycorrhizal associations of some plant species and vegetation types. In this review we focused on the vegetation growing in Minas Gerais State, in southeastern Brazil (Figure 1).

Table 1. Summary of evidence on AMF in Dry forest, Brazil

Source	Location/ forest type	Spore number	AMF Species Richness	Mycorrhizal colonization[#]
This study, Pagano et al. (2008)	Dry forest, Minas Gerais, Restored and disturbed sites	227.40*	18	50%
Mergulhão et al. (2009)	Native preserved caatinga, Araripina, Pernambuco State	344	19	73%
Lima et al. (2007)	Dry forest, Paraíba and Pernambuco States; Undisturbed, cultivated and disturbed sites	133 to 149; (7 to 8.5 viable spores)	ND	26 to 50%
Maia et al. (2006)	Dry forest, Brazil	ND	20	ND
Souza et al. (2003)	Caatinga vegetation, Alagoas State	4.15	24	20%
Santos et al. (2000)	Pernambuco and Bahia States	ND	ND	77%
Silva et al. (2001, 2005)	Preserved and mining disturbed caatinga, Bahia State	151	15	ND

ND = not determined in the study, # Maximal AM colonization reported, *Spore number $100g^{-1}$ soil.

Figure 1. Location of the study area, in the northern of Minas Gerais, Brazil.

REPORTS FOR MINAS GERAIS STATE BRAZIL

Soils and Climatic Conditions

In the Northwest region of Minas Gerais, Brazil, soils are sandy loam, acid, and nutrient-poor (Rizzini 1997). Soil texture is sandy with lower levels of clay and silt. Soils present low calcium, Mg and P concentration, and Soil organic matter (OM) (1.36%) a high base saturation (Pagano et al. 2009). Soil analysis, from undisturbed sites and from disturbed sites (forest is cut for wood purposes) showed more Ca, K and OM content in the preserved area than in the disturbed one, as well as more macropores in the disturbed site (Pagano et al. 2008).

The climate in the Northwest region of Minas Gerais, Brazil is semi-arid BSh according to Köppen's classification. Annual precipitation is of 1 mm (July) to 217 mm (December), mean temperature of 14.8 °C (dry season) to 34 °C (rainy season), and an annual mean of potential evaporation of 4.8 mm/day. Annual mean precipitations (871 mm) are concentrated in the November – March months (Pagano et al. 2009).

**Table 2. Soil characteristics and plants species studied
in the dry forest from southeast of Brazil**

Vegetal cover type	Soil pH	SOM	C	P	Sand[b] (%)	Plant species/ Cover	Reference
Natural	5.7	1.36	0.78	5	80	Remnant dry deciduous forest. Common tree species: *Anadenanthera colubrina, Anadenanthera peregrina, Balfourodendron molle, Myracrodruon urundeuva, Plathymenia reticulata, Schinopsis brasiliensis, Cavannilesia arborea*	Rizzini (1997)
Mixed forests	5.8	0.82	0.47	5.7 2	84	*Anadenanthera peregrina, Plathymenia reticulata, Tabebuia heptaphylla, eucalyptus camaldulensis*	Pagano et al. (2008) Pagano et al. (2009)
Disturbed	5.9	0.9	0.52	6	78	Herbs, shrubs and trees	Pagano et al. (2008)

[a]Mean of two measures from one composite sample. pH (H_2O) 1:1; SOM = Soil organic matter (%); C - carbon (%); Available P (mg dm^{-3}); [b]Particle size distribution: sand 2-0.02 mm.

Spores from fresh soil samples at three different sites (elevation 516 m above sea level) in the semiarid region in the State of Minas Gerais, Brazil, namely: Biological Reserve, Mixed forests and Disturbed site, were compared.

Rainy season

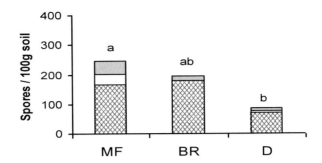

Dry season

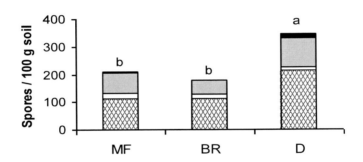

Figure 2. Spore abundance for the members of Acaulosporaceae, Gigasporaceae, Glomeraceae and Entrophosporaceae in the Biological reserve (BR), mixed forest (MF) and disturbed vegetation (D). Data are means of rainy and dry season sampling. Glomeraceae (*hatched*), Gigasporaceae (*white*), Acaulosporaceae (*gray*), Entrophosporaceae (*black*). Bars with the same letter do not differ significantly ($P < 0.05$) by Tukey's test.

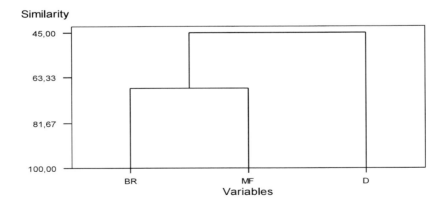

Figure 3. Complete link cluster analysis of 3 sampling sites in Minas Gerais, Brazil, based on similarities of AMF species composition at the dry (A) and rainy period (B) (*BR* Biological reserve; *D* Disturbed site; *MF* Mixed forest)//nudist dry for.

Mycorrhizas, an Important Component in Semiarid Sites

The undisturbed area, located into a Biological Reserve presenting preserved forest (PF) (15° 09'S, 43°56' W), supported a diverse flora, which fits the phytosociological description of woody Caatinga (Rizzini 1997). The Mixed forest (MF) (15°09'03"S, 43°49'26"W) were agroforestry systems of *Plathymenia reticulata* Benth (Leguminosae), *Anadenanthera peregrina*, noduliferous tree species that belongs to the pioneer vegetation, *Tabebuia heptaphylla* (Vell.) Tol. (Bignoniaceae), a non-legume native tree, and *Eucalyptus camaldulensis* Dehnh (Mirtaceae), selected for wood provision for the local populations. The disturbed (D) area adjacent to the mixed forest (15°09'03"S, 43°49'26"W) presented scrub vegetation (Table 2).

Spore Distribution and Diversity

In the Northwest region of Minas Gerais, the diversity of AMF, analyzed based on spore morphology, revealed higher species richness and at least eighteen AM fungal known species present in the natural dry environments. The AM spore diversity is illustrated in Figure 3, including those AMF which cannot be described at species level. This agrees with reports from Maia et al. (2006) who showed twenty AMF species in natural caatinga from Brazil.

Pagano et al. (2009) found *A. bireticulata* and *Glomus brohultii* (or possibly *Glomus macrocarpum*) for a natural dry forest vegetation type, and possibly new species of AMF for the disturbed site. However, other reports didn't show the occurrence of new species of AMF (Mergulhão et al. 2009) for a natural preserved site in Araripina, Pernambuco State.

The number of AM fungal spores detected in field-collected soils was relatively high, ranging from 186 to 227 spores/100 g of soil (Table 5).

All recovered spores found in the three types of vegetation sampled in Minas Gerais belonged to Acaulosporaceae, Entrophosporaceae, Gigasporaceae and Glomeraceae. In the rainy season, Glomeraceae occurrence was higher in PF and MF than in the D. However, in the dry season the disturbed soil presented a higher occurrence of this Glomalean family (Figure 5). In the dry season an increase of Acaulosporaceae were found in the three types of vegetation. In the disturbed site a pronounced increase also of Glomeraceae and Gigasporaceae was found, and notably, Entrophosporaceae was present (Figure 1).

Eighteen different morphotypes were detected overall, after the analysis of soil samples. Only fourteen isolates were morphologically identified at the species level: *Acaulospora bireticulata* Rothwell & Trappe, *A. delicata* Walker, Pfeiffer & Bloss, *A. laevis* Gerdemann & Trappe, *A. mellea* Spain & Schenck, *A. scrobiculata* Trappe, *A. rehmii* Sieverding & Toro, *Entrophospora infrequens* Ames & Schneid., *Gigaspora albida* Schenck & Smith, *Gi. margarita* Becker & Hall, *Glomus brohultii* Sieverd. & Herrera, *G. macrocarpum* Tul. & Tul., *G. taiwanense* (Wu & Chen) Almeida & Schenck, *S. cerradensis* Spain & Miranda and *Racocetra gregaria*. Among the rest there were one *Acaulospora*, and two *Racocetra* species. They were denominated as follows: *Acaulospora* sp. 1, *Racocetra* sp. 1, and *Racocetra* sp. 2 (Figure 2). Of the 18 AM fungal types obtained in this work, seven belong to Acaulosporaceae, one to Entrophosporaceae, seven to Gigasporaceae, and three to Glomeraceae.

Acaulospora scrobiculata and *Acaulospora* sp. 1 were the most frequent species in MF (dry period) and in D (rainy period), respectively. In general, *A. rhemii* and *S. cerradensis* presented lower isolation frequency.

Figure 4. Partial view of the preserved (a), disturbed (b) sites and mixed forest (c), in Minas Gerais, Brazil.

Nine species were found in both MF and in the Disturbed area. In the disturbed soil, species that proved almost exclusive was *R. gregaria* (Table 2). *Racocetra* sp. 1 and sp. 2 as well as *S. cerradensis* were also recovered from Disturbed area but with a lesser frequency. The preserved area showed fewer species and surprisingly, *Gigaspora albida* and *Gi. ramisporophora* were prominent (Table 2). In the three sites *G. brohultii* was the most frequent found species (~100% of samples). On the other hand, five species isolated from PF can be considered rare species, because they presented < 20% frequency (Table 2).

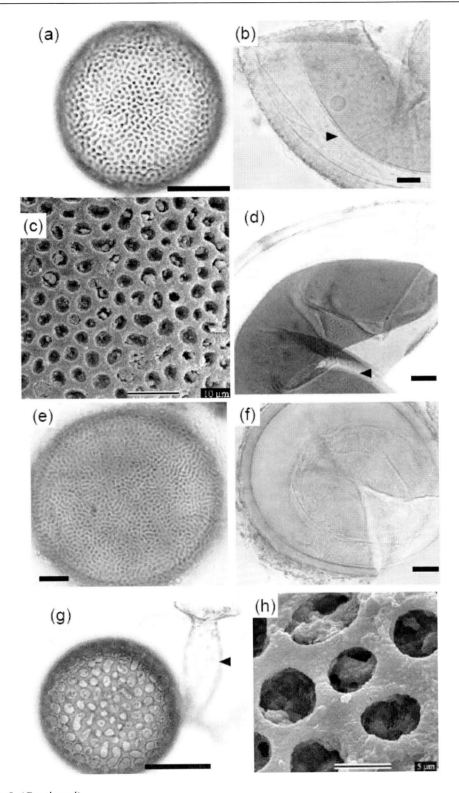

Figure 5. (Continued)

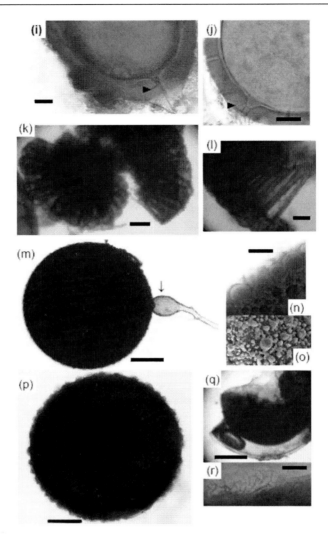

Figure 5. Some spores of species of Glomeromycota found in Minas Gerais, Brazil: 1- Acaulosporaceae (a-c) *Acaulospora scrobiculata*, (b) crushed spore in Melzer's reagent, germinative wall 2 (arrow), (c) detail of ovoid concave depressions on the surface, scanning electron micrograph (SEM), (d) *Acaulospora mellea* broken spore in Melzer's reagent showing germinal wall 2 with beads (arrow), (e) *Acaulospora rehmii*; (f) *Acaulospora* sp. 1 crushed spore, (g) *Acaulospora bireticulata* spore with soporiferous collapsed saccule (arrow head) in PVLG; (h) detail of double reticulum on outer wall of *A. bireticulata*; 2- Glomeraceae: (i) *Glomus brohultii* detail of spores and hypha attachment (arrow head); (j) *Glomus macrocarpum*; (k-l) *Glomus taiwanense*, (k) sporocarp and (l) detailed of spores; 3- Gigasporaceae: (m-o) *Racocetra gregaria* (m) General view of a spore with its subtending hypha (arrow), (n) Detail of surface ornamentations on outer wall consisting of warts, (o) SEM; (p) *Racocetra* sp. 1, (q-r) *Racocetra* sp. 2 (q) broken spore with germination shield and (r) detailed of wall ornamentation; Scale bar: a, e, g, k, p = 50 μm; b, c, d, f, i, o = 10 μm; j, h, j, n, r = 5 μm and m, q= 100 μm.

Description of unidentified AM species: *Acaulospora* sp. 1: spore pale yellow, 90- 180 μm, wall outer layer, 4.54 μm, smooth. *Racocetra* sp. 1: spore dark red brown, globose, 490 x 500 μm, wall 2.2 μm thick. Warts pale yellow, rounded dome-shaped warts (13.6 - 14 μm diam) 4 - 5 μm high, 6 - 7 μm wide, surrounded by small (4 μm diam) warts 2 – 2.5 μm high, 3.5 - 3.67 μm wide. *Racocetra* sp. 2: spore pale, globose, 463 x 463.6 μm, wall 2.2 μm thick. Two spore walls. Outer spore pale wall ornamented on the upper surface with a peculiar pattern. Inner wall brown.

In the Northwest region of Minas Gerais, there was considerable overlap (six species) with the species reported by Maia et al. (2006). Moreover, we found five species in common with Albuquerque (2008), who reported 14 to 21 species for natural sites of the caatinga biome in Pernambuco State. Furthermore, six AM species were in common with the list of species reported by Mergulhão et al. (2009).

With regard to Acaulosporaceae, species found in the studied areas such as *Acaulospora rehmii* and *A. scrobiculata* have been cited to occur in the Brazilian semiarid (Maia and Trufem 1990, Silva et al. 2001, Souza et al. 2003). *A. scrobiculata* has also been found in preserved and mining sites of the Bahia State semiarid, Brazil (Silva et al. 2005) and also in a natural preserved site in Araripina, Pernambuco State (Mergulhão et al. 2009).

Moreover, *A. mellea*, *A. rehmii* and *A. scrobiculata* have been cited for plantations in the State of Pernambuco (Silva et al. 2007), but no reports have been filed in the semiarid of Brazil for *A. bireticulata*, reported here as new to Brazilian semiarid species. In the present study Glomeraceae was dominant regardless of the vegetation cover of the site, *Glomus brohultii* Sieverd. & Herrera being the most representative species. *Glomus* was previously found in the Brazilian dry forest of Pernambuco (Silva et al. 2001) and Alagoas (Souza et al. 2003) States. *Glomus brohultii* has been previously reported for Cuba, tropical South America and Africa (Herrera-Peraza et al. 2003), and China (Wu et al. 2007, Chen et al. 2008). However, due to the difficulties of identification of some *Glomus* species, *Glomus brohultii* and *Glomus macrocarpum* possibly could be identified as the same AMF.

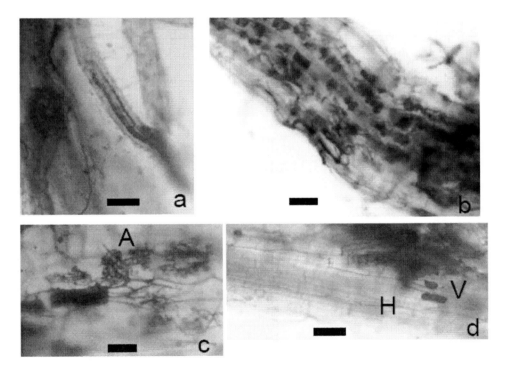

Figure 6. AM colonization in fine roots of the studied plants in Minas Gerais, Brazil; fine root of *Plathymenia reticulata* (a); arbuscules in root cells (b); Arbuscules in root cells of *S. brasiliensis* (c); (d); AM colonization in roots of *E. contortisiliquum* showing hyphae (H) and terminal vesicles (V). Bars for figures a, d = 100 μm; b, c = 50 μm.

Albuquerque (2008) reported *Glomus etunicatum* and *G. macrocarpum* to be common in the natural dry forest of Pernambuco State, Brazil; however, Mergulhão et al. (2009) reported the last species only for the disturbed site studied.

Glomus taiwanense (a sporocarpic species) was previously reported in the Atlantic Forest of the same Brazilian State (Goto and Maia 2005). Many sporocarpic species also have been reported to be specialists for grasslands (Oehl et al. 2003), which could explain their presence only in the scrub vegetation in the present study.

On the other hand, *S. cerradensis* has been frequently cited for Cerrado soils (Miranda and Miranda 1997), as well as *R. gregaria* was previously reported for non-salinized semiarid areas or with low salinity for the State of Pernambuco (Yano-Melo et al. 2003).

In the Northwest region of Minas Gerais we found *Gi. albida* and *Gi. margarita* in BR. The same two species have been cited to occur in the Brazilian semiarid (Maia and Trufem 1990, Silva et al. 2001), and in the caatinga vegetation of Alagoas State (Souza et al. 2003). Coincidentally, *Gi. albida* was the only *Gigaspora* species mentioned in arid zones of the Indian Thar Desert (Panwar and Tarafdar 2006) and also occurring in dry sites in Africa (Diallo et al. 1999). *Gigaspora margarita* was observed in a preserved caatinga site at Bahia State (Silva et al 2005) and in semiarid Sahelian areas (Dalpé et al. 2000), being mentioned for various countries (worldwide distribution). The other *Gigaspora* species (*G. ramisporophora*), presented in both D and BR, was also reported for the semiarid of Brazil (Albuquerque 2008).

The AM fungal richness in NW Minas Gerais was comparable to that found by Silva et al. (2005) in natural caatinga vegetation for the State of Bahia, and to that found by Souza et al. (2003) for the State of Alagoas. Moreover, this species richness was similar to the cultivated areas of caatinga for the State of Pernambuco (Maia and Trufem 1990). On the other hand, AM fungal richness was higher than that showed by Silva et al. (2001) for native caatinga in State of Bahia (Table 1).

The high AM species richness found in the disturbed site reflected in a higher sporulation in the dry period, bearing in mind the facultative mycotrophism of semiarid plants (Allen et al. 1995) and the presence of herbaceous and scrub vegetation (Table 2).

The spore number recovered from the soil samples in NW Minas Gerais was relatively high at the three sites. Most of the AMF spores obtained were small, corroborating the dominance of small-spored species of Glomeromycota, mainly fall into the genera *Acaulospora* and *Glomus* (Morton 1988), as a selective adaptation to water stress (Brundrett et al. 1999, Boddington and Dodd 2000). Some *Scutellospora* (with larger spores), though less dominant, were more common in the dry season at the disturbed site.

Moreover, Gai et al. (2006) for semiarid Tibetan grasslands reported *Glomus* and *Acaulospora* comprising 97% of the species, whereas in Minas Gerais we found that accounted for 50%. On the other hand, Stutz et al. (2000) did not find *Scutellospora* species in three arid regions of North America and Africa. However, Ulhmann et al. (2004) reported three species of *Scutellospora* in a forest savannah used for cattle grazing in Africa.

Lovera and Cuenca (2007) showed that there is a high AM diversity in the natural savannah from Venezuela and many new AM species occurring there. They pointed out that spores belonging to *Scutellospora* and *Gigaspora* are highly vulnerable to disturbance, related with the slow recovery and the type of plant community that finally will establish in the degraded areas.

In NW Minas Gerais, the morphotyping allowed us to identify the AMF spores in the field. Five AMF isolates (three *Racocetra*) were exclusive to semiarid disturbed area composed by scrub plants, suggesting a vegetation preference, and the importance of the native AMF for the type of plant community established in the area. It is opportune to mention the recently reorganization of *Gigasporaceae* with a new genera: *Racocetra* (Oehl et al. 2008, Morton and Msiska 2010).

It is known that most *Scutellospora* species have been described from warmer climates characterized by pronounced rainfall and a dry season (Tchabi et al. 2008). In our study, we recovered four previously named *Scutellospora* species, one in the mixed forest and the rest in the disturbed site. *Racocetra gregaria* (previously named *Scutellospora gregaria*, Figure 5) seems to be a common AM species in the disturbed site. Two different morphotypes of *Racocetra* were also present with less abundance. *Scutellospora cerradensis* was also present in the disturbed site (Table 3).

Table 3. Frequency of occurrence (%), of the AMF in the three sites at different seasons

N°	AMF species	Rainy period			Dry period	Rainy period	Dry period
DD		PF[†]	MF	D	PF	MF	D
	Acaulosporaceae						
1	*Acaulospora bireticulata*	0	33.3	0	0	16.6	50
2	*A. delicata*	0	33.3	0	0	33.3	0
3	*A. laevis*	0	33.3	0	50	0	0
4	*A. scrobiculata*	0	66.6	66.6	75	100	23.9
5	*Acaulospora* sp. 1	33.3	77.7	100	0	16.6	0
6	*A. mellea*	0	22.2	33.3	0	50	0
7	*A. rhemii*	0	33.3	0	25	16.6	25
	Entrophosporaceae						
8	*Entrophospora infrequens*	0	11.1	33.3	0	0	50
	Gigasporaceae						
9	*Gigaspora albida*	33.3	0	0	50	0	0
10	*Gigaspora margarita*	44.4	44.4	33.3	75	71.4	25
11	*G. ramisporophora*	0	0	100	50	0	0
12	*Scutellospora cerradensis*	0	0	0	0	0	25
	Racocetraceae						
13	*Racocetra gregaria*	0	0	0	0	16.6	75
14	*Racocetra* sp. 1	0	0	0	0	0	25
15	*Racocetra* sp. 2	0	0	0	0	0	33.3
	Glomeraceae						
16	*Glomus brohultii*	90.9	100	81.4	100	100	100
18	*Glomus macrocarpum*	0	0	0	0	0	33.3
17	*Glomus taiwanense*	0	0	0	0	0	33.3

RA (%); *PF* Preserved dry forest; *D* Disturbed; *MF* Mixed forest. [†]Site code (soil samples collected in 2003 and 2004).

Root Colonization by AMF

Microscopic analysis of the mycorrhizal status of roots of the studied trees in NW Minas Gerais revealed that all samples formed only AM, and no ectomycorrhizal fungi were detected.

The mycorrhizal colonization level was generally high (more than 48%, Pagano et al. 2010), thus reflecting the mycotrophic nature of the studied species. An "*Arum*-type" mycorrhizal structure was noted in all the studied species (Pagano et al. 2009, 2010). Moreover, the higher frequency of oval and elongated vesicles compared to irregular and lobed vesicles highlighted the dominance and diversity of *Glomus* species over *Acaulospora* species.

It has been showed that the vegetal species informed here are recommended for land restoration (Lorenzi 1992, Carvalho 2003), and the additional fact that they are AM dependent in natural and restored semiarid ecosystems supplies important information for restoration programs. The presence of the commonly *Glomus* spores, together with the observed *Glomus* type of colonization, suggest that this AM genus could be a potential inoculum for the trees.

AMF AND LAND DISTURBATION

In La Gran Sabana, southeastern Venezuela, disturbed sites, planted with *Brachiaria decumbens* and fertilized, presented a higher spore number; however, the AM diversity did not attain the level of natural ecosystem after 7 years (Cuenca et al. 1998). Moreover, the restoration efforts were not very successful (Lovera and Cuenca 2007).

In NW Minas Gerais, the comparison of mixed plantations and disturbed areas with the undisturbed one showed a change in AM species composition. Five AMF species occurred in both caatinga and MF, two were exclusive to PF (*A. delicata*, *G. albida*), and five occurred only in Disturbed area (*G. macrocarpum*, *G. taiwanense*, *S. cerradensis*, *Scutellospora* sp. 1 and *Scutellospora* sp. 2). A more related AM composition was found between the undisturbed vegetation and the mixed forest. When changes in vegetation type occur the AM species are also modified due to the preference of AM by plants in natural environments (van der Heijden et al. 2003), which could explain the more related AMF composition found between the BR and MF in comparison to D (Pagano et al. 2011). This agrees with suggestions that vegetation types with long-lived trees may have distinctive AM fungal biotas (Helgason et al. 1998, Merryweather and Fitter 1998).

Glomus brohultii was found in D, PF and MF sites (Table 3). *Glomus macrocarpum and G. taiwanense* were found in D, but were scarcely represented.

The results showed that *A. scrobiculata* was dominant in soil presenting scrub vegetation, which may be attributed to the high competitiveness of this species (Muthukumar and Udaiyan 2002).

Based on similarities of AMF species composition, the spore communities in MF and in PF were more similar to each other (72.43% similarity) in their composition than to the spore community of the disturbed vegetation. The complete link cluster analysis (Figure 3) of the sampling sites suggests a successful restoration, revealed by the higher AM species similarity, compared with those found in the native vegetation cover.

Timber extraction, burning and monocultures may reduce the fungal species found in the soil as well as change AM colonization percentage of roots, number of spores and AMF infectivity. Selective logging (thinning) can affect AM propagules increasing the number of viable spores, and showing more intense root colonization and a lower proportion of vesicles (Lima et al. 2007); however, the spores could not remain viable due to the severe climatic conditions.

We can conclude that the studied plant species contains natural AM fungal species richness that can be affected by land use, and that the restoration using these plants can be facilitated with the presence of AMF, as can be provided by inoculation (spores). Thus, AMF are an alternative for restoration of this biome. The potential uses of microbiological inoculants need more studies, and the AM formation must be included in a complete understanding of plant life histories.

Table 4. Distribution of AMF species at the 3 studied sites

Species	PF[†]	MF	D
Acaulospora bireticulata			
A. delicata			
A. laevis			
A. scrobiculata			
Acaulospora sp. 1			
A. mellea			
A. rhemii			
Entrophospora infrequens			
Gigaspora albida			
Gigaspora margarita			
G. ramisporophora			
Glomus brohultii			
Glomus macrocarpum			
Glomus taiwanense			
S. cerradensis			
Racocetra gregaria			
Racocetra sp. 1			
Racocetra sp. 2			
Spore number‡	186.73	227.40	217.31
Species richness	9	10	15

[†] Sites. ‡mean spores per 100 g-1 soil (samples collected in 2003 and 2004).

PROBLEMS ENCOUNTERED IN THE STUDIES AND FUTURE DIRECTIONS

The main problems facing AMF studies in Brazil are low financial support and the detrimental state of the highways used for field trips for collection of samples. What is lacking is a suitable body of research that can be used to undertake continuous field studies and the use of new methods derived from the molecular evaluation are also relevant. Complications due to the few time of the field surveys and few technical support provide extra problems. Moreover, a conditioned sample is critical, because some samples must be processed rapidly in order to do not be detrimental by fungal action.

More detailed studies involving nursery experiments under different conditions and seedlings development in the field are needed for most native species used for restoration purposes.

Table 5. Mycorrhizal status of the trees studied in southeast of Brazil

Family	Plant species	AM type	Previous report
Mimosaceae	*A. peregrina*	ND	AM[1]
	E. contortisiliquum	ND	AM[2]
Bignoniaceae	*T. heptaphylla*	*Arum*	AM[2]
Leguminosae	*P. reticulata*	*Arum*	AM[2,3]
Anacardiaceae	*S. brasiliensis*	*Arum*	AM[2]

Note: Indicated are the families. AM type: colonization type: *Arum* type, *Paris* type. Arbuscular mycorhizal (AM) colonization report. [1]Wang and Qiu (2006), [2]Pagano et al. (2010), [3]Pagano et al. (2009). ND= Not determined.

CONCLUSION

In the introduction to this chapter, we briefly described the available reports for Brazil and for the semiarid region in Minas Gerais State.

Throughout the chapter, we have showed that the analyzed tree species in Minas Gerais State, Brazil, maintain higher root colonization and spore numbers, confirming the benefic role of AM, and highlighting the importance of mycorrrhizae as an essential component for establishment and sustainability of plant communities in this biome. Nonetheless, further studies are required to achieve maximum benefits from these microorganisms and their associations.

Studying the presence and abundance of mycorrhizal symbionts in the dry forest is an important step in assessing the diversity and richness of the AM fungal community in this area. We thus focused on identifying AM fungi in soils according to the morphological characteristics of the fungi. However, this question must be further investigated since the diversity of spores in arid soils could not reflect the diversity of AMF in roots (Uhlmann et al. 2006).

As expected, AMF are a common and important component in semiarid vegetation in Brazil, which supplies important information for restoration programs. Since this chapter is primarily concerned with AMF occurrence in Brazil, we have refrained from discussing most studies in dry forests from other countries.

The evidence presented here suggests that the choice of tree species would have great implication in the conservation of AMF species, and that the ability of native AMF to colonize plants in natural conditions and the loss of these fungi with disturbance require more studies, due to the fact that highly dependent tree hosts should be selected over mycorrhizal-independent hosts.

ACKNOWLEDGMENTS

Marcela Pagano is grateful to CAPES (Council for the Development of Higher Education at Graduate Level, Brazil) for scholarships granted from 2003 to 2007, and for Post-doctoral scholarships granted from 2009 to 2010. Marta Noemí Cabello is researcher from Comisión de Investigaciones Científicas de la Provincia de Buenos Aires (CIC), Argentina.

REFERENCES

Albuquerque, PP. *Diversidade de Glomeromycetes e atividade microbiana em solos sob vegetação nativa do Semi-Árido de Pernambuco,* Doctoral dissertation, Recife: Federal University of Pernambuco; 2008 (in Portuguese).

Allen, EB; Allen, MF; Helm, DJ; Trappe, JM; Molina, R; Rincon, E. Patterns and regulation of mycorrhizal plant and fungal diversity. *Plant Soil*, 1995, 170, 47-62.

Boddington, CL; Dodd, JC. The effect of agricultural practices on the development of indigenous arbuscular mycorrhizal fungi. I. Field studies in an Indonesian ultisol. *Plant Soil*, 2000, 218,1–2.

Brundrett, MC; Abbot, LK; Jasper, DA. Glomalean fungi from tropical Australia I. Comparison of the effectiveness of isolation procedures. *Mycorrhiza*, 1999, 8, 305-314.

Carvalho, PER. *Espécies florestais brasileiras. Recomendações Silviculturais, potencialidades e uso da madeira.* Brasília: EMBRAPA-CNPF, 2003 (in Portuguese).

Chen, Y; Yuan, J; Yang, Z; Xin, G; Fan, L. Associations between arbuscular mycorrhizal fungi and *Rhynchrelyrum repens* in abandoned quarries in southern China. *Plant Soil,* 2008, 304, 257-266.

Cuenca, G; De Andrade, Z; Escalante, G. Arbuscular mycorrhizae in the rehabilitation of fragile degraded tropical lands. *Biol. Fertil. Soils,* 1998, 26, 107-111.

Diallo, AT; Samb, PI; Ducousso, M. Arbuscular mycorrhizal fungi in the semi-arid areas of Senegal. *Europ. J. Soil Biol.*, 1999, 35, 65-75.

Gai, JP; Feng, G; Cai, XB; Christie, P; Li, XL. A preliminary survey of the arbuscular mycorrhizal status of grassland plants in southern Tibet. *Mycorrhiza*, 2006, 16,191-196.

Gange, AC; West, HM. Interactions between arbuscular mycorrhizal fungi and foliar-feeding insects in *Plantago lanceolata* L. *New Phytol.*, 1994, 128,79-87.

Gentry, AH. Diversity and floristic composition of neotropical dry forests. In: Bullock SH, Money HA, Medina E, editors. *Seasonally dry tropical forests.* Cambridge: Cambridge University Press; 1995; 146-194.

Goto, BT; Maia, LC. Sporocarpic species of arbuscular mycorrhizal fungi (Glomeromycota), with a new report from Brazil. *Acta Bot. Bras.*, 2005, 19,633-637.

Helgason, T; Daniel, TJ; Husband, R; Fitter, A; Young, JPY. Ploughing up the wood-wide web. *Nature*, 1998, 394, 431.

Herrera-Peraza, RA; Ferrer, RL; Sieverding, E. *Glomus brohultii*: A new species in the arbuscular mycorrhiza forming Glomerales. *J. Appl. Bot.*, 2003, 77, 37-40.

Klironomos, JN; McCune, J; Hart, M; Neville, J. The influence of arbuscular mycorrhizae on the relationship between plant diversity and productivity. *Ecol. Letters*, 2000, 3,137-141.

Lima, RLFA; Salcedo, IH; Fraga, VS. Propagules of Arbuscular Mycorrhizae in P deficient soils under different land uses, in Semiarid NE Brazil. *Bras. Ci. Solo*, 2007, 31, 257-268.

Lorenzi, H. *Árvores brasileiras: manual de identificação e cultivo de plantas arbóreas nativas do Brasil.* Nova Odessa: Plantarum; 1992 (in Portuguese).

Lovera, M; Cuenca, G. Diversity of arbuscular mycorrhizal fungi (AMF) and mycorrhizal potential of the soil from a natural and a disturbed savannah from La Gran Sabana, Venezuela. *Interciencia,* 2007, 32, 108-114 (in Spanish).

Maia, LC; Trufem, SFB. Fungos micorrízicos vesículo-arbusculares em solos cultivados no Estado de Pernambuco, Brasil. *Rev. Bras. Bot.*, 1990, 13,89-95 (in Portuguese).

Maia, LC; Yano Melo, AM; Goto, BT. Filo Glomeromycota. In: Gusmão LFP, Maia LC, editors. *Diversidade e caracterização dos fungos do Semi-Árido Brasileiro.* Recife: Associação de Plantas do Nordeste (APNE); 2006; 109-126 (in Portuguese).

Maia, LC; Silva, G.A., Yano Melo, AM; Goto, BT. Fungos micorrízicos arbusculares no bioma Caatinga. In: Siqueira, JO, Souza, FA, Cardoso EJBN, Tsai SM. Micorrizas: 30 anos de pesquisa no Brasil, 2010, 311-339 (in Portuguese).

Mergulhão, ACES; Figueiredo, MVB; Burity, HA; Maia, LC. Host and successive cycles of multiplication affect the detection of arbuscular mycorrhizal fungi in gypsum mining impacted areas. *R. Árvore*, 2009, 33, 227-236 (in Portuguese).

Merryweather, JW; Fitter, AH. The arbuscular mycorrhizal fungi of *Hyacinthoides non-scripta* I. Diversity of fungal taxa. *New Phytol.*, 1998, 138,117-129.

Miranda, JCC; Miranda, LN. Micorriza arbuscular. In: Vargas M T, Hungria M, editors. *Biologia dos solos dos cerrados.* Planaltina, DF: Embrapa- CPAC; 1997; 69-123 (in Portuguese).

Morton, JB. Taxonomy of VA mycorrhizal fungi: classification, nomenclature and identification. *Mycotaxon,* 1988, 32, 267-324.

Morton JB., Msiska Z. Phylogenies from genetic and morphological characters do not support a revision of Gigasporaceae (Glomeromycota) into four families and five genera. *Mycorrhiza*, 2010, 20, 483-496.

Muthukumar, T; Udaiyan, K. Seasonality of vesicular arbuscular mycorrhizae in sedges in a semi-arid tropical grassland. *Acta Oecologica*, 2002, 23, 337-347.

Nouaim, R; Chaussod R. *Rôle des mycorhizes dans l'alimentation hydrique et minérale dês plantes, notamment des ligneux de zones arides.* CIHEAM - Options Mediterraneennes; 1996.

Oehl, F; Sieverding, K; Ineichen, K; Mader, P; Boller, T; Wiemken, A. Impact of land use intensity on the species diversity of arbuscular mycorrhizal fungi in agroecosystems of Central Europe. *Appl. Environ. Microbiol.*, 2003, 69, 2816-2824.

Oehl, F; de Souza, FA; Sieverding, E. Revision of *Scutellospora* and description of five new genera and three new families in the arbuscular mycorrhiza-forming *Glomeromycetes*. *Mycotaxon*, 2008, 106, 311-360.

Pagano, MC. *Characterization of Glomalean mycorrhizal fungi and its benefits on plant growth in a semi-arid region of Minas Gerais (Jaíba Project), Brazil.* Doctoral thesis, Belo Horizonte: Federal University of Minas Gerais; 2007.

Pagano, MC; Cabello, MN; Bellote, AF; Sá, NM; Scotti, MR. Intercropping system of tropical leguminous species and *Eucalyptus camaldulensis*, inoculated with rhizobia and/or mycorrhizal fungi in semiarid Brazil. *Agrofor. Syst.*, 2008, 74, 231-242.

Pagano, MC; Scotti, MR; Cabello, MN. Effect of the inoculation and distribution of mycorrhizae in *Plathymenia reticulata* Benth under monoculture and mixed plantation in Brazil. New Forests, 2009, 38,197-214.

Pagano, MC; Cabello, MN; Scotti, MR. Agroforestry In Dry Forest, Brazil: Mycorrhizal Fungi Potential. In: Kellymore LR, editor. *Handbook on Agroforestry: Management Practices and Environmental Impact.* Nova Science Publishers, New York, 2010, pp.367-388.

Pagano MC, Utida MK, Gomes EA, Marriel IE, Cabello MN, Scotti MR. Plant-type dependent changes in arbuscular mycorrhizal communities as soil quality indicator in semi-arid Brazil. *Ecological Indicators*, 2011, 11, 643-650.

Panwar, J; Tarafdar, JC. Distribution of three endangered medicinal plant species and their colonization with arbuscular mycorrhizal fungi. *J. Arid Environ.*, 2006, 65,337-350.

Prado, D; Gibbs, PE. Patterns of species distributions in the dry seasonal forest of South America. *Ann. Missouri Bot. Gard.*, 1993, 80,902-927.

Santos, BA; Silva, GA; Maia, LC; Alves, MV. Mycorrhizae in Monocotyledonae of Northeast Brazil: subclasses Alismatidae, Arecidae and Zingiberidae. *Mycorrhiza*, 2000, 10,151-153.

Schüßler, A; Gehrig, H; Schwarzott, D; Walker, C. Analysis of partial Glomales SSU rRNA gene sequences: implications for primer design and phylogeny. *Mycol. Res.*, 2001, 105,5-15.

Silva, GA; Maia, LC; Silva, FSB; Lima, PCF. Potencial de infectividade de fungos micorrízicos arbusculares oriundos de área de caatinga nativa e degradada por mineração, no Estado da Bahia, Brasil. *Rev. Brasil. Bot.*, 2001, 24,135-143 (in Portuguese).

Silva, GA; Trufem, SFB; Saggin Junior, O; Maia, LC. Arbuscular mycorrhizal fungi in a semiarid copper mining area in Brazil. *Mycorrhiza*, 2005, 15,47-53.

Silva, LX; Figueiredo, MVB; Silva, GA; Goto, BT; Oliveira, JP; Burity, HA. Fungos micorrízicos arbusculares em áreas de plantio de leucena e sábia no estado de Pernambuco. *R. Árvore*, 2007, 31,427-435 (in Portuguese).

Smith, SE; Read, DJ. *Mycorrhizal symbiosis*. 3rd ed. London: Academic Press; 2008.

Souza, RG; Maia, LC; Sales, MF; Trufem, SFB. Diversidade e potencial de infectividade de fungos micorrízicos arbusculares em área de caatinga, na Região de Xingo, Estado de Alagoas, Brasil. *Rev. Bras. Bot.*, 2003, 26,49-60 (in Portuguese).

Stutz, JC; Copeman, R; Martin, CA; Morton, JB. Patterns of species composition and distribution of arbuscular mycorrhizal fungi in arid regions of southwestern North America and Namibia, Africa. *Can. J. Bot.*, 2000, 78, 237-245.

Tchabi, A; Coyne, D; Hountondji, F; Lawouin, L; Wiemken, A; Oehl, F. Arbuscular mycorrhizal fungal communities in sub-Saharan Savannas of Benin, West Africa, as affected by agricultural land use intensity and ecological zone. *Mycorrhiza*, 2008, 18,181-195.

Uhlmann, E; Görke, C; Petersen, A; Oberwinker, F. Arbuscular mycorrhizae from semiarid regions of Namibia. Can. J. Bot., 2004, 82,645-653.

Uhlmann, E; Görke, C; Petersen, A; Oberwinkler, F. Arbuscular mycorrhizae from arid parts of Namibia. J. Arid Environ., 2006, 64,221-237.

van der Heijden, MGA. Arbuscular mycorrhizal fungi as a determinant of plant diversity: in search of underlying mechanisms and general principles. In: van der Heidjen MGA and Sanders IR, editors. *Mycorrhizal Ecology*. Berlin: Springer; 2003.

Varma A. Arbuscular mycorrhiza fungi: the state of the art. *Crit. Rev. Biotechnol.*, 1995, 15, 179-199.

Wang, B; Qiu YL. Phylogenetic distribution and evolution of mycorrhizas in land plants *Mycorrhiza,* 2006, 16, 299-363.

Wu, FY; Ye, ZH; Wu, SC; Wong, MH. Metal accumulation and arbuscular mycorrhizal status in metallicolous and nonmetallicolous populations of *Pteris vittata* L. and *Sedum alfredii* Hance. *Planta*, 2007, 226,1363-1378.

Yano-Melo, AM; Maia, LC; Morgado, LB. Fungos micorrízicos arbusculares em bananeiras cultivadas no vale do submédio São Francisco. *Acta Botânica Brasílica*, 1997, 11, 115-121 (in Portuguese).

Yano-Melo, AM; Trufem, SFB; Maia, LC. Arbuscular mycorrhizal fungi in salinized and surrounded areas at the São Francisco Submedium Valley, Brazil. *Hoehnea,* 2003, 30, 79-87.

In: Mycorrhiza: Occurrence in Natural and Restored…
Editor: Marcela Pagano

ISBN: 978-1-61209-226-3
© 2012 Nova Science Publishers, Inc.

Chapter 8

MYCORRHIZAL STATUS AND RESPONSIVENESS OF EARLY SUCCESSIONAL COMMUNITIES FROM CHAQUEAN REGION IN CENTRAL ARGENTINA

Carlos Urcelay, Paula A. Tecco, Marisela Pérez, Gabriel Grilli, M. Silvana Longo and Romina Battistella

Instituto Multidisciplinario de Biología Vegetal (UNC-CONICET) and FCEFyN,
Universidad Nacional de Córdoba, Casilla de Correo 495, 5000 Córdoba, Argentina

ABSTRACT

Theoretical models suggest that early successional communities are dominated by non-mycorrhizal plant species and thus, assume that mycorrhiza interactions do not play an important role in structuring these plant communities. Here we test these classic models gathering data from Chaquean region in central Argentina. The evidence shows that, in contrast to model predictions, most of the studied ruderal species in these communities harbor arbuscular mycorrhizal fungi (AMF) and dark septate endophytes (DSE) in their roots. In addition, data from synthetic experiment show that AMF have antagonistic effects on three abundant ruderal fast-growing pioneer species (growing alone and interacting with each other) and promote evenness in this simplified plant community. The effects of AMF on ruderal pioneers observed here mirror those attributed to herbivores and pathogens, suggesting a previous unrecognised mechanism by which AMF might promote secondary succession. Altogether, the evidence suggests that AMF and DSE should be included in models that predict the effects of fungal root symbionts in early secondary succession communities. The evidence so far suggests that AMF-ruderal species symbiosis has potential implications in agroecosystems management, ecological restoration and phytoremediation.

Keywords: plant strategies – ruderals – arbuscular mycorrhizas – dark septate endophytes – Chaquean region – Argentina

INTRODUCTION

Arbuscular mycorrhizal fungi (AMF) (Phylum Glomeromycota) are obligatory symbionts that depend on plant photosynthetic carbon and in turn, provide plants with soil nutrients or other benefits (Smith and Read 2008). The mycotrophic status of host plant and the effects of mycorrhizal fungi on plant growth depend on several factors, chief among them: phylogenetic lineage of host plant, plant strategy and abiotic properties of ecosystems. These aspects of the plant-fungus interaction impact on the structure of plant communities (Smith and Read 2008 and references therein).

It has been proposed that the mycorrhizal status and responsiveness of plant communities change during succession (Read 1989). According to classic ecological theory, after a disturbance, early secondary successional communities will be dominated by ruderal plants (R-species *sensu* Grime 1979). Based on evidence mostly gathered in temperate regions from the northern hemisphere, these ruderal plants are predicted to belong to the non-mycorrhizal families *Chenopodiaceae, Brassicaceae* and *Polygonaceae*. The reduction of mycorrhizal inoculum after disturbance has been proposed as a pre-requisite for their success (Smith and Read 2008).

It is now widely accepted that AMF influence plant communities structure and that the direction and magnitude of the effect is related to the mycorrhizal dependency of the dominants and subordinates plant species (Urcelay and Díaz 2003). Because fast-growing ruderals with short life-cycles frequently dominate in recently disturbed sites which generally undergo nutrient pulses and competitive release, it has been suggested that they may be less able to benefit from mycorrhizal colonisation (Grime 2001). Therefore, some theoretical models have predicted that in early successional communities, there will be no effect of AMF on plant composition since they are dominated by non-mycorrhizal species (Read 1989, Urcelay and Díaz 2003). Instead, in these communities AMF would facilitate the establishment of late successional species (i.e., mycorrhizal dependent hosts), promoting plant succession (Janos 1980, Allen and Allen 1990, Gange et al. 1993).

Altogether, the evidence and heuristic models suggest that early secondary successional communities would be mainly dominated by ruderal non-mycorrhizal plant families and that the role of AMF in these communities, if any, would be related to the establishment of late successional mycorrhizal species. However, whether this conceptual framework can be generalized to several plant communities in different ecosystems remains poorly known, particularly in Chaquean domain from southern South America.

In this chapter we review published and unpublished evidence on the mycorrhizal status of early successional plant species from the Chaco Serrano district ('*Chaquean region*' hereafter) and examine the possible role of AMF in these communities. Specifically, we aim to answer the following questions: (1) which is the taxonomic affiliation and the mycorrhizal status of the most abundant early successional species in the Chaquean region? (2) Do mycorrhizal fungi have a role in structuring these communities? and (3) Does the evidence fit previous models on the role of mycorrhizal fungi in early successional communities? (4) Which are the implications of the evidence on land management?

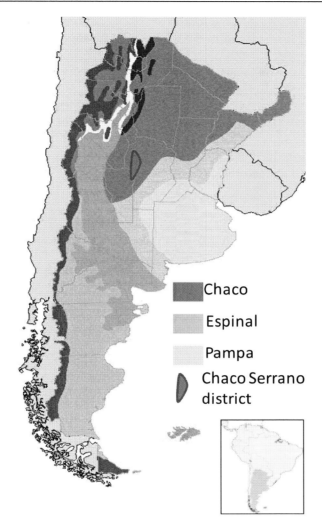

Figure 1. Biogeographic regions of Argentina. Chaco, Espinal and Pampa constitute the Chaquean Domain. Data of this chapter were mainly gathered in Chaco Serrano District within Chaco but extend to the entire Chaquean domain since species studied here are abundant in most of the ruderal communities present in this biogeographic domain (Zuloaga & Morrone 1996). Source of the map: www.ambiente.gov.ar.

MYCORRHIZAL STATUS OF RUDERAL SPECIES IN THE CHAQUEAN REGION

The dominant vegetation of mountains from central Argentina (from 29° 00' to 33° 30' S) corresponds to the woodland belt of the Chaco Serrano district that is included in the Chaquean region. This region together with the Espinal and Pampa constitute the Chaquean domain (figure 1) (Luti et al. 1979, Cabido et al. 1998). Historic mean annual precipitation in Chaco Serrano district is 720 mm (de Fina 1992), with most of the rainfall occurring in the warm season. Historic mean annual temperature is 14.5 ° C (de Fina 1992). Soils are sandy, well drained and shallow (litic Ustorthents Entisols). Chemical properties in representative area (31°30' N, 64°35' W, 880 masl) average: 7.01 % $C_{organic}$, 2.85 mg g^{-1} N_{total}, 0.17 mg g^{-1} P_{total} and pH = 6.6 (Urcelay et al. 2009). Due to human activities such as livestock grazing, logging and frequent fires, the original woodland has been almost completely transformed into a mosaic of grassland, shrubland and open woodland (Gavier and Bucher 2004, Zak and Cabido 2002, Zak et al. 2004). As a consequence of this scenario early secondary successional plant communities are widely extended along the regional landscape.

Table 1. Mycorrhizal status of ruderal early successional plant species from the "Chaquean region" in central Argentina

Family and species	Mycorrhizal status
Asteraceae	
Bidens pilosa L. var. Pilosa	AM-DSE
Bidens subalternans DC.	AM-DSE
Carduus thoermeri Weinm	AM-DSE
Galinsoga parviflora Cav.	AM-DSE
Tagetes minuta L.	AM-DSE
Taraxacum officinale G. Weber ex F. H. Wigg.	AM-DSE
Xanthium cavanillesii Schouw	AM-DSE
Zinnia peruviana (L.) L.	AM-DSE
Chenopodiaceae	
Chenopodium ambrosioides L.	NM
Euphorbiaceae	
Euphorbia acerensis Boiss.	AM-DSE
Euphorbia dentata Michx	AM-DSE
Fabaceae	
Vicia graminea Sm.	AM-DSE
Lamiaceae	
Leonurus japonicus Houtt.	AM-DSE
Poaceae	
Bromus catharticus Vahl var. Catharticus	AM-DSE
Cenchrus spinifex Cav.	AM-DSE
Muhlenbergia peruviana (P. Beauv.) Steud.	AM
Melinis repens (Willd.) Zizka	AM-DSE
Solanaceae	
Petunia axillaris (Lam.) Britton, Stern & Poggenb.	AM-DSE

AM: arbuscular mycorrhizas, DSE: dark septate endophytes, NM: non-mycorrhizal.

As in many other systems, Asteraceae and Poaceae are by far the families with highest number of early successional species (i.e. ruderals) in the Chaquean region, particularly in Chaco Serrano district (Giorgis et al. in press). By reviewing published evidence (Roumet et al. 2006, Urcelay and Battistella 2007, Fracchia et al. 2009) and unpublished data from the authors, we found out that the majority of the ruderal species with known mycorrhizal status harbour arbuscular mycorrhizal fungi and dark septate endophytes (DSE) in their roots (Table 1). Only one species, *Chenopodium ambrosioides*, lacks fungal symbionts in roots.

The fact that most of these species with AM belong to Asteraceae, Poaceae, Euphorbiaceae, Fabaceae, Lamiaceae and Solanaceae (Table 1), do support the proposed relationships between the phylogenetic affiliation of host plants and their mycorrhizal status (Trappe 1987, Brundrett 2009). However, the evidence of AMF in the roots of most of the ruderal species in these communities do not support the theoretical models discussed above. In other words, early successional communities in central Argentina are dominated by AM plant families, and this could be a sign of an important influence of AMF on growth and competition on these ruderal species.

It is well known that AMF generally affect plant growth. If the effect of AMF is not neutral, and is either positive or negative (Johnson et al. 1997), AMF could affect the plant composition and the dominance hierarchies as seen for some plant communities (Urcelay and Díaz 2003), hence influence plant succession. It is worth asking then, how do AMF affect ruderal species in early successional communities from the Chaquean region? If the effect on dominant early ruderals is positive, AMF would favour dominance and enhance productivity probably inhibiting plant succession. On the contrary, if the effect is negative, AMF would weigh down dominance hierarchies and plant productivity, and in turn promote plant succession as has been observed for other soil organisms. This second alternative resembles the effect of some invertebrates and pathogens, which have been shown to negatively affect early successional species, in particular those who dominates, promote shifts in plant composition, leading then to plant succession (Brown and Gange 1989, 1992, van der Putten et al. 1993, Bradford et al. 2002, De Deyn et al. 2003). In the next section we explore these alternative scenarios through an experimental approach on three abundant ruderal species of the Chaquean region.

Models on mycorrhizal type distribution and their role in ecosystems generally do not consider DSE (Read 1989, 1991). These fungi are a widespread group of root symbionts (Mandyam and Jumpponen 2005) generally belonging to Helotiales (Ascomycota) (Upson et al. 2009) that range from mutualism to parasitism (Jumpponen 2001, Mandyam and Jumpponen 2005). Whether they should be considered mycorrhizal or not, still remains controversial (Smith and Read 2008). Nonetheless, it has been shown that this type of symbionts might be nutritionally important to plants in adverse environments (Haselwandter and Read 1982, Newsham 1999, Barrow and Osuna 2002). Data from the Chaquean region suggest that they are widely distributed in these ecosystems and that they might have a role in early successional communities. Despite this, no study have examined their effects on plant growth in this region therefore it would be clearly desirable to evaluate the role of these symbionts in structuring these plant communities.

**Table 2. ANOVA for the effects of mycorrhizal fungi and Neighbour on *Bidens pilosa,*
Tagetes minuta and *Zinnia peruviana* shoot biomass**

Effect	F	df	P
Bidens Shoot biomass			
Mycorrhiza	696.81	1	<0.0001
Neighbour	16.55	3	<0.0001
M x N	5.22	3	0.0026
Tagetes Shoot biomass			
Mycorrhiza	405.08	1	<0.0001
Neighbour	18.59	3	<0.0001
M x N	0.64	3	0.5950
Zinnia Shoot biomass			
Mycorrhiza	40.58	1	<0.0001
Neighbour	2.97	3	0.0376
M x N	0.42	3	0.7420

THE ROLE OF AMF IN RUDERAL COMMUNITIES

The evidence regarding the effects of mycorrhizal responsiveness in Chaquean region is very scarce. It has been recently shown that ruderal *Bidens pilosa* is slightly negatively or not affected by AMF, either growing alone or subjected to intraspecific competition while the deciduous shrub *Acacia caven* (Fabaceae) is positively affected by AMF growing alone and at low densities of intraspecific competition (Pérez and Urcelay 2009). This evidence would be in line with the assumption that species related to contrasting ecological strategies are differentially affected by AMF in this region. Given the framework presented in the introduction of this chapter and considering the mycorrhizal status of ruderal species shown in previous section, three specific questions arise: (1) which is the effect of AMF on growth of ruderal fast growing species that frequently succeed in disturbed sites? (2) How do AMF affect interespecific competition between ruderal neighbours? and (3) How do AMF effects on growth and neighbour competition affect the overall productivity?

These questions were assessed by carrying out a greenhouse experiment considering three abundant species that frequently dominate disturbed sites in the Chaquean domain (figure 1): *Bidens pilosa, Tagetes minuta* and *Zinnia peruviana* (figure 2). Specifically, the study examined the effect of AMF on the growth of these locally abundant ruderal fast-growing species. The three species belong to Asteraceae and, as showed above, are known to be highly colonised by AMF in the field (Urcelay and Battistella 2007). These species generally coexist in disturbed sites and are then replaced by perennial forbs and grasses together with shrubs being finally dominated by Chaquean woodland-shrublands.

The source of mycorrhizal inoculums used to examine the effect of AMF on plant growth included spores communities and root fragments as in Pérez and Urcelay (2009) because it has been clearly recognised that the net effect of a set of AMF isolates on plant growth is different from the effect of each fungal isolates (van der Heijden et al.1998). The selected inoculum have all the isolates (with their inherent proportions) that occur in the field, having in this way, a more realistic picture of the effect of mycorrhizal fungi on plant species in the natural community (for further details see Pérez and Urcelay 2009).

Figure 2. Ruderal species (Asteraceae) grown in greenhouse with (+M) and without (-M) arbuscular mycorrhizal fungi.

EFFECT OF AMF ON GROWTH AND INTERACTIONS BETWEEN RUDERAL SPECIES

When growing alone, the total biomass of *Bidens* and *Tagetes* were negatively affected ($T= -16.3716$, $P<0.00001$; $T= -10.4430$, $P<0.00001$, respectively) while *Zinnia* was not significantly affected ($T=-1.8196$, $P= 0.0876$) by AMF. With AMF, *Bidens* grew three-fold and *Tagetes* two-fold less than without the presence of fungal symbionts. These results together with previous evidence (e.g. Francis and Read 1995, Klironomos 2003), suggest that antagonistic or parasitic effects of AMF on ruderal plants species is frequent in these type of communities. Rapid plant growth, high seed set together with the occurrence in sites in which arbuscular mycorrhizal fungi has been reduced, such as disturbed sites (Jasper et al. 1991), could explain the success of these species in the early successional stages after disturbance in natural communities.

The shoot biomass of *Bidens*, *Tagetes* and *Zinnia* were negatively affected by the mycorrhizal fungi and the neighbour presence (Table 2, figure 3). In the presence of AMF, *Bidens* was significantly affected when growing with *Zinnia* and both neighbours together (Figure 3a). Without AMF, *Bidens* was mainly affected by both neighbours together. *Tagetes* shoot biomass was negatively affected by both neighbours, either alone or together, with and without AMF (Figure 3b). In turn, *Zinnia* shoot biomass was significantly affected by the presence of neighbours but only significantly in presence of AMF (Figure 3c). These results agree with previous evidence in which specific plant combinations altered the way by which AMF affect plant interactions (Marler et al. 1999, Callaway et al. 2003).

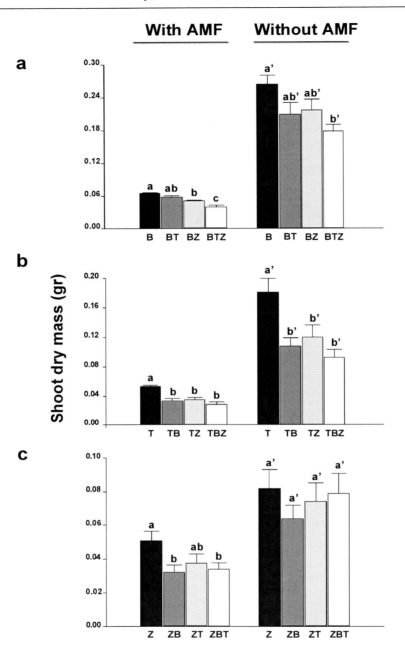

Figure 3. Shoot biomass of a) *Bidens pilosa*, grown with and without AMF in isolation and in neighbor presence: (BT) with *Tagetes minuta*, (BZ) with *Zinnia peruviana*, and (BTZ) with *Tagetes* and *Zinnia* together. b) *Tagetes minuta*, grown with and without AMF in isolation and in neighbor presence: (TB) with *Bidens pilosa*, (TZ) with *Zinnia peruviana*, and (TBZ) with *Bidens* and *Zinnia* together. c) *Zinnia peruviana*, grown with and without AMF in isolation and in neighbor presence: (ZB) with *Bidens pilosa*, (ZT) with *Tagetes minuta*, and (ZBT) with *Bidens* and *Tagetes* together. Error bars indicate + 1 SE ($n = 10$ replicates). Bars with the same letters are not significantly different within each treatment (i.e. with and without AMF) (Tukey's HSD test, $P<0.05$).

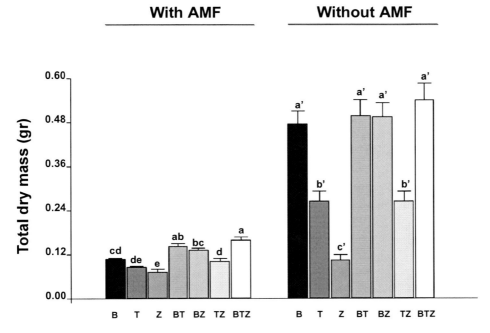

Figure 4. Total biomass per pot of species grown with and without AMF alone and species combination: (B) *Bidens pilosa*, (T) *Tagetes minuta*, (Z) *Zinnia peruviana*, (B-T) *Bidens* and *Tagetes*, (BZ) *Bidens* and *Zinnia*, (TZ) *Tagetes* and *Zinnia*, (BTZ) *Bidens*, *Tagetes* and *Zinnia*. Error bars indicate + 1 SE ($n = 10$ replicates). Bars with the same letters are not significantly different within each treatment (i.e. with and without AMF) (Tukey's HSD test, $P<0.05$).

With the exception of *Zinnia* growing alone, the total biomass of all neighbour combinations were markedly lower when growing with mycorrhizal fungi, indicating a strong overall negative effect of arbuscular mycorrhizal on these ruderal species (Figure 4). In general, *Bidens* showed the highest biomass followed by *Tagetes* and *Zinnia* respectively. These differences were most evident when growing without AMF (figure 2).

It has been postulated that the negative effect of AMF on certain herbaceous species could be attributed to a soil fungi-mediated parasitism in which fixed carbon or other resources are transferred from one plant to another (beneficiary) via common mycorrhizal network (Callaway et al. 2001, Callaway et al. 2003). In the present study no species benefited either by AMF or neighbour presence, suggesting direct adverse effects of AMF on these ruderal forbs.

Table 3. ANOVA for the effect of arbuscular mycorrhizal fungi and neighbours composition on shoot, root and total biomass per pot

	F	Df	P
Total biomass			
Mycorrhiza	620.18	1	<0.00001
Combination	62.00	6	<0.00001
M x N	6.62	6	<0.00001

EFFECTS OF AMF ON DOMINANCE IN RUDERAL COMMUNITY

Through changes in the competitive abilities of plant species, above- and belowground biotic interactions are known to determine plant community structure (Grime et al. 1987, Crawley 1997, Olff and Ritchie, 1998, van der Heijden et al. 1998, Hartnett and Wilson 1999, Klironomos 2002). As mentioned above, AMF are known to promote plant succession by having no effects on early successional pioneer species and improving nutrient uptake and growth on later successional species (Janos 1980, Allen and Allen 1990, Gange et al. 1993). On the other hand, it has been demonstrated that herbivores (Crawley 1997, Olff and Ritchie 1998), soil pathogens (van der Putten et al. 1993, Bever 1994, Packer and Clay 2000), and root-feeding insects (Brown and Gange 1989, 1992, Wardle and Barker 1997, Bradford 2002) influence plant species composition leading to successional changes by selective suppression of early stage plant species. The negative effect of AMF on pioneers shown here, mirror those attributable to herbivores and pathogens, suggesting an additional mechanism by which AMF might promote plant succession. Accordingly, the Simpson dominance index in the three-species treatment was significantly higher when AMF was absent (T test, $P<0.05$), indicating that together with the overall negative effects on plant species, the fungal symbionts promoted evenness in these synthetic communities. These results are consistent with other findings in which invertebrate soil fauna promoted secondary grassland succession stages, by selectively suppressing early dominants, and enhanced evenness in the plant community (De Deyn et al. 2003). Moreover, our results suggest that AMF not only promote secondary plant succession through positive effects on species from later stages of succession (Perez and Urcelay 2009) but also by antagonistic effect on ruderal pioneers, in particular those which dominate.

EFFECTS OF AMF ON PRODUCTIVITY IN RUDERAL COMMUNITY

It has been previously demonstrated that AMF could enhance productivity (Grime et al. 1987; Gange et al. 1990, van der Heijden et al. 1998). However, the effect of AMF on productivity of a community might be variable because it depends on plant mycorrhizal dependency and the cost and benefits given by the associated fungal communities to the plants (Klironomos et al. 2000). Previous study in an early successional community from England, show that AMF enhanced plant productivity because they positively affected dominant species (Gange et al. 1990). In contrast, the evidence from Chaquean region shows that, due to negative effects on growth of dominant ruderal species, AMF decreased productivity in these early successional plants. Specifically, total yield of species growing together without AMF was similar to that of the dominant species (of the species combination) growing alone, indicating stronger competitive interactions between species when growing without the fungal symbionts (Figure 4). Through these effects, AMF might be responsible not only of changes in plant community composition as discussed above, but also of changes in productivity among different successional stages.

All in all, the evidence from Chaquean region strongly suggests that mycorrhizal communities have an important role in structuring early secondary plant communities.

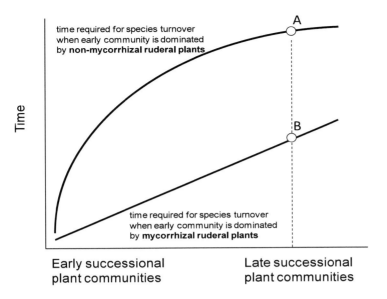

Figure 5. Time required for late successional communities to replace early successional communities depending on the mycorrhizal status and responsiveness of dominant ruderal plant species at the beginning of the succession. Communities originally dominated by non-micorrhizal ruderal plants would undergo slow species turnover (A), while communities originally dominated by micorrhizal ruderal species would require less time to be replaced by late successional communities (B).

MODEL ON THE EFFECT OF AMF ON PLANT SUCCESSION

According to the evidence discussed in previous sections, early successional communities are dominated by mycorrhizal ruderal plants. However, the taxonomic affiliation of these plant species varies among different ecosystems from distinct geographic regions. Given the strong relationship between taxonomic affiliation of host plants and the type of mycorrhizal association that they establish (Brundrett 2009), it could be hypothesized that the role of mycorrhizal fungi in early successional communities would depend on plant community composition. Based on evidence discussed in this chapter, early successional communities could be dominated by non-mycorrhizal or by mycorrhizal ruderal plants that may be adversely affect by AMF. This beget at least two possible scenarios regarding the role of mycorrhizas in plant succession (Figure 5).

In communities dominated by non-mycorrhizal plants, AMF would promote the establishment of late successional mycorrhizal species generating a positive feedback between plant and AMF communities. As a consequence, these interactions will promote, with time, the replacement of the non-mycorrhizal species by the mycorrhizal ones (Read 1989, Smith and Read 2008). At the beginning of the successional process, changes in plant communities are predicted to be slow but then would accelerate as mycorrhizal species begin to establish and can positively feedback on AMF communities (see A in figure 5).

In communities dominated by mycorrhizal host plants, the successional replacement would be much faster since AMF not only may promote the establishment of later species but also accelerate the process by negatively affecting the early successional species. That is, less time will be required for later successional communities to replace earlier ones. Certainly, a completely depauperated fungal community in soils following a great disturbance, would delay successional changes in both type of situations, but as AMF communities begin to re-establish, the successional changes would be faster in those communites dominated by mycorrhizal ruderals. The later prediction seems to better fit the situation in the Chaquean region (see B in figure 5).

IMPLICATIONS FOR MANAGEMENT

Most of the ruderal species depicted in this chapter are common weeds beyond the Chaquean region extending to most of the Chaquean domain plain in central Argentina including Espinal and Pampa (Zuloaga and Morrone 1996) (figure 1). This extensive area is traditionally characterized by agricultural practices, being mainly covered by soybean but also corn and wheat (Aizen et al. 2009).

It is well known that agricultural practices have strong consequences on biodiversity (McLaughlin and Mineau 1995). The application of herbicides, pesticides and fertilizers has significant impacts on the environment by negatively affecting biodiversity, increasing eutrophication of waters and accumulating pesticides (Tilman et al. 2002). Agricultural practices strongly affect belowground organisms that may directly affect ecosystem services such as crop yield and indirectly impact on the carbon and nutrient cycling, soil structure, food web interactions that ultimately affect productivity (Barrios 2007). Nonetheless, different agricultural practices may differentially affect soil microorganisms, in particular mycorrhizal fungi. Several examples, mainly from the northern hemisphere, have shown that AM fungal diversity is negatively affected by intensive agricultural management such as high-nutrient and pesticide input and/or physical soil disturbance (An et al. 1993; Oehl et al. 2003; Hijri et al. 2006, Verbruggen et al. 2010). In the chaquean domain (Chaco, Espinal and Pampa in figure 1), it has been recently shown that tillage and mono-cropping reduce AMF spore density in comparison with non-tillage practices (Cofré et al. 2010). The negative effects of some agricultural practices on AMF communities undermine the recognized key role that AMF have in agroecology, such as their potential role in weed management (Jordan et al. 2000).

It has been generally accepted that AMF would have beneficial effects in agroecosystems by positively affecting the growth of mycorrhizal crops (Gianinazzi et al. 2010). It has been also suggested that AMF would enhance the competitive abilities of these crops when they interact with non-mycorrhizal weeds (Jordan et al. 2000). There is now increasing evidence showing the antagonistic effects of AMF on several weedy species (Vatovec et al. 2005, Jordan and Huerd 2008). Accordingly, the evidence gathered in this chapter so far, suggests that AMF colonize roots of some of the most widespread weeds in central Argentina and directly decreased their performance. In this way, the interaction between AMF and weeds may have at least two beneficial consequences for mycorrhizal crops: on the one hand, the root of the weeds may act as sources of inoculums for the persistence of AMF communities in soils allowing for subsequent mycorrhizal colonization of crop roots; on the other hand, the fungal symbionts may directly control weed growth and in turn enhance competitive abilities of crops (Jordan and Vatovec 2004).

As a consequence of all of the evidence observed here and from other regions, it seems that agricultural practices that negatively affect AMF communities would, not only negatively affect crop productivity and other ecosystem services provided by the fungi but also favour the competitive abilities and spread of weeds in these agroecosystems.

In addition, the interactions between AMF and pioneer species suggest the possibility of the potential role of this symbiosis in restoration ecology. This assertion is based on the fact that the mechanisms by which the combination of AMF-ruderal species enhances plant succession also fit the situation of ecosystem restoration. Accordingly, it has been shown that disturbed areas that initially harbour mycorrhizal species enhance the establishment and growth of other mycorrhizal species accelerating the recovery of natural communities (Carrillo-García et al. 2002; Rosales et al. 2008). This possibility also extends to phytoremediation of contamined soils since it has been demonstrated that AMF may protect plants from toxicity caused by heavy metal concentrations (Mathur et al. 2007, Wong et al. 2007). These issues certainly deserve further attention.

CONCLUSION

One of the most important aspects of plant ecology is to elucidate factors that determine succession in ecosystems, in particular how they influence diversity and productivity.

This chapter shows that unlike the predictions of classic theoretical models, in the Chaquean region, most early secondary successional species are mycorrhizal, showing AMF and DSE in their roots. This suggests an important role of these fungal symbionts in structuring these plant communities. Through their antagonistic effects, AMF differentially affect plant interactions, decrease productivity, counteract dominance in early successional ruderal species and, in this way, may accelerate secondary plant succession. In the case of DSE, despite their widespread occurrence in plant roots, we still do not know their effects on plant growth and competition. Certainly this issue deserves experimental testing.

Altogether, the evidence presented here challenges previous models on mycorrhizal distribution and functioning in early successional plant communities and suggests that AMF and DSE should be included in future theoretical models regarding their role in early successional communities, including the possibility of their antagonistic effects.

The interactions between AMF and ruderal species depicted here also highlight the potential role of these ruderal mycorrhizal species in agroecosystems management, ecological restoration and phytoremediation.

REFERENCES

Aizen, MA; Garibaldi, LA; Dondo, M. Expansión de la soja y diversidad de la agricultura en Argentina. *Ecol. Austral,* 2009, 19, 45-54.

Allen, EB; Allen, MF. The mediation of competition by mycorrhizae in successional and patchy environments. In: *Perspective in plant competition* (eds. Grace, J. B. & Tilman, D.). Academic Press, Inc. San Diego; 1990; pp. 367-390.

An, Z-Q; Hendrix, JW; Hershman, DE; Ferris, RS; Henson, GT. The influence of crop rotation and soil fumigation on a mycorrhizal fungal community associated with soybean. *Mycorrhiza*, 1993, 3, 171-182

Barrios, E. Soil biota, ecosystem services, and land productivity. *Ecol. Economics*, 2007, 64, 269-285.

Barrow, JR; Osuna, P. Phosphorus solubilization and uptake by dark septate fungi in four wing saltbrush *Atriplex canescens* (Pursh) Nutt. *Journal of Arid Environment*, 2002, 51, 449-459.

Bever, JD. Feedback between plants and their soil communities in an old field community. *Ecology*, 1994, 75, 1965-1977.

Bradford, MA; Jones, TH; Bardgett, RD; Black, HIJ; Boag, B; Bonkowski, M; Cook, R; Eggers, T; Gange, AC; Grayston, SJ; Kandeler, E; McCaig, AE; Newington, JE; Prosser, JI; Setala, H, Staddon, PL; Tordoff, GM; Tscherko, D; Lawton, JH. Impacts of Soil Faunal Community Composition on Model Grassland Ecosystems. *Science*, 2002, 298, 615-618.

Brown, VK; Gange, AC. Differential effects of above- and below-ground insect herbivory during early plant succession. *Oikos*, 1989, 54, 67-76.

Brown, VK; Gange, AC. Secondary plant succession: how is it modified by insect herbivory? *Vegetatio*, 1992, 101, 3-13.

Brundrett, MC. Mycorrhizal associations and other means of nutrition of vascular plants: understanding the global diversity of host plants by resolving conflicting information and developing reliable means of diagnosis. *Plant Soil*, 2009, 320, 37-77.

Cabido, M; Funes, G; Pucheta, E; Vendramini, F; Díaz, S. A chorological analysis of the mountains from Central Argentina. Is all what we call Sierra Chaco really Chaco? Contribution to the study of the flora and vegetation of the Chaco. XII. *Candollea*, 1998, 53, 321-331.

Callaway, RM; Newingham, B; Zabinsky, C; Mahall, BR. Compensatory growth and competitive ability of an invasive weed are enhanced by soil fungi and native neighbours. *Ecol. Lett.*, 2001, 4, 429-433.

Callaway, RM; Mahall, BR; Wicks, C; Pankey, J; Zabinski, C. Soil fungi and the effects of an invasive forb on grasses: neighbor identity matters. *Ecology*, 2003, 84, 129-135.

Carrillo-García, A; León De La Luz, J-L; Bashan, Y; Bethlenfalvay, GJ. Nurse Plants, mycorrhizae, and plant establishment in a disturbed area of the Sonoran desert. *Restor. Ecol.*, 2002, 7, 321-335.

Crawley, MJ. The structure of plant communities. In: *Plant ecology* (ed. Crawley, M. J.) 2 edn. Blackwell Science, Inc. Oxford; 1997; pp. 475-531.

Cofré, N; Becerra, A; Urcelay, C; Domínguez, L; Wall, L. Densidad y composición de esporas micorrícico-arbusculares en suelos bajo diferentes prácticas agrícolas. *Actas del XXII Congreso Argentino de la Ciencia del Suelo*, 2010, 4pp.

De Deyn, GB; Raaijmakers, CE; Zoomer, HR; Berg, MP; de Ruiter, PC; Verhoef, HA; Bezemer, TM; van der Putten, WH. Soil invertebrate fauna enhances grassland succession and diversity. *Nature*, 2003, 422, 711-713.

de Fina, AL. Aptitud agroclimática de la República Argentina. Academia Nacional de Agronomía y Veterinaria, 1992.

Fracchia, S; Aranda A; Gopar, A; Silvani V; Fernandez, L; Godeas, A. Mycorrhizal status of plant species in the Chaco Serrano Woodland from central Argentina. *Mycorrhiza*, 2009, 19, 205-214.

Francis, R; Read, DJ. Mutualism and antagonism in the mycorrhizal symbiosis, with special reference to impacts on plant community structure. *Can. J. Bot.*, 1995, 73, 1301-1309.

Gange, AC; Brown, VK; Farmer, LM. A test of mycorrhizal benefit in an early successional plants community. *New Phytol.*, 1990, 115, 85-91.

Gange, AC; Brown, VK; Sinclair, S. Vesicular-arbuscular mycorrhizal fungi: a determinant of community structure in early succession. *Funct. Ecol.*, 1993, 7, 616-622.

Gavier, GI; Bucher, EH. Deforestación de las Sierras Chicas de Córdoba (Argentina) en el período 1970-1997. *Misceláneas N° 101*, Acad. Nac. Cienc.Córdoba, Argentina; 2004.

Gianinazzi, S; Gollotte A; Binet M-N; van Tuinen, D; Redecker, D; Wipf, D. Agroecology: the key role of arbuscular mycorrhizas in ecosystem services. *Mycorrhiza*, 2010, DOI: 10.1007/s00572-010-0333-3

Giorgis, MA; Cingolani AM; Chiarini, F; Chiapella, J; Barboza, G; Espinar LA; Morero, R; Gurvich DE; Tecco, PA; Subils, R; Cabido, M. Composición florística del Bosque Chaqueño Serrano de la provincia de Córdoba, Argentina. *Kurtziana*, (In press).

Grime, JP. *Plant Strategies and Vegetation Processes*. John Wiley, New York, USA; 1979.

Grime, JP; Mackey, JM; Miller, SH; Read, DJ. Floristic diversity in a model system using experimental microcosm. *Nature*, 1987, 328, 420-422.

Grime, JP. *Plant strategies, vegetation processes, and ecosystem properties*. John Wiley & Sons, LTD, New York.; 2001.

Haselwandter, K; Read, DJ. The significance of rootfungus association in two *Carex* species of high-alpine plant communities. *Oecologia*, 1982, 53, 352-354.

Hartnett, DC; Wilson, GWT. Mycorrhizae influence plant community structure and diversity in tallgrass prairie. *Ecology*, 1999, 80, 1187–1195.

van der Heijden, MGA; Klironomos, JN; Ursic, M; Moutoglis, P; Stritwolf-Engel, R; Boller, T; Wiemken, A; Sanders, I. Mycorrhizal fungal diversity determines plant biodiversity, ecosystem variability and productivity. *Nature*, 1998, 396, 69-72.

Hijri, I; Sykorova, Z; Oehl, F; Ineichen, K; Mäder, P; Wiemken, A; Redecker, D. Communities of arbuscular mycorrhizal fungi in arable soils are not necessarily low in diversity. *Mol. Ecol.*, 2006, 15, 2277-2289.

Janos, DP. Vesicular-arbuscular mycorrhizae affect lowland tropical rain forest plant growth. *Ecology*, 1980, 61, 151-162.

Jasper, DA; Abbot, LK; Robson, AD. The effect of soil disturbance on vesicular-arbuscular mycorrhizal fungi, in soils from different vegetation types. *New Phytol.*, 1991, 112, 101-107.

Johnson, NC; Graham, JH; Smith, FA. Functioning of mycorrhizas along the mutualism parasitism continuum. *New Phytol.*, 1997, 135, 1-12.

Jordan, NR; Vatovec, C. Agroecological benefits from weeds. In: *Weed biology and management* (ed. Inderjit). Kluwer Academic Publishers, The Netherlands, 2004, pp.137-158.

Jordan, NR; Huerd, S. Effects of soil fungi on weed communities in a corn–soybean rotation. *Renew. Agric. Food Syst.*, 2008, 23, 108-117.

Jordan, NR; Zhang, J; Huerd, S. Arbuscular mycorrhizal fungi: potential roles in weed management. *Weed Res.*, 2000, 40, 397-410.

Jumpponen, A. Dark septate endophytes - are they mycorrhizal? *Mycorrhiza,* 2001, 11, 207-211.

Klironomos, JN. Feedback to soil biota contributes to plant rarity and invasiveness in communities. *Nature*, 2002, 417, 67-70.

Klironomos, JN. Variation in plant response to native and exotic arbuscular mycorrhizal fungi. *Ecology*, 2003, 84, 2292-2301.

Klironomos, JN; McCune, J; Hart, M; Neville, J. The influence of arbuscular mycorrhizae on the relationship between plant diversity and productivity. *Ecol. Lett.*, 2000, 3, 137-141.

Luti, R; Solís, M; Galera, FM; Muller, N; Berzal, N; Nores M; Herrera, M; Barrera, JC. Vegetación, En: Vázquez, J; Miatello, R & Roque, M (eds.). Geografía Física de la provincia de Córdoba; 1979, pp. 297-368.

Mandyam, K; Jumpponen, A. Seeking the elusive function of the root-colonising dark septate endophytic fungi. *Studies in Mycology,* 2005, 53, 173-189.

Marler, MJ; Zabinski, CA; Callaway, RM. Mycorrhizae indirectly enhance competitive effects of an invasive forb on a native bunchgrass. *Ecology*, 1999, 80, 1180-1186.

Mathur, N; Singh, J; Bohra, S; Quaizi, A; Vyas, A. Arbuscular mycorrhizal fungi: a potential tool for phytoremediation. *J. Pl. Sci.*, 2007, 2, 127-140.

McLaughlin A, Mineau P. The impact of agricultural practices on biodiversity. *Agric., Ecosyst. & Environ.*, 1995, 55, 201-212.

Newsham, KK. *Phialophora graminicola*, a dark septate fungus, is a beneficial associate of the grass *Vulpia ciliata* ssp. *ambiqua. New Phytologist,* 1999,144, 517-524.

Oehl, F; Sieverding, E; Ineichen, K; Mäder, P; Boller, T; Wiemken, A. Impact of land use intensity on the species diversity of arbuscular mycorrhizal fungi in agroecosystems of Central Europe. *Appl. Environ. Microbiol.*, 2003, 69, 2816-2824.

Olff, H; Ritchie, M. Effects of herbivores on grassland plant diversity. *Trends Ecol Evol.*, 1998, 13, 261-265.

Packer, A; Clay, K. Soil pathogens and spatial patterns of seedling mortality in a temperate tree. *Nature*, 2000, 404, 278-281.

Perez, M; Urcelay, C. Differential growth response to arbuscular mycorrhizal fungi and plant density in two wild plants belonging to contrasting plant functional types. *Mycorrhiza*, 2009, 19, 517-523.

van der Putten, WH; Van Dijk, C; Peters, BAM. Plant-specific soil-borne diseases contribute to succession in foredune vegetation. *Nature*, 1993, 362, 53-56.

Read, DJ. Mycorrhizas and nutrient cycling in sand dune ecosystems. Proceedings of the Royal Society of Edinburgh; 1989; 96B, 89-110.

Read, DJ. Mycorrhizas in ecosystems. *Experienta*, 1991, 47, 376-391.

Rosales, J; Cuenca, G; Ramírez, N; De Andrade, Z. Native colonizing species and degraded land restoration in La Gran Sabana, Venezuela. *Restor. Ecol.*, 2008, 5, 147-155.

Roumet, C; Urcelay, C; Díaz, S. Suites of root traits differ between annual and perennial species growing in the field. *New Phytol.*, 2006, 170, 357-368.

Smith, SE; Read, DJ. Mycorrhizal symbiosis, 3rd edn. Academic Press, Amsterdam. Boston; 2008.

Tilman D, Cassman KG, Matson PA, Naylor R, Polasky S. Agricultural sustainability and intensive production practices. *Nature*, 2002, 418, 671-677.

Trappe, JM. Phylogenetic and ecological aspects of mycotrophy in the angiosperms from an evolutionary standpoint. In: *Ecophysiology of VA mycorrhizal plants* (ed. Zafir, GR) CRC Press, Boca Raton, Fl.; 1987, pp. 5-25.

Upson, R; Newsham, KK; Bridge, PD; Pearce, DA; Read, DJ. Taxonomic affinities of dark septate root endophytes of *Colobanthus quitensis* and *Deschampsia antarctica*, the two native Antarctic vascular plant species. *Fungal Ecol.*, 2009, 2, 184-196.

Urcelay, C; Díaz, S. The mycorrhizal dependence of subordinates determines the effect of arbuscular mycorrhizal fungi on plant diversity. *Ecol. Lett.*, 2003, 6, 388-391.

Urcelay, C; Battistella, R. Colonización micorrícica en distintos tipos funcionales de plantas herbáceas del centro de Argentina. *Ecol. Austral*, 2007, 17, 179–188.

Urcelay C; Díaz, S; Gurvich, DE; Chapin III, FS; Cuevas, E; Domínguez, LS. Mycorrhizal community resilience in response to experimental plant functional type removals in a woody ecosystem. *J. Ecol.*, 2009, 97, 1291-1301.

Vatovec, C; Jordan, NR; Huerd, S. Responsiveness of certain agronomic weed species to arbuscular mycorrhizal fungi. *Renew. Agric. Food Syst.*, 2005, 20, 181-189.

Verbruggen, E; Röling WFM; Gamper, HA; Kowalchuk, GA; Verhoef, HA; van der Heijden, MGA. Positive effects of organic farming on below-ground mutualists: large-scale comparison of arbuscular mycorrhizal communities in agricultural soils. *New Phytol.*, 2010, 186, 968-979.

Wardle, DA; Barker, GM. Competition and herbivory in establishing grassland communities: implications for plant biomass, species diversity and soil microbial activity. *Oikos*, 1997, 80, 470-480.

Wong, CC; Wu, SC; Kuek, C; Khan, AG; Wong, MH. The role of mycorrhizae associated with vetiver grown in Pb-/Zn-contaminated soils: greenhouse study. *Restor. Ecol.*, 2007, 15, 60-67.

Zak, M; Cabido, M. Spatial patterns of the Chaco Vegetation of central Argentina: integration of remote sensing and phytosociology. *Appl. Veg. Sci.*, 2002, 5, 213-226.

Zak, MR; Cabido, M; Hodgson, JG. Do subtropical seasonal forests in the Gran Chaco, Argentina, have a future? *Biol. Cons.*, 2004, 120, 589-598.

Zuloaga, FO; Morrone, O. Catálogo de las Plantas Vasculares de la República Argentina. *Monographs in Systematic Botany from the Missouri Botanical Garden*, 1996, 60, 1-323.

In: Mycorrhiza: Occurrence in Natural and Restored...
Editor: Marcela Pagano

ISBN: 978-1-61209-226-3
© 2012 Nova Science Publishers, Inc.

Chapter 9

MEASURING AND ESTIMATING ECTOMYCORRHIZAL FUNGAL DIVERSITY: A CONTINUOUS CHALLENGE

Ornella Comandini[1], Andrea C. Rinaldi[1] and Thomas W. Kuyper[2]
[1] Department of Biomedical Sciences and Technologies,
University of Cagliari, I-09042 Monserrato (CA), Italy
[2] Department of Soil Quality, Wageningen University, Box 47,
NL-6700 AA Wageningen, The Netherlands

ABSTRACT

Thousands of ectomycorrhizal (ECM) fungal species exist, but estimates of global species richness of ECM fungi differ widely. Many genera have been proposed as being ECM, but in a number of studies evidence for the hypothesized ECM habit is lacking. Progress in estimating ECM species richness has therefore been slow. Recently, we have published evidence for the ECM habit of fungal species and for the identification of the mycobiont(s) in specific ECM associations, using published and web-based mycorrhiza literature. The identification methods considered were morpho-anatomical characterization of naturally occurring ECMs, pure culture synthesis, and molecular identification. In addition, stable isotope data of C and N, and phylogenetic information were also considered as relevant criteria to assess ECM habit. Our survey indicated that for 343 fungal genera an ECM status has been alleged, and for about two thirds (236 genera) of these supportive evidence of ECM status exists or can at least be entertained as the more reasonable hypothesis. On the basis of our literature search we conservatively estimated ECM species richness around 7750 species. However, on the basis of estimates of knowns and unknowns in macromycete diversity, we suggested that a final estimate or ECM species richness between 20,000 and 25,000 would be more realistic. Recent updates, taking into consideration evidence that became available after our study was released, have not changed figures substantially (234 genera, 7950 species), confirming that current knowledge of ECM fungal diversity, as supported by experimental evidence, is only partly complete, and that inclusion of many fungal genera in this trophic and ecological category is not verified at this stage. Care must thus be used when compiling lists of ECM and saprotrophic species on the basis of published information only. We reflect on the status of the various sources of evidence. We also discuss interesting avenues of future research, including a wider assessment and understanding of the ubiquity and diversity of the secondary root-associated fungi (root endophytes, especially Ectomycorrhiza-Associated Ascomycetes).

ECTOMYCORRHIZAL SYMBIOSIS ON THE STAGE!

Just a few data are needed in order to let even the non-specialist appreciate the ecological importance played by the ectomycorrhizal (ECM) symbiosis. It is estimated that 2-3% of all vascular plants (some 6000-10,000 plant species) are ECM (Smith and Read, 2008; Brundrett, 2009). Although this number might sound very low compared to the diversity of arbuscular mycorrhizal plants – estimated at over 200,000 species, or two thirds of all vascular plants) – it should be recalled that most ECM plants are trees, which form the core structure of boreal forests in the northern hemisphere (e.g., Pinaceae, Fagaceace, Betulaceae), and are also often dominant elements in many temperate, sub-tropical, and tropical areas of the southern hemisphere (e.g., Nothofagaceae, Myrtaceae, Dipterocarpaceae, several cesalpinioid genera in Fabaceae). In a recent account of plant mycorrhizal diversity on a global scale, Brundrett (2009) listed 26 families of angiosperm and gymnosperm ECM plants, hosting 145 genera in total. These figures are most likely bound to be updated (i.e., increased) in the future, with new ECM plants identified in particular in the tropics. For example, Sarcolaenaceae and Asteropeiaceae have been shown to be ECM only very recently (Ducousso *et al.*, 2004; Ducousso *et al.*, 2008), and our knowledge of the ECM associations of a key family such as the Dipterocarpaceae (with about 500 species) can be considered preliminary. As an example we refer to widely divergent suggestions about the age of the ECM symbiosis in that family. Based on the sistergroup relation between Sarcolaenaceae and Dipterocarpaceae, Ducousso *et al.* (2004) suggested an age of the ECM habit of at least 88 million years, and Moyersoen (2006) describing the South American ECM dipterocarp *Pakaraimaea* suggested a Gondwanaland evolution of the ECM habit at least 135 million years ago (m.y.a.). Such a scenario seemed also plausible because of the parallel situation in Fabaceae, tribe Macrolobieae, with ECM species in South America (*Dicymbe* – Henkel *et al.*, 2002) and Africa (Alexander, 1989; Onguene and Kuyper, 2001). Alexander (2006) already critically reflected on alternative explanations (independent acquisition of the ECM habit, long-distance dispersal). These alternative explanations fit much better with the phylogeny of and molecular clock for the Cistaceae (the sister group of Dipterocarpaceae + Sarcolaenaceae) which dates that latter split at 25 m.y.a. and that of Dipterocarpaceae and Sarcolaenaceae at 5-6 m.y.a. (Guzmán and Vargas, 2009). It may therefore come as no big surprise that the Angiosperm Phylogeny Group (2009: 115) even suggests that Dipterocarpaceae and Sarcolaenaceae (even with Cistaceae) are confamilial. On the basis of data on angiosperm and gymnosperm phylogenies the ECM habit seems to have arisen around 25-26 times.

Shifting attention from plants to the fungal ECM partners, one first notes that the ECM habit in the fungi evolved more frequently. Tedersoo *et al.* (2010) mentioned the possibility that the ECM habit had evolved at least 66 times. This estimate substantially enlarges earlier estimates (see below). Still the data do not provide any well-substantiated suggestion about secondary reversals to a saprotrophic mode (see below).

At a closer inspection, however, when it comes to a detailed description of which fungi have an ECM trophic status, the view gets more blurry, for a number of reasons. For many years, reports of ECM associations have been based only, or prevalently, on field observations of sporocarps in the vicinity of potential hosts, and this has led to the accumulation of a number of misidentifications and contradictory reports on the ECM status of many fungal (and also plant!; cf *Cupressus*, Brundrett, 2009) genera. Since these claims, several of which are almost certainly erroneous, have been perpetuated in more recent publications, this spurious burden has slowed down progress in estimating ECM fungal richness. It was only since mid 1980's, with the standardisation of the morpho-anatomical descriptions of ECMs followed by the introduction of molecular tools for the identification of root tips, that reliable data have started to accumulate in great number (Agerer, 1986; Horton and Bruns, 2001).

However, new methods do not always make things less complicated, let alone unambiguous. The classification of a plant-fungus association as ECM can still be problematic. Such problems arise from the definition of ectomycorrhiza *per se* – morphological or otherwise. Morphological definitions include the presence of a mantle (which, however, can be very thin or even almost absent), or of a Hartig net which is considered the symbiotic interface. Other definitions that are based on outcomes as mutualistic (mutually beneficial for both plant and fungus) also have their problems, and mutualistic root-associations in ascomycetous groups have also been reported as Dark Septate Endophytic fungi (DSE; Summerbell, 2005) or Ectomycorrhiza-Associated Ascomycetes (EAA; Tedersoo *et al.*, 2009). Furthermore the border between biotrophic root-associated fungi and certain ECM groups gets blurred, as in the genus *Entoloma* and possibly *Morchella*, where ECM genera occur in the sistergroup *Leucangium* and *Fischerula* (Læssøe and Hansen, 2007). Finally, the debate about so-called mixotrophic life styles (a combination of ectomycorrhizal and saprotrotrophic nutritional modes, also known as facultative ECM) is still going on (see below). As a result of these various factors, the picture is by no means resolved, and estimates of global species richness of ECM fungi differ widely.

We have recently gathered and critically analysed available evidence for the ECM habit of fungal species and for the identification of the mycobiont(s) involved in specific ECM associations worldwide, using published and web-based mycorrhiza literature (Rinaldi *et al.*, 2008). Morpho-anatomical characterization of naturally occurring ECMs, pure culture synthesis, and molecular identification were considered as the main identification methods; in addition, phylogenetic information and stable isotope data of Carbon and Nitrogen were used as relevant criteria to assess ECM habit. Following this rationale, we have indicated that out of 343 genera for which ECM status has been alleged, evidence supporting this conclusion exist, or ECM habit can be at least inferred, for 236 fungal genera, hosting about 7750 species (Rinaldi *et al.*, 2008). We also argued that 'true' species richness may fall rather between 20,000 and 25,000 species. Recently, Tedersoo *et al.* (2010) also critically analysed the situation. Their paper suggests ECM richness on generic level to be between 214 and 247, hence fairly similar to our analysis. Their list of cases where they did deviate from our suggestion is also relatively small, and a critical reflection would show that disagreement is only minor (see below). We comment briefly on that list, because addressing the discrepancies could help us to pinpoint those cases where more research is important. Almost complete agreement also indicates that the 'hard' criteria to infer ECM status and the 'secondary' criteria (phylogeny, stable isotopes) provide a robust assessment. We here present and discuss updates to the 'catalogue' of ECM fungi that became available after our study was released.

PHYLOGENETIC EVIDENCE

Tedersoo *et al.* (2010) proposed that the ECM habit had arisen at least 66 times. The largest number of ECM lineages was in the Pezizales (16). That large number is certainly surprising, considering that the ECM diversity of the Pezizales was only recently recognised (Tedersoo *et al.*, 2006). Certainly, with more and more independent lineages of ECM fungi, the question becomes more pressing whether reversals (from ECM to saprotrophic habit) have occurred. For instance, Tedersoo *et al.* (2010) treated the /tarzetta and /pulvinula clade as two separate ECM lineages, excluding the non-ECM genera *Geopyxis* and *Stephensia*. In our earlier study both genera were treated as ECM (Rinaldi *et al.*, 2008), which implies one only origin in this clade. Tedersoo *et al.* (2010) also suggested that the ECM habit arose in 5 lineages in the Thelephorales. That conclusion is based on the (putative) non-ECM status of several new genera of that order, a part of *Tomentella*, and the genera *Amaurodon* and *Lenzitopsis* for which the authors listed arguments as to their treatment as non-ECM fungi. Rinaldi *et al.* (2008) postulated a monophyletic origin of the ECM habit in the Thelephorales, and, on that basis only, treated those genera as ECM. Such discrepancies invite further research (*Amaurodon* is the basal lineage of the Thelephorales in Larsson *et al.*, 2004, and its placement as ECM or non-ECM is on that basis only ambiguous) and a novel testing whether reversals in such more narrowly delimited groups have occurred. The discrepancy also invites the question of accuracy of names that are attached to sequences that form the basis for such phylogenies. For instance, Wei *et al.* (2010) listed a species of *Anthracobia* (*A. subatra*) in the ECM clade with *Sphaerosporella* and one of the two *Trichophaea* subclades and noted that it will be likely that future studies will reveal *A. subatra* and possibly further *Anthracobia* species to form ECM. However, the phylogeny by Perry *et al.* (2007) makes it more plausible that either *Anthracobia* is paraphyletic, or that the species *A. subatra* was assigned to the wrong genus or the specimen was misidentified.

Eremiomyces was treated as non ECM by Tedersoo *et al.* (2010). A phylogeny by Læssøe and Hansen (2007) suggests placement in the *Peziza depressa* clade, together with the ECM genera *Ruhlandiella*, *Terfezia*, *Cazia, Tirmania*, and non-ECM *Plicaria*. The non-ECM habitat (Ferdman *et al.*, 2005) suggests it is better considered non-ECM. The phylogeny by Læssøe and Hansen (2007) also makes it likely that *Kalaharituber* is non-ECM. This conclusion fits with its geography in a non-ECM habitat (Ferdman *et al.*, 2005).

Several studies obtained root tip sequences that were assigned to novel ECM genera. However, not all of them have been listed as ECM, in some cases because the best BLAST match was still too low to be considered as reliable identification. Morris *et al.* (2008b) listed from a cloud forest in Guerrero, Mexico the putative genera *Hyphodiscus* and *Cheilymenia* with 93 and 91% BLAST match respectively, which we consider as too low to allow a generic placement of these sequences.

ROOT ENDOPHYTIC AND ECM ASCOMYCETES

Several records of new ECM genera refer to Ascomycetes. In some cases the ECM morphotype has been described and molecularly characterised (*Pinirhiza sclerotia* = *Acephala macrosclerotiorum*, Münzenberger *et al.*, 2009; *Piceirhiza bicolorata* =

Meliniomyces vraolstadiae, M. bicolor, Hambleton and Sigler, 2005), while in other cases the only evidence is their molecular identification on (pooled) ectomycorrhizal root tips or older descriptions as ECM, albeit with remarks that the symbiosis was not always mutualistic, that mantles were thin and irregular and/or that the Hartig net was absent. Such cases apply to *Phialocephala fortinii* (a close relative to *Acephala*), and *Cadophora finlandica*, *Chloridium paucisporum* and even the ericoid mycorrhizal fungus *par excellence Rhizoscyphus ericae* (close relatives of *Meliniomyces*). However, as noted by Tedersoo *et al.* (2009) not all fungi reported as ECM are truly mycorrhizal and root endophytes are likely also included in the list. For that reason the authors coined the term ECM-Associated Ascomycota (EAA). In several cases, however, the distinction between EAA and ECM fungi is not clear to us, especially because several taxa are ecologically heterogeneous, containing root endophytes, ericoid mycorrhizal and ECM fungi, sometimes even in the same fungal species (see discussions in Grünig *et al.,* 2008; Tedersoo *et al.,* 2009; Alberton *et al.,* 2010). In that regard it is surprising to read that Tedersoo *et al.* (2010) claimed that, except for *Meliniomyces bicolor*, the ECM lifestyle is phylogenetically distinct from the endophytic and ericoid mycorrhizal life style. For us, these cases suggest that both life styles are closer, at least in the helotialean fungi, and even question the definition of ectomycorrhiza in relation to other mutualistic associations (Alberton *et al.,* 2010).

Among EAA we include a group of Ascomycetes that are only known as environmental sequences (Porter *et al.,* 2008). These fungi have been isolated from ECM root tips (Rosling *et al.,* 2003) but we do not consider them as ECM. It is not clear to us whether the ECM Sordariales (Trowbridge and Jumpponen, 2004; Tedersoo *et al.,* 2007) have to be reclassified as EAA. Other Ascomycetes that might need reconsideration (and renewed reflection on how to define ECM in relation to other root symbioses) are *Arachnopeziza* in mesh bags in a conifer forest in Sweden (Hedh *et al.,* 2008) and *Lachnum* (and a number of other ascomycete sequences that suggest secondary colonizers at best) on *Kobresia* in the Himalaya (Gao and Yang, 2010).

ECM FUNGI GUILD: THE NEW MEMBERS ARE...

A small number of new fungal genera have been described that might prove to be ECM. In all instances, to date no evidence exist to support this conclusion, with the exclusion of field observations of potential host plants and phylogenetic relationships (Table 2). The new monotypic genus *Imaia* has been recently introduced to accommodate the peculiar hypogeous ascomycete *Terfezia gigantea*, which occurs in moist, temperate forests in Japan and eastern North America (Kovács *et al.,* 2008). Molecular analysis shows that *Imaia* belongs to the Morchellaceae, with ECM *Leucangium*. Trappe and colleagues have reported a new truffle-like genus from the Australian Northern Territory, namely *Ulurua* (Trappe *et al.,* 2010). Although no detailed information was given as for the habitat of this monotypic taxon, phylogenetic analyses indicated a close relationship between *Ulurua* and the putatively ECM *Mycoclelandia* (Table 2).

Table 1. Genera of proven ectomycorrhizal fungi[§]

Genus	Reference [†]	Root tip morphology	Root tip molecular	Synthesis	Isotopic	Phylogeny	Species number (estim.d)[‡]	Notes
ASCOMYCOTA								
Amylascus	Maia *et al.*, 1996					Hansen *et al.*, 2005	2	
Balsamia	Trappe, 1969	Palfner and Agerer, 1998	Palfner and Agerer, 1998		Hobbie *et al.*, 2001	Hansen and Pfister, 2006	6	
Barssia	Trappe, 1969				Hobbie *et al.*, 2001	Hansen and Pfister, 2006)	2	
***Cadophora*[#]**	Wang and Wilcox, 1985[#]	Wang and Wilcox, 1985[#]	Vrålstad *et al.*, 2002[#]	Wilcox and Wang, 1987[#]		Hambleton and Sigler, 2005[#]	2	1
Cazia	Agerer, 2006					Hansen and Pfister, 2006	1	
Cenococcum	Trappe, 1962	Harniman and Durall, 1996	Mahmood *et al.*, 1999	Godbout and Fortin, 1983			1	
Chloridium	Wang and Wilcox, 1985	Wang and Wilcox, 1985		Wilcox and Wang, 1987a			1	2
Choiromyces	Trappe, 1969		Izzo *et al.*, 2005a		Hobbie *et al.*, 2001	Hansen and Pfister, 2006	5	
Delastria	Agerer, 2006					? (Lumbsch and Huhndorf, 2007)	1	
Dingleya	Trappe, 1969				Hobbie *et al.*, 2001	Hansen and Pfister, 2006	7	

Elaphomyces	Trappe, 1969	Agerer, 1999a	Tedersoo *et al.*, 2003	Miller and Miller, 1984			25	
Fischerula	Molina *et al.*, 1992				Hobbie *et al.*, 2001	Hansen and Pfister, 2006	2	
Genea	Trappe, 1969	Jakucs *et al.*, 1998; Tedersoo *et al.*, 2006a	Tedersoo *et al.*, 2006a; Smith *et al.*, 2006			Hansen and Pfister, 2006	32	
Genabea[#]	Trappe, 1969[#]		Izzo *et al.*, 2005; Smith *et al.*, 2006[#]			Perry *et al.*, 2007[#]	1	3
Geopora	Trappe, 1969	Tedersoo *et al.*, 2006a	Tedersoo *et al.*, 2006a			Hansen and Pfister, 2006	13	
Geopyxis	Vrålstad *et al.*, 1998		Vrålstad *et al.*, 1998			Hansen and Pfister, 2006	7	4
Gilkeya[#]	Smith *et al.*, 2006	Moser *et al.*, 2009[#]	Moser *et al.*, 2009[#]			Smith *et al.*, 2006	1	
Glischroderma[#]	Tedersoo *et al.*, 2006b		Norman and Egger, 1999[#]; Tedersoo *et al.*, 2006b; Kjøller *et al.*, 2006				1	5
Gyromitra	Trappe, 1969				- (Hobbie *et al.*, 2001)	Hansen and Pfister, 2006	18	6
Helvella	Trappe, 1969	Tedersoo *et al.*, 2006a	Tedersoo *et al.*, 2006a; Kjøller *et al.*, 2006		Högberg *et al.*, 1999; Hobbie *et al.*, 2001	Hansen and Pfister, 2006	52	

Table 1 (Continued).

Genus	Reference †	Root tip morphology	Root tip molecular	Synthesis	Isotopic	Phylogeny	Species number (estim.d)‡	Notes
Humaria	Molina *et al.*, 1992	Ingleby *et al.*, 1990; Tedersoo *et al.*, 2006a	Tedersoo *et al.*, 2006			Hansen and Pfister, 2006	16	
Hydnobolites	Molina *et al.*, 1992		Tedersoo *et al.*, 2008a				3	
Hydnoplicata[#]	Jumpponen *et al.*, 2004[#]	Warcup, 1990[#]	Morris *et al.*, 2008a[#]				2	
Hydnotrya	Trappe, 1969	Tedersoo *et al.*, 2006a	Tedersoo *et al.*, 2006a	Lu *et al.*, 1998	+/- (Hobbie *et al.*, 2001)	Hansen and Pfister, 2006	15	
Hydnotryopsis	Brundrett *et al.*, 1996					Hansen *et al.*, 2005	2	
Labyrinthomyces	Molina *et al.*, 1992			Brundrett *et al.*, 2005		Hansen and Pfister, 2006	7	
Lachnum[#]			Gao and Yang, 2010[#]					
Leptodontidium[#]	Fernando and Currah, 1995	Fernando and Currah, 1996	Diedhiou *et al.*, 2009[#]	Fernando and Currah, 1995			10 ?	7
Leucangium[#]	Trappe, 1969[#]	Palfner and Agerer, 1998[#]	Palfner and Agerer, 1998[#]		+/- (Hobbie *et al.*, 2001)[#]	Hansen and Pfister, 2006[#]	1	8
Loculotuber	Agerer, 2006					Læssøe and Hansen, 2007	1	
Marcelleina[#]	Tedersoo *et al.*, 2010[#]		Morris *et al.*, 2008a[#]			Hansen *et al.*, 2005[#]	8	

Meliniomyces	Tedersoo *et al.*, 2006b	Brand *et al.*, 1992	Tedersoo *et al.*, 2006b; Twieg *et al.*, 2007				3	
Muciturbo	Maia *et al.*, 1996			Warcup, 1990b			3	
Nothojafnea	Maia *et al.*, 1996			Warcup, 1990b			2	
Otidea[#]	Trappe, 1969	Moser *et al.*, 2009[#]	Kennedy *et al.*, 2003; Moser *et al.*, 2009[#]		Hobbie *et al.*, 2001; 2002	Hansen and Pfister, 2006	23	
Pachyphloeus	Molina *et al.*, 1992	Tedersoo *et al.*, 2006a	Tedersoo *et al.*, 2006a			Hansen *et al.*, 2005	6	
Paurocotylis	Brundrett *et al.*, 1996				+/- (Hobbie *et al.*, 2001; 2002)	Hansen and Pfister, 2006	1	4
Peziza	Molina *et al.*, 1992	Valentine *et al.*, 2004; Tedersoo *et al.*, 2006a	Valentine *et al.*, 2004; Tedersoo *et al.*, 2006a	Warcup, 1990b	- (Hobbie *et al.*, 2002)	Hansen *et al.*, 2005	104	9
Phialocephala	Wang and Wilcox, 1985	Wang and Wilcox, 1985	Menkis *et al.*, 2005	Wilcox and Wang, 1987b			6	10
Picoa[#]	Agerer, 2006[#]			Gutiérrez *et al.*, 2003[#]			4	
Pseudotulostoma	Miller *et al.*, 2001	Henkel *et al.*, 2006	Henkel *et al.*, 2006				2	

Table 1 (Continued).

Genus	Reference †	Root tip morphology	Root tip molecular	Synthesis	Isotopic	Phylogeny	Species number (estim.d)‡	Notes
Pulvinula	Maia *et al.*, 1996	Amicucci *et al.*, 2001	Amicucci *et al.*, 2001	Warcup, 1990b		Hansen and Pfister, 2006	27	
Reddellomyces	Molina *et al.*, 1992			Brundrett *et al.*, 2005		Hansen and Pfister, 2006	4	
Ruhlandiella	Molina *et al.*, 1992			Warcup, 1991		Hansen *et al.*, 2005	1	
Sarcosphaera	Molina *et al.*, 1992					Hansen *et al.*, 2005	1	
Scabropezia[#]	Tedersoo *et al.*, 2006[#]		Morris *et al.*, 2008a[#]			Hansen *et al.*, 2005[#]	3	
Sowerbyella	Hobbie *et al.*, 2001				Hobbie *et al.*, 2001; 2002		16	
Sphaerosporella	Molina *et al.*, 1992	Meotto and Carraturo, 1988; de Román and de Miguel, 2005b		Danielson, 1984b		Hansen and Pfister, 2006	2	
Sphaerozone	Molina *et al.*, 1992	Brand, 1988					4	
Stephensia	Molina *et al.*, 1992			Warcup, 1990b	Hobbie *et al.*, 2001		7	4
Tarzetta	Tedersoo *et al.*, 2006a	Tedersoo *et al.*, 2006a	Tedersoo *et al.*, 2006a			Hansen and Pfister, 2006	9	
Terfezia	Trappe, 1969	Gutiérrez *et al.*, 2003	Walker *et al.*, 2005	Kovács *et al.*, 2003; Gutiérrez *et al.*, 2003		Hansen *et al.*, 2005	12	

Tirmania	Trappe, 1969		Walker *et al.*, 2005	Fortaz and Chevalier, 1992		Hansen *et al.*, 2005	3	
Tricharina	Agerer, 2006		Trocha *et al.*, 2006			Hansen and Pfister, 2006	2?	11
Trichophaea	Molina *et al.*, 1992	Tedersoo *et al.*, 2006°	Tedersoo *et al.*, 2006a			Hansen and Pfister, 2006	25	12
Tuber	Trappe, 1969	Comandini and Pacioni, 1997	Giomaro *et al.*, 2002	Giomaro *et al.*, 2002	Hobbie *et al.*, 2001; 2002	Hansen and Pfister, 2006	86	
Underwoodia	Agerer, 2006					Hansen and Pfister, 2006	3	
Wilcoxina	Molina *et al.*, 1992	Ingleby *et al.*, 1990; Tedersoo *et al.*, 2006a	Tedersoo *et al.*, 2006a	Scales and Peterson, 1991		Hansen and Pfister, 2006	3	
Wynnella	Agerer, 2006				Hobbie *et al.*, 2001	Hansen and Pfister, 2006	1	
BASIDIOMYCOTA								
Albatrellus	Molina *et al.*, 1992	Agerer, 1996	Bidartondo *et al.*, 2000		Högberg *et al.*, 1999	Miller *et al.*, 2006	16	13
Alnicola	Trappe, 1962	Moreau *et al.*, 2006a	Moreau *et al.*, 2006a			Matheny *et al.*, 2006	~30	
Alpova[#]	Molina *et al.*, 1992	Wiedmer *et al.*, 2001	Kennedy and Hill, 2010[#]	Molina, 1981		Binder and Hibbett, 2006	20	
Amanita	Trappe, 1962	Mleczko, 2004	Bidartondo *et al.*, 2000	Cripps and Miller, 1995	Högberg *et al.*, 1999; Hobbie *et al.*, 2001; 2002	Zhang *et al.* 2004; Tulloss, 2008	~500	14

Table 1 (Continued).

Genus	Reference †	Root tip morphology	Root tip molecular	Synthesis	Isotopic	Phylogeny	Species number (estim.d)‡	Notes
Amphinema	Molina *et al.*, 1992	Montecchio et al., 2002	Montecchio et al., 2002	Weiss, 1991			6	
Anamika	Agerer, 2006					Matheny et al., 2006	1	
Andebbia	Agerer, 2006					Hosaka et al., 2006	1	
Astraeus[#]	Molina *et al.*, 1992	Giraud, 1988	Dickie and Reich, 2005[#]	Molina, 1981		Binder and Hibbett, 2006	2	
Aureoboletus	Molina *et al.*, 1992					Binder and Hibbett, 2006	5	
Auritella	Matheny and Bougher, 2006					Matheny and Bougher, 2006	7	
Austroboletus	Molina *et al.*, 1992					Binder and Hibbett, 2006	~30	
Austrogautieria	Molina *et al.*, 1992	Thoen and Ba, 1989		Lu *et al.*, 1998		Hosaka et al., 2006	6	
Austropaxillus	Agerer, 2006	Palfner 2001				Binder and Hibbett, 2006	9	
Bankera	Molina *et al.*, 1992	Agerer and Otto, 1997		Danielson, 1984a			2	
Boletellus	Trappe, 1962					Binder and Hibbett, 2006	~50	
Boletopsis	Trappe, 1962	Agerer, 1992	Izzo *et al.*, 2005°				5	

Boletus	Trappe, 1962	Hahn, 2001; Jakucs and Beenken, 2001	Jonsson *et al.*, 1999a; Jakucs and Beenken, 2001	Brunner *et al.*, 1992; Pera and Alvarez, 1995	Högberg *et al.*, 1999; Hobbie *et al.*, 2001	Binder and Hibbett, 2006	~300	15
Bothia	Halling *et al.*, 2007					Halling *et al.*, 2007	1	
Byssocorticium	Molina *et al.*, 1992	Brand, 1991	Horton *et al.*, 2005				9	
Byssoporia	Molina *et al.*, 1992	Zak and Larsen, 1978		Kropp, 1982		Miller *et al.*, 2006	1	
Calostoma	Agerer, 2006	Wilson *et al.*, 2007	Wilson *et al.*, 2007		Wilson *et al.*, 2007	Binder and Hibbett, 2006	15	
Cantharellus	Trappe, 1962	Mleczko, 2004b	Countess and Goodman, 2000	Danell, 1994	Högberg *et al.*, 1999; Hobbie *et al.*, 2001; 2002	Moncalvo *et al.*, 2006	65	
Castoreum	Molina *et al.*, 1992			Brundrett *et al.*, 2005			3	
Catathelasma[#]	Trappe, 1962			+/- (Hutchinson, 1992)[#]	+/- (Kohzu *et al.*, 1999)[#]	- (Matheny *et al.*, 2006)	4	16
Ceratobasidium[#]	Tedersoo *et al.*, 2010[#]		Yagame *et al.*, 2008[#]	Warcup, 1991[#]		see Tedersoo *et al.*, 2010[#]	11 (?)	
Chalciporus	Brundrett *et al.*, 1996				- (Högberg *et al.*, 1999)	Binder and Hibbett, 2006	25	17
Chamonixia	Molina *et al.*, 1992	Raidl, 1999				Binder and Hibbett, 2006	8	
Chondrogaster	Molina *et al.*, 1992					Hosaka *et al.*, 2006	2	

Table 1 (Continued).

Genus	Reference †	Root tip morphology	Root tip molecular	Synthesis	Isotopic	Phylogeny	Species number (estim.d)‡	Notes
Chroogomphus	Molina *et al.*, 1992	Agerer, 1990	Cullings *et al.*, 2000		Högberg *et al.*, 1999	+/- (Binder and Hibbett, 2006)	18	
Clavariadelphus[#]	Molina *et al.*, 1992	Iosifidou and Raidl, 2006	Morris *et al.*, 2008a[#]			- (Hosaka *et al.*, 2006; Giachini *et al.*, 2010[#])	19	18
Clavulina	Brundrett *et al.*, 1996	Tedersoo *et al.*, 2003	Tedersoo *et al.*, 2003		- (Hobbie *et al.*, 2002)	Moncalvo *et al.*, 2006	~40	
Coltricia	Agerer, 2006	Thoen and Ba, 1989; Tedersoo *et al.*, 2007b	Tedersoo *et al.*, 2007b	Danielson, 1984°		Larsson *et al.* 2006	20	
Coltriciella	Tedersoo *et al.*, 2007b	Tedersoo *et al.*, 2007b	Tedersoo *et al.*, 2007b			Larsson *et al.* 2006	7	
Cortinarius	Trappe, 1962	Kuss *et al.*, 2004	Kuss *et al.*, 2004	Godbout and Fortin, 1983	Högberg *et al.*, 1999; Hobbie *et al.*, 2001	Matheny *et al.*, 2006	~2000	19
Craterellus	Trappe, 1962	Fransson, 2004	Fransson, 2004		Högberg *et al.*, 1999	Moncalvo *et al.*, 2006	20	
Descolea	Trappe, 1962	Palfner, 1997		Lu *et al.*, 1998		- ? (Matheny *et al.*, 2006)	~15	
Descomyces	Brundrett *et al.*, 1996	Agerer *et al.*, 2001	Agerer *et al.*, 2001	Lu *et al.*, 1998			5	
Destuntzia	Molina *et al.*, 1992					Albee-Scott, 2007b	5	

Diplocystis	Louzan *et al.*, 2007					Louzan *et al.*, 2007	1	
Entoloma	Trappe, 1962	Agerer, 1997	Montecchio *et al.*, 2006	Antibus *et al.*, 1981	- (Högberg *et al.*, 1999)	+/- (Matheny *et al.*, 2006)	~25 (?)	20
Fistulinella	Brundrett *et al.*, 1996					Binder and Hibbett, 2006	15	
Gallacea[#]	McKenzie, 2006					Hosaka *et al.*, 2006; Giachini *et al.*, 2010[#]	6	21
Gastroboletus	Molina *et al.*, 1992			Molina and Trappe, 1982		Binder and Hibbett, 2006	13	
Gautieria	Molina *et al.*, 1992	Palfner, 2001	Walker *et al.*, 2005	Duñabeitia *et al.*, 1996		Hosaka *et al.*, 2006	25	
Gloeocantharellus	Lee *et al.*, 2002					Hosaka *et al.*, 2006	3	
Gomphidius	Trappe, 1962	Agerer, 1991b	Olsson *et al.*, 2000	Ohga and Wood, 2000	Högberg *et al.*, 1999; Hobbie *et al.*, 2001; 2002	+/- (Binder and Hibbett, 2006)	10	
Gomphus	Molina *et al.*, 1992	Agerer *et al.*, 1998a	Agerer *et al.*, 1998a			Hosaka *et al.*, 2006	10	
Gummiglobus	Brundrett *et al.*, 1996					Hosaka *et al.*, 2006	2	
Gymnopaxillus	Agerer, 2006					Binder and Hibbett, 2006	4	

Table 1 (Continued).

Genus	Reference †	Root tip morphology	Root tip molecular	Synthesis	Isotopic	Phylogeny	Species number (estim.d)‡	Notes
Gyrodon	Trappe, 1962	Becerra *et al.*, 2005	Becerra *et al.*, 2005			Binder and Hibbett, 2006	10	
Gyroporus	Trappe, 1962	Agerer, 1999b				Binder and Hibbett, 2006	10	
Hallingea	Rinaldi *et al.*, 2008					Hosaka *et al.*, 2006	3	
Hebeloma	Trappe, 1962	Jakucs *et al.*, 1999	Jakucs *et al.*, 1999	Brunner *et al.*, 1991	Clemmensen *et al.*, 2006	Matheny *et al.*, 2006	~150	
Heimioporus[#]	Halling, 2007			see Tedersoo *et al.*, 2010[#]	Kohzu *et al.*, 1999[#]		16	22
Hydnellum	Molina *et al.*, 1992	Agerer, 1993; Kernaghan, 2001	Kernaghan, 2001		- (Högberg *et al.*, 1999)		38	
Hydnum	Trappe, 1962	Agerer *et al.*, 1996	Agerer *et al.*, 1996	Lu *et al.*, 1998	Högberg *et al.*, 1999	Moncalvo *et al.*, 2006	5 (?)	23
Hygrophorus	Trappe, 1962	Treu, 1990	Cullings *et al.*, 2000	Kropp and Trappe, 1982	Högberg *et al.*, 1999	Matheny *et al.*, 2006	~100	
Hymenogaster	Trappe, 1962	Donnini and Bencivenga, 1995		Brundrett *et al.*, 2005			~100	
Hysterangium	Trappe, 1962	Raidl and Agerer, 1998		Molina and Trappe, 1982		Hosaka *et al.* 2006	~50	24

Genus							Species	Ref
Inocybe	Trappe, 1962	Magyar *et al.*, 1999	Magyar *et al.*, 1999	Cripps and Miller, 1995	Högberg *et al.*, 1999; Hobbie *et al.*, 2001	Matheny *et al.*, 2006	~500	
Laccaria	Trappe, 1962	Torres *et al.*, 1995	Buée *et al.*, 2005	Godbout and Fortin, 1983	Högberg *et al.*, 1999; Hobbie *et al.*, 2001	Matheny *et al.*, 2006	~75	25
Lactarius	Trappe, 1962	Eberhardt *et al.*, 2000	Nuytinck *et al.*, 2004	Flores *et al.*, 2005	Högberg *et al.*, 1999; Hobbie *et al.*, 2001	Miller *et al.*, 2006	~450	26
Leccinum	Trappe, 1962	Müller and Agerer, 1990	den Bakker *et al.*, 2004	Molina and Trappe, 1982	Högberg *et al.*, 1999	Binder and Hibbett, 2006	~85	27
Lenzitopsis	Agerer, 2006					Stalpers, 1993	1	28
Leucogaster[#]	Molina *et al.*, 1992		Izzo *et al.*, 2005b			Miller *et al.*, 2006; Albee-Scott, 2007a,b[#]	20	
Leucopaxillus[#]	Trappe, 1962			Lu *et al.*, 1998		Matheny *et al.*, 2006; Moncalvo *et al.* 2002[#]	~15	29
Leucophleps	Molina *et al.*, 1992		Izzo *et al.*, 2005b			Albee-Scott, 2007a	4	
Lindtneria	Rinaldi *et al.*, 2008					Martin *et al.*, 2004	11	

Table 1 (Continued).

Genus	Reference †	Root tip morphology	Root tip molecular	Synthesis	Isotopic	Phylogeny	Species number (estim.d) ‡	Notes
Lyophyllum[#]	Trappe, 1962	Agerer and Beenken, 1998	Agerer and Beenken, 1998	Parladé *et al.*, 1996a; Yamada *et al.*, 2001[#]		+/- (Matheny *et al.*, 2006)	~40	30
Malajczukia	Molina *et al.*, 1992					+ ? (Hosaka *et al.*, 2006)	8	
Melanogaster	Molina *et al.*, 1992	Wiedmer *et al.*, 2004	Cline *et al.*, 2005	Parladé *et al.*, 1996b		Binder and Hibbett, 2006	25	
Membranomyces	Tedersoo *et al.*, 2003	Tedersoo *et al.*, 2003	Tedersoo *et al.*, 2003			Moncalvo *et al.*, 2006	2	
Mesophellia	Molina *et al.*, 1992			Lu *et al.*, 1998		Hosaka *et al.*, 2006	~15	
Multifurca	Buyck *et al.*, 2008					Buyck *et al.*, 2008	5	
Mycolevis	Molina *et al.*, 1992					Miller *et al.*, 2006; Albee-Scott, 2007b	1	
Nothocastoreum	Brundrett *et al.*, 1996					Hosaka *et al.*, 2006	1	
Octaviania	Molina *et al.*, 1992	Chilvers, 1968				Binder and Hibbett, 2006	15	
Paragyrodon	Binder and Hibbett, 2006					Binder and Hibbett, 2006	1	
Paxillus	Trappe, 1962	Mleczko, 1997	Lilleskov *et al.*, 2002	Molina, 1981	Högberg *et al.*, 1999	Binder and Hibbett, 2006	15	

Genus	Col2	Col3	Col4	Col5	Col6	Col7	Col8	Col9
Phellodon	Molina *et al.*, 1992	Agerer, 1992b			+/- (Högberg *et al.*, 1999)		16	
Phylloboletellus[#]	Agerer, 2006[#]					Binder and Hibbett, 2006[#]	1	
Phylloporus	Trappe, 1962					Binder and Hibbett, 2006	~50	
Piloderma	Molina *et al.*, 1992	Goodman and Trofymow, 1996	Dahlberg *et al.*, 1997	Baxter and Dighton, 2001			6	
Pisolithus	Trappe, 1962	de Román and de Miguel, 2005	Moyersoen *et al.*, 2003	Baxter and Dighton, 2001		Binder and Hibbett, 2006	~12	
Polyozellus	Kropp and Trappe, 1982					Stalpers, 1993	1	
Polyporoletus[#]	Agerer, 2006	Agerer *et al.*, 1998b	Agerer *et al.*, 1998b			Miller *et al.*, 2006; Albee-Scott, 2007a[#]	1	
Protoglossum[#]	Tedersoo *et al.*, 2010[#]			Burgess *et al.*, 1993[#]		Peintner *et al.*, 2001[#]	8	31
Protubera[#]	Tedersoo *et al.*, 2010[#]					Giachini *et al.*, 2010[#]	8	
Pseudotomentella	Agerer, 2006	Di Marino *et al.*, 2007	Kõljalg *et al.*, 2000; Di Marino *et al.*, 2007				12	

Table 1 (Continued).

Genus	Reference †	Root tip morphology	Root tip molecular	Synthesis	Isotopic	Phylogeny	Species number (estim.d)‡	Notes
Pulveroboletus	Molina *et al.*, 1992					Binder and Hibbett, 2006	25	
Quadrispora#	Tedersoo et al., 2010#					Peintner et al., 2001#	2	
Ramaria[#]	Trappe, 1962	Nourha *et al.*, 2005	Nourha *et al.*, 2005		+/- (Hobbie *et al.*, 2001; 2002)	Hosaka *et al.*, 2006; Giachini *et al.*, 2010[#]	~220 (?)	32
Retiboletus	Binder and Hibbett, 2006					Binder and Hibbett, 2006)	5	
Rhizopogon	Trappe, 1962	Jakucs *et al.*, 1998b	Jakucs *et al.*, 1998b	Massicotte *et al.*, 1999	Högberg *et al.*, 1999	Binder and Hibbett, 2006	~150	
Rhodactina	Yang *et al.*, 2006					Yang *et al.*, 2006	1	
Rhopalogaster[#]	Rinaldi *et al.*, 2008					Hosaka *et al.*, 2006	1	
Riessia	Lee *et al.*, 1997	Lee *et al.*, 1997					4	
Riessiella	Lee *et al.*, 1997	Lee *et al.*, 1997					2	
Royoungia[#]	Brundrett *et al.*, 1996		Bidartondo *et al.*, 2009[#]			Binder and Hibbett, 2006	1	
Russula	Trappe, 1962	Beenken, 2001a; 2001b	Beenken, 2001a; 2001b	Taylor and Alexander, 1989	Högberg *et al.*, 1999; Hobbie *et al.*, 2001	Miller *et al.*, 2006	~750	33

Genus	Ref 1	Ref 2	Ref 3	Ref 4	Ref 5	Ref 6	Count	
Sarcodon	Agerer, 2006	Agerer, 1991c	Izzo et al., 2005a				36	
Scleroderma	Trappe, 1962	Ingleby, 1999	Valentine et al., 2004	Mohan et al., 1993b		Binder and Hibbett, 2006	~30	
Sebacina	Selosse et al., 2002	Urban et al., 2003	Urban et al., 2003			Weiß et al., 2004[#]	9	
Serendipita[#]	Tedersoo et al., 2010[#]		Ryberg et al., 2009[#]	Warcup, 1988[#]		Ryberg et al., 2009[#]	7 (?)	34
Setchelliogaster	Molina et al., 1992			Brundrett et al., 2005			5	
Sistotrema[#]	Nilsson et al., 2006a	Di Marino et al., 2008	Nilsson et al., 2006a; Dunham et al., 2007[#]			Moncalvo et al., 2006	46 (?)	
Stephanopus	Molina et al., 1992	Palfner, 2001					5	
Strobilomyces	Trappe, 1962	Matsuda and Hijii, 1999	Matsuda and Hijii, 1999			Binder and Hibbett, 2006	~20	
Suillus	Trappe, 1962	Treu, 1990	Horton et al., 2005	Samson and Fortin, 1988	Högberg et al., 1999	Binder and Hibbett, 2006	50	
Thelephora	Trappe, 1962	Agerer and Weiss, 1989	Mahmood et al., 1999	Mohan et al., 1993a			~50	
Tomentella	Agerer, 2006	Jakucs et al., 2005	Jakucs et al., 2005	Kõljalg, 1992			80	
Tomentellopsis	Agerer, 2006	Agerer, 1998	Kõljalg et al., 2002	Kõljalg et al., 2002			7	
Tremellodendron	Selosse et al., 2002		Walker, 2003; Tedersoo et al., 2006b				10	

Table 1 (Continued).

Genus	Reference †	Root tip morphology	Root tip molecular	Synthesis	Isotopic	Phylogeny	Species number (estim.d)‡	Notes
Tremellogaster[#]	Tedersoo et al., 2010[#]					Binder and Hibbett, 2006[#]	1	
Tremelloscypha	Selosse et al., 2002					Weiß and Oberwinkler, 2001	2	
Tricholoma	Trappe, 1962	Comandini et al., 2004	Comandini et al., 2004	Brunner et al., 1992	Högberg et al., 1999; Hobbie et al., 2001	Matheny et al., 2006	~200	
Truncocolumella	Trappe, 1962	Eberhart and Luoma, 1996	Horton et al., 2005	Massicotte et al., 2000		Binder and Hibbett, 2006	3	
Tulasnella	Bidartondo et al., 2003		Bidartondo et al., 2003	Bidartondo et al., 2003		Moncalvo et al., 2006	~50	
Turbinellus[#]	Hosaka et al., 2006					Hosaka et al., 2006; Giachini et al., 2010[#]	5	
Tylopilus	Trappe, 1962	Uhl, 1989; Raidl and Hahn, 2006	Jonsson et al., 1999b; Burke et al., 2005; 2006		Högberg et al., 1999	Binder and Hibbett, 2006	~75	35
Tylospora	Agerer, 2006	Eberhardt et al., 1999	Eberhardt et al., 1999	Taylor and Alexander, 1990			2	

Xanthoconium	Brundrett *et al.*, 1996					Binder and Hibbett, 2006	7	
ZYGOMYCOTA								
Endogone	Molina *et al.*, 1992	Chu-Chou and Grace, 1983	Chu-Chou and Grace, 1984	Warcup, 1990a			~20	
Densospora[#]	McGee, 1996[#]			Warcup, 1985[#]			4	
Sclerogone	Brundrett *et al.*, 1992			Warcup, 1990a			1	

[§] The nomenclature used in this paper is that adopted by Kirk *et al.* (2008) in the *Dictionary of the Fungi*. To improve readability, taxa are listed alphabetically within their phylum, regardless of the class, order and family. Genera for which two or more different lines of evidence support an ECM habit are reported in boldface, whereas those for which a single type of experimental evidence is available are in boldface and underlined. In both these cases, the number of currently recognized species is reported, as retrieved from the *Dictionary of Fungi*, or from more recent taxonomic monographs of specific genera. An entry does not necessarily imply that all species of that genus are ECM. Synonymized genera are not listed, but in some cases are indicated in the notes to specific taxa;

[†] a reference (not necessarily the first one in chronological terms) alleging or quoting the ECM status of each listed genus, is reported. The quoted references are listed in Rinaldi *et al.* (2008) (freely available at www.fungaldiversity.org). Only articles with more recent evidence (marked [#], as specified below) are listed in the references section of this chapter;

[‡] for those genera where both mycorrhizal and non mycorrhizal species are thought to occur, only the species believed or proved to form ECM are reported;

[#] newly reported evidence supporting proved or putative ECM associations, and/or entries that were otherwise missing in Rinaldi *et al.* (2008) or that have been modified.

Notes:

1. as *Phialophora*, in Rinaldi *et al.* (2008);
2. *Chloridium paucisporum* has been considered as ectomycorrhizal by Wang and Wilcox (1985). The ITS sequence of the type strain differs only 1 bp of that of *Cadophora finlandica* with which it may be conspecific (Harney *et al.*, 1997). Other species of that genus (26) are not ECM.;
3. as *Myrmecocystis*, in Trappe (1969);
4. *Geopyxis, Paurocotylis* and *Stephensia* were considered non-ECM by Tedersoo *et al.* (2010). The phylogeny in Læssøe & Hansen (2007) puts ECM *Tarzetta* in a clade with *Geopyxis, Paurocotylis* and *Stephensia*. Stable isotope data reported in Rinaldi *et al.* (2008) for *Stephensia* and *Paurocotylis* tend to support an ECM status;

Table 1 (Continued).

5   Norman and Egger (1999) had the anamorphic *Glischroderma* in a clade with *Scabropezia* and *Pachyphloeus* (both ECM);

6   *Gyromitra* was considered as non-ECM by Tedersoo *et al.* (2010). The published phylogenies put it in Discinaceae, with ECM *Hydnotrya* and non-ECM *Discina* and *Pseudodiscina*;

7   listed by Tedersoo *et al.* (2010) as non-ECM, but reported as ECM by Diedhiou *et al.* (2009). See also main text discussion on ECM and EAA;

8   as *Picoa*, in Trappe (1969);

9   *Peziza* is highly polyphyletic, with some clades being ECM. The phylogeny in Hansen *et al.* (2005) and Læssøe and Hansen (2007) suggests that at least 3 clades form ECM (*P. depressa*, *P. succosa* [*Galactinia*] and *P. gerardii* [plus *Marcelleina*]. The fire-place species of *Plicaria* are likely non-ECM. The anamorphic *Chromelosporium* has, according to Norman and Egger (1999), teleomorphs in *Plicaria* (non-ECM) and possibly *Peziza* (ECM);

10   according to Tedersoo *et al.* (2009, 2010) *Phialocephala* is EAA rather than ECM. See text for discussion. Considering that *Acephala* is ECM, the status of more taxa of this clade as ECM is likely;

11   listed as non-mycorrhizal/ECM by Tedersoo *et al.* (2010). *Tricharina* is likely polyphyletic, with *T. ochroleuca*, being part of the ECM *Geopora* clade (Perry *et al.*, 2007), confirmed as ECM. Other species may be non-ECM;

12   *Trichophaea* is polyphyletic, but both clades are ECM (Tedersoo *et al.*, 2010);

13   *Albatrellus* includes *Scutiger* (Kirk *et al.*, 2008);

14   *Amanita* includes *Amarrendia* and *Torrendia* (Kirk *et al.*, 2008);

15   *Boletus* is grossly polyphyletic (Binder and Hibbett, 2006) and according to Kirk *et al.* (2008) includes *Pseudoboletus* (*B. parasiticus*) and *Xerocomus* (which almost certainly is polyphyletic, and has to be split in *Xerocomus* and *Xerocomellus*). All taxa are considered ECM;

16   *Catathelasma* has been reported as ECM in Tedersoo *et al.* (2010) although, admittedly, no conclusive proof exists to support this classification. In vitro synthesis with *Pinus* resulted in poorly developed ECM-like structures with a bi-seriate Hartig net (Hutchinson, 1992). Assessment of stable isotopes revealed that *Catathelasma* was enriched in ^{15}N as expected for ECM fungi, but displayed a surprisingly high ^{13}C signal (Kohzu *et al.*, 1999). In the multilocus phylogenetic overview of Matheny *et al.* (2006) *Catathelasma* formed a distinct, ecologically heterogeneous, tricholomatoid clade composed of the saprotrophic *Clitocybe subvelosa* and *Callistosporium*, and *Catathelasma*. We accept *Catathelasma* as ECM on a provisional basis.

17   *Chalciporus* listed as non-ECM by Tedersoo *et al.* (2010). The status of *Chalciporus piperatus* has been disputed (Rinaldi *et al.*, 2008) and no new data have come to light. Tedersoo *et al.* (2010) suggested that the taxon is sister to *Buchwaldoboletus* (another bolete with a likely association, in this case *Phaeolus schweinizii*). As *Pseudoboletus* (*Boletus*) *parasiticus* is listed as ECM by Tedersoo *et al.* (2010), the combination of ECM with a biotrophic relation to other fungi (as in Gomphidiaceae) is possible. Phylogenetic data (Binder and Hibbett, 2006) suggest that ECM habit in the crown group of Boletales must have evolved more than once (because *Phlebopus* and *Boletinellus* are non-ECM). Accepting *Chalciporus* and *Buchwaldoboletus* as non-ECM involves the assumption of one more ECM origin. The position in the phylogeny (http://www.clarku.edu/faculty/dhibbett/boletales_stuff/Global_Boletales_2004_28S.gif) does support ECM status. Following

the Dictionary of the Fungi (Kirk *et al.*, 2008) we treat *Rubinoboletus* as synonym, and Tedersoo *et al.* (2010) also accepted *Rubinoboletus* as ECM;

18 *Clavariadelphus* is polyphyletic with some species being ECM and other non-ECM;

19 *Cortinarus* includes *Cuphocybe, Rapacea, Thaxterogaster, Dermocybe* and *Rozites* (Kirk *et al.*, 2008);

20 The phylogeny published by Co-David *et al.* (2009) suggests one origin of ECM in the genus, only in north-temperate species. The species of *Rhodogaster* and *Richoniella* (treated as part of *Entoloma*) in that phylogeny are considered non-ECM;

21 The phylogeny by Giachini *et al.* (2010) suggests that *Gallacea* and *Protubera*, as sistergroup to *Hysterangium* are ECM. Tedersoo *et al.* (2010) accepted *Gallacea* as ECM and listed *Protubera* as non-mycorrhizal/ECM. The latter genus was listed as non-ECM by Rinaldi *et al.* (2008);

22 in Kohzu *et al.* (1999) as *Heimiella*;

23 The Dictionary of the Fungi lists 120 species for *Hydnum*, but the description includes wood-decomposing fungi and therefore refers to a polyphyletic assemblage of tooth-fungi. *Hydnum* sensu stricto contains only a few species;

24 *Hysterangium* includes *Aroramyces* (Kirk *et al.*, 2008);

25 *Laccaria* includes *Hydnangium* and *Podohydnangium* (Kirk *et al.*, 2008);

26 *Lactarius* includes *Arcangeliella* and *Zelleromyces* (Kirk *et al.*, 2008). *Lactarius* is polyphyletic and will in the future be split (*Lactarius* and *Lactifluus*) – see also Albee-Scott, 2007b;

27 *Leccinum* includes *Leccinellum* (den Bakker *et al.*, 2004);

28 since *Lenzitopsis* was not included in Larsson *et al.* (2004), and no new data have become available, we continue to treat this genus as ECM;

29 Moncalvo *et al.* (2002) have *Leucopaxillus* (plus *Porpoloma*, also sometimes mentioned as ECM but without substantial support) as sister to *Tricholoma*. Matheny *et al.* (2006) have a *Leucopaxillus* nested in ECM *Tricholoma*, but one may wonder whether the collection has been correctly named. The name *Leucopaxillus* is used for two groups, *Leucopaxillus* sensu stricto, which we consider ECM, and *Aspropaxillus*, which is saprotrophic is grasslands. Kohzu *et al.* (1999) studied *A. giganteus*, which indeed is a saprotroph;

30 *Lyophyllum* is polyphyletic, but a consistent new taxonomy has not been published. Several species form ECM (see Tedersoo *et al.*, 2010);

31 in Burgess *et al.* (1993) as *Hymenogaster*;

32 *Ramaria* is polyphyletic (Giachini *et al.*, 2010) and not all species are ECM. However, the non-ECM species are a minority;

33 *Russula* includes *Cystangium, Gymnomyces, Macowanites* (Kirk *et al.*, 2008) – see also Albee-Scott, 2007b;

34 Tedersoo *et al.* (2010) consider *Serendipita* (syn. *Sebacina vermifera*) as ECM, on the basis of the molecular phylogeny provided by Weiß *et al.* (2004) and its ability to form ECM structures in vitro (Warcup, 1988). We agree with this vision;

35 *Tylopilus* includes *Porphyrellus* (Kirk *et al.*, 2008).

Table 2. Genera of putative ectomycorrhizal fungi, with newly reported cases [§]

Genus	Reference [†]	Root tip morphology	Root tip molecular	Synthesis	Isotopic	Phylogeny	Species number (estim.d)[‡]	Notes
ASCOMYCOTA								
Aleurina[#]	Tedersoo *et al.*, 2010[#]						10?	1
Acephala[#]	Münzenberger *et al.*, 2009[#]	Münzenberger *et al.*, 2009[#]	Münzenberger *et al.*, 2009[#]	Münzenberger *et al.*, 2009[#]		Münzenberger *et al.*, 2009[#]	1	
Anthracobia[#]	Wei *et al.*, 2010[#]					Wei *et al.*, 2010[#]	15 ?	2
Arachnopeziza[#]	Hedh *et al.*, 2008[#]		Hedh *et al.*, 2008[#]				15	3
Gelinipes[#]	Tedersoo *et al.*, 2010[#]						1	1
Gymnohydnotrya	Brundrett *et al.*, 1996						3	
Hydnocystis	Molina *et al.*, 1992						2	
Imaia[#]	Kovács *et al.*, 2008[#]					Kovács *et al.*, 2008[#]	1	
Mycoclelandia	Brundrett *et al.*, 1996						2	
Oidiodendron[#]	Diedhiou *et al.*, 2009[#]		Diedhiou *et al.*, 2009[#]				28	3
Paradoxa	Molina *et al.*, 1992						1	
Phaeangium	Maia *et al.*, 1996						1	
Pseudaleuria	Maia *et al.*, 1996						1	
Sphaerosoma	Molina *et al.*, 1992						3	
Trichophaeopsis[#]	Tedersoo *et al.*, 2010[#]					Perry *et al.*, 2007[#]	2	
Ulurua[#]	Tedersoo *et al.*, 2010[#]					Trappe *et al.*, 2010[#]	1	
Unicava[#]	Tedersoo *et al.*, 2010[#]							1
BASIDIOMYCOTA								
Afroboletus[#]	Onguene, 2000[#]; Agerer, 2006						7	
Austrogaster	Agerer, 2006						3	
Boletochaete	Brundrett *et al.*, 1996						3	
Chlorogaster[#]	Rinaldi *et al.*, 2008[#]						1	
Corditubera	Agerer, 2006						5	
Cystogomphus	Molina *et al.*, 1992						1	
Duraniella[#]	Desjardin *et al.*, 2008[#]					Desjardin *et al.*, 2008[#]	1	
Fevansia	Trappe and Castellano, 2000						1	
Gastroleccinum	Agerer, 2006						1	

Genus	Reference						Count 1	Count 2
Gigasperma	Brundrett et al., 1996						2	4
Gomphogaster	Molina et al., 1992						1	
Gummivena[#]	Trappe and Bougher, 2002						1	
Heliogaster[#]	Orihara et al., 2010[#]					Orihara et al., 2010[#]	1	
Horakiella	Brundrett et al., 1996						1	
Maccagnia	Agerer, 2006						1	5
Mayamontana	Castellano et al., 2007						1	
Mackintoshia	Agerer, 2006						1	
Mycoamaranthus	Bougher, 1995						3	
Neopaxillus[#]	Tedersoo et al., 2010[#]						5	
Paxillogaster	Agerer, 2006						1	
Phyllobolites[#]	Tedersoo et al., 2010[#]						1	
Protogautieria[#]	Tedersoo et al., 2010[#]						2	
Psiloboletinus	Trappe, 1962						1	6
Pterygellus	Verbeken and Walleyn, 1999						5	
Setogyroporus	Buyck et al., 1996						1	
Singeromyces[#]	Tedersoo et al., 2010[#]						1	
Sinoboletus	Dell et al., 2005						10	
Spongiforma[#]	Desjardin et al., 2009[#]					Desjardin et al., 2009[#]	1	
Timgrovea	Brundrett et al., 1996						5	7

Table 2 (Continued).

Genus	Reference [†]	Root tip morphology	Root tip molecular	Synthesis	Isotopic	Phylogeny	Species number (estim.d)[‡]	Notes
Tubosaeta[#]	Onguene, 2000[#]; Agerer, 2006						5	
Veloporphyrellus	Agerer, 2006						2	
ZYGOMYCOTA								
Peridiospora	Brundrett, 2008						4	
Youngiomyces	Brundrett, 2008						4	

[§] For the adopted nomenclature and other details, see legend to Table 1. In this Table, genera for which no experimental or phylogenetic evidence is currently available, but for which ECM status can be hypothesized based on habitat features and/or personal experience, are in boldfaced small caps. All other notations as in Table 1;

[†] a reference (not necessarily the first one in chronological terms) alleging or quoting the ECM status of each listed genus, is reported. The quoted references are listed in Rinaldi *et al.*, 2008 (freely available at www.fungaldiversity.org). Only articles with more recent evidence (marked [#], as specified below) are listed in the references section of this chapter;

[‡] for those genera where both mycorrhizal and non mycorrhizal species are thought to occur, only the species believed or proved to form ECM are reported;

[#] newly reported cases of putative or proved ECM associations, and/or entries that were otherwise missing in Rinaldi *et al.* (2008) or that have been modified.

Notes:

1 *Gelinipes* and *Unicava* (both provisional genera) cluster with *Aleurina* (which is polyphyletic) in Perry *et al.*, 2007. According to Tedersoo *et al.* (2010), *A. argentina*, which occurs in *Nothofagus* forests in Chile, clusters with ECM taxa;

2 see main text for discussion;

3 possibly secondary root colonizers, i.e. EAA;

4 the genus is heterogeneous (see Matheny *et al.*, 2006);

5 likely a synonym of *Laccaria* (Kirk *et al.*, 2008);

6 very closely related (possibly a synonym) to *Suillus*;

7 taxonomic position unclear.

Switching to Basidiomycetes, two other interesting 'newcomers' are *Duraniella* and *Spongiforma*, gasteroid boletes described from Malaysia and Thailand, respectively (Table 2). *Duraniella* occurs under *Shorea* spp. (Dipterocarpaceae) in open secondary forests (Desjardin *et al.*, 2008). A similar habitat and taxonomic placement among the Boletineae has been inferred for *Spongiforma*, which occurred solitarily in soil under *Dipterocarpus* sp. and *Shorea* sp. in primary forests (Desjardin *et al.*, 2009). Within the Boletales, the new gasteroid genus *Heliogaster* has been introduced by Orihara *et al.* (2010) to accommodate the sequestrate *Octaviania columellifera*, an endemic of Japan, where it occurs primarily associated with *Abies firma* (Orihara *et al.*, 2010). A molecular phylogeny placed *Heliogaster* in a distinct position with regard to authentic *Octaviania* species, close to the *Xerocomus* (= *Boletus*) *chrysenteron* complex.

In addition to newly described, potential ECM genera, Table 2 also lists several already existing taxa for which an ECM status has been either alleged or proved recently. Münzenberger *et al.* (2009) have proved – by a mix of synthesis, morpho-anatomical, and molecular data – that the ECM morphotype *Pinirhiza sclerotia* is formed by the dark septate fungus *Acephala macrosclerotiorum*, a new species and close relative of *Phialocephala fortinii*, as also discussed above.

In total, 234 fungal genera are reported in Table 1 and 2 as – at least potentially – ECM. This value is very close to that presented in Rinaldi *et al.* (2008), but the list has somewhat changed. Besides the changes and additions noted above, several sequestrate genera (e.g., *Arcangeliella*, *Zelleromyces*, *Gymnomyces*, *Macowanites*, *Cystangium* in the Russulaceae; *Amarrendia* and *Torrendia* in the Amanitaceae; *Hydnangium* and *Podohydnangium* in the Laccariaceae) have been eliminated because their synonymic status with agaricoid genera has been established, according to the last edition of the Dictionary of the Fungi (Kirk *et al.*, 2008). In the case of *Entoloma* the sequestrate genera *Richoniella* and *Rhodogaster* have also been synonymised (Co-David *et al.*, 2009) but we consider these latter taxa as non-ECM. The phylogeny published by Co-David *et al.* (2009) suggests a monophyletic origin of the ECM habit; likewise we exclude *E. nitidum* from the ECM species. In several other instances, genera quoted by Tedersoo *et al.* (2010) have been added (e.g., *Gelinipes*, *Neopaxillus*, *Phyllobolites*, *Singeromyces*, *Unicava*), although no direct experimental evidence exists so far to support this conclusion. Similarly, the new count yields around 7,950 species of ECM fungi. Consequently, we do not see a need to revise our estimate of global ECM fungal species richness between 20,000 and 25,000 species.

SAPROTROPHIC LIFE STYLES OF ECM FUNGI

In our previous review we discussed, as part of our reflection on the value of phylogenetic supportive evidence for an ECM status, the possibility of mixotrophy, a nutritional mode whereby species are both ECM in combination with a saprotrophic life style (with regard to the possibility that secondary reversals from an ECM nutritional mode to a saprotrophic one affect the placement of genera in our list and that of Tedersoo *et al.* (2010), see above). Mixotrophy as a concept is widely used in ECM research, when it comes to the status of certain chlorophyllous orchids that associate with ECM fungi (Selosse and Roy, 2009) and whose ^{13}C signal is intermediate between autotrophic plants and achlorophyllous mycoheterotrophic plants. Assigning such status on the basis of stable isotopes only is not without its problems (as signals are also affected by changes in water use efficiency, and differential provenance of carbon for the build-up of leaves and stems), but is a topic better left out of this review. The relevant question is whether the concept of mixotrophy is also applicable to (putative) ECM fungi. In one (trivial) sense, the answer would be yes: in seasonal climates that are characterized by leaf fall the C nutrition of the fungus must either be dependent on C stored and used later (so uncoupled from recent photosynthate), or on carbon that remains available after the root tip died. Only in the latter case could a claim be made for a saprotrophic life style, although these ECM fungi are unlikely to be strongly competitive compared to saprotrophic fungi. The suggestion by Talbot *et al.* (2008) that ECM fungi are decomposers in disguise (the so-called plan B hypothesis), with a substantial ability to decompose soil organic carbon under conditions that carbon supply to the plant is low, is still speculative – with various field observations that are consistent with such a hypothesis but without any experimental demonstration. Observations that ECM tips show expression levels of hydrolytic and oxidative enzymes through activity profiling (Courty *et al.,* 2005) and some upregulation of enzyme activities during bud break (Courty *et al.,* 2007) are also consistent with such a hypothesis, but Baldrian (2009) proposed several other explanations for increased enzyme activities. An earlier claim that truffles can become independent from the tree for carbon supply and live saprotrophically in later stages, was shown to be incorrect by Zeller *et al.* (2008).

The occurrence of class II peroxidases (lignin peroxidases; manganese peroxidases) in various groups of ECM fungi was recently described by Bödeker *et al.* (2009). They noted such enzymes in *Russula*, *Lactarius* and *Cortinarius*. Their data show that these ECM fungi do not contain lignin peroxidases, but manganese peroxidases. Manganese peroxidases also play a role in detoxification of phenolic compounds and are also especially expressed under conditions of oxidative stress and at high manganese levels, which may happen as a function of decreasing pH in deeper soil layers (Sinsabaugh, 2010).

In our opinion claims that ECM fungi do not only possess substantial saprotrophic potential, but also show substantial saprotrophic activity in the field, has never received unambiguous support. Suggestions to the contrary by Talbot *et al.* (2008), Cullings and Courty (2009) and Courty *et al.* (2010) rather reflect the wish to come up with a new paradigm for mycorrhizal interactions. It might be significant that Cullings and Courty (2009) also explicitly refer to numerous reversals from an ECM to a saprotrophic nutritional habit; a claim for which, in our opinion, no good evidence exists either.

CONCLUDING REMARKS

This is certainly an exciting moment for the study of global ECM diversity. An increasing amount of reliable, taxonomically checked molecular data, both translating into direct identification of mycobionts in ECM tips and in more accurate and comprehensive phylogenetic studies, coupled to a better understanding of the ECM symbiosis, is giving us the tools to explore this intriguing, lively liaison at an unprecedented level of detail. Although a lot of work still remains to be done, the recent studies critically assessing ECM diversity at a global scale (Rinaldi *et al.*, 2008; Tedersoo *et al.*, 2010; Comandini *et al.*, this paper) will hopefully help further exploring uncharted territories.

REFERENCES

Agerer, R. (1986) Studies on Ectomycorrhizas II. Introducing remarks on characterization and identification. *Mycotaxon* 26: 473-492.

Agerer, R. (2006) Fungal relationships and structural identity of their ectomycorrhizae. *Mycological Progress* 5: 67-107.

Albee-Scott, S.R. (2007a) The phylogenetic placement of the Leucogastrales, including *Mycolevis siccigleba* (Cribbeaceae), in the Albatrellaceae using morphological and molecular data. *Mycological Research* 111: 653-662.

Albee-Scott, S.R. (2007b) Does secotioid inertia drive the evolution of false-truffles? *Mycological Research* 111: 1030-1039.

Alberton, O., Kuyper, T.W. and Summerbell, R.C. (2010) Dark septate root endophytic fungi increase growth of Scots pine seedlings under elevated CO_2 through enhanced nitrogen use efficiency. *Plant and Soil* 328: 459-470.

Alexander, I.J. (1989) Systematics and ecology of ectomycorrhizal legumes. *Monographs in Systematic Botany of the Missouri Botanical Garden* 29: 607-624.

Alexander, I.J. (2006) Ectomycorrhizas – out of Africa? *New Phytologist* 172: 589-591.

Angiosperm Phylogeny Group (2009) An update of the Angiosperm Phylogeny group classification for the orders and families of flowering plants: APG III. *Botanical Journal of the Linnaean Society* 161: 105-121.

den Bakker, H.C., Zuccarello, G.C., Kuyper, T.W. and Noordeloos, M.E. (2004) Evolution and host specificity in the ectomycorrhizal genus *Leccinum*. *New Phytologist* 163: 201-215.

Baldrian, P. (2009) Ectomycorrhizal fungi and their enzymes in soils: is there enough evidence for their role as facultative soil saprotrophs? *Oecologia* 161: 657-660.

Bidartondo, M.I., Ameri, G. and Döring, H. (2009) Closing the mycorrhizal DNA sequence gap. *Mycological Research* 113: 1025-1026.

Binder, M. and Hibbett, D.S. (2006) Molecular systematics and biological diversification of Boletales. *Mycologia* 98: 971–981.

Bödeker, I.T.M., Nygren, C.M.R., Taylor, A.F.S., Olson, Å. and Lindahl, B.D. (2009) ClassII peroxidase-encoding genes are present in a phylogenetically wide range of ectomycorrhizal fungi. *ISME Journal* 3: 1387-1395.

Brundrett, M.C. (2009) Mycorrhizal associations and other means of nutrition of vascular plants: understanding the global diversity of host plants by resolving conflicting information and developing reliable means of diagnosis. *Plant and Soil* 320: 37-77.

Burgess, T.I., Malajczuk, N. and Grove, T.S. (1993) The ability of 16 ectomycorrhizal fungi to increase growth and phosphorus uptake of *Eucalyptus globulus* and *E. diversicolor*. *Plant and Soil* 153: 155–164.

Co-David, D., Langeveld, D. and Noordeloos, M.E. (2009) Molecular phylogeny and spore evolution of Entolomataceae. *Persoonia* 23: 147-176.

Courty, P.-E., Pritsch, K., Schloter, M., Hartmann, A. and Garbaye, J. (2005) Activity profiling of ectomycorrhiza communities in two forest soils using multiple enzymatic tests. *New Phytologist* 167: 309-319.

Courty, P.-E., Breda, N. and Garbaye, J. (2007) Relation between oak tree phenology and the secretion of organic matter degrading enzymes by *Lactarius quietus* ectomycorrhizas before and during bud break. *Soil Biology & Biochemistry* 39: 1655-1663.

Courty, P.-E., Buée, M., Diedhiou, A.G., Frey-Klett, P., Le Tacon, F., Rineau, F., Turpault, M.-P., Uroz, S. and Garbaye, J. (2010) The role of ectomycorrhizal communities on forest ecosystem processes: new perspectives and emerging concepts. *Soil Biology & Biochemistry* 42: 679-698.

Cullings, K. and Courty, P.-E. (2009) Saprotrophic capabilities as functional traits to study functional diversity and resilience of ectomycorrhizal community. *Oecologia* 161: 661-664.

Desjardin, D.E., Wilson, A.W. and Binder, M. (2008) *Duraniella*, a new gasteroid genus of boletes from Malaysia. *Mycologia* 100: 956-961.

Desjardin, D.E., Binder, M., Roekring, S. and Flegel, T. (2009) *Spongiforma*, a new genus of gasteroid boletes from Thailand. *Fungal Diversity* 37: 1-8.

Dickie, I.A., and Reich, P.B. (2005) Ectomycorrhizal fungal communities at forest edges. *Journal of Ecology* 93: 244–255.

Diedhiou, A.G., Dupouey, J.-L., Buée, M., Dambrine, E., Laüt, L. and Garbaye, J. (2009) Response of ectomycorrhizal communities to past Roman occupation in an oak forest. *Soil Biology & Biochemistry* 41: 2206–2213.

Ducousso, M., Bourgeois, C., Buyck, B., Eyssartier, G., Vincelette, M., Rabevohitra, R., Béna, G., Randrihasipara, L., Dreyfus, B. and Prin, Y. (2004) The last common ancestor of Sarcolaenaceae and Asian dipterocarp trees was ectomycorrhizal before the India-Madagascar separation, about 88 million years ago. *Molecular Ecology* 13: 231–236.

Ducousso, M., Ramanankierana, H., Duponnois, R., Rabévohitra, R., Randrihasipara, L., Vincelette, M., Dreyfus, B. and Prin,Y. (2008) Mycorrhizal status of native trees and shrubs from eastern Madagascar littoral forests with special emphasis on one new ectomycorrhizal endemic family, the Asteropeiaceae. *New Phytologist* 178: 233–238.

Dunham, S.M., Larsson, K.-H. and Spatafora, J.W. (2007) Species richness and community composition of mat-forming ectomycorrhizal fungi in old- and second-growth Douglas-fir forests of the HJ Andrews Experimental Forest, Oregon, USA. *Mycorrhiza* 17: 633-645.

Ferdman, Y., Aviram, S., Roth-Bejerano, N., Trappe, J.M. and Kagan-Zur, V. (2005) Phylogenetic studies of *Terfezia pfeilii* and *Choiromyces echinulatus* (Pezizales) support new genera for southern African truffles: *Kalaharituber* and *Eremiomyces*. *Mycological Research* 109: 237-245.

Gao, Q. and Yang, Z.L. (2010) Ectomycorrhizal fungi associated with two species of *Kobresia* in an alpine meadow in the eastern Himalaya. *Mycorrhiza* 20: 281-287.

Giachini, A.J., Hosaka, K., Nouhra, E., Spatafora, J. and Trappe, J.M. (2010) Phylogenetic relationships of the Gomphales based on nuc-25S-rDNA, mit-12S-rDNA, and mit-*atp*6-rDNA combined sequences. *Fungal Biology* 114: 224-234.

Grünig, C.R., Queloz, V., Sieber, T.N. and Holdenrieder, O. (2008) Dark septate endophytes (DSE) of the *Phialocephala fortinii* s.l. – *Acephala applanata* species complex in tree roots: classification, population biology, and ecology. *Botany* 86: 1355–1369.

Gutiérrez, A., Morte, A. and Honrubia, M. (2003) Morphological characterization of the mycorrhiza formed by *Helianthemum almeriense* Pau with *Terfezia claveryi* Chatin and *Picoa lefebvrei* (Pat.) Maire. *Mycorrhiza* 13: 299–307.

Guzmán, B. and Vargas, P. (2009) Historical biogeography and character evolution of Cistaceae (Malvales) based on analysis of plastid *rbcL* and *trnL-trnF* sequences. *Organisms, Diversity & Evolution* 9: 83-99.

Hambleton, S. and Sigler, S. (2005) *Meliniomyces*, a new anamorph genus for root-associated fungi with phylogenetic affinities to *Rhizoscyphus ericae* (= *Hymenoscyphus ericae*), *Leotiomycetes*. *Studies in Mycology* 53: 1-27.

Hansen, K., LoBuglio, K.F. and Pfister, D.H. (2005) Evolutionary relationships of the cup-fungus genus *Peziza* and Pezizaceae inferred from multiple nuclear genes: RPB2, β-tubulin, and LSU rDNA. *Molecular Phylogenetics and Evolution* 36: 1–23.

Hansen, K. and Pfister, D. H. (2006) Systematics of the Pezizomycetes – the operculate discomycetes. *Mycologia* 98: 1031–1041.

Harney, S.K., Rogers, S.O and Wang, C.J.K. (1997) Molecular characterization of dematiaceous root endophytes. *Mycological Research* 101: 1397-1404.

Hedh, J., Wallander, H. and Erland, S. (2008) Ectomycorrhizal mycelial species composition in apatite amended and non-amended mesh bags buried in a phosphorus-poor spruce forest. *Mycological Research* 112: 681-688.

Henkel, T.W., Terborg, J. and Vilgalys R.J. (2002) Ectomycorrhizal fungi and their leguminous hosts in the Pakaraima Mountains of Guyana. *Mycological Research* 106: 515-531.

Hobbie, E.A., Weber, N.S. and Trappe, J.M. (2001) Mycorrhizal vs saprotrophic status of fungi: the isotopic evidence. *New Phytologist* 150: 601–610.

Horton, T.R. and Bruns, T.D. (2001) The molecular revolution in ectomycorrhizal ecology: peeking into the black-box. *Molecular Ecology* 10: 1855-1871.

Hutchison, L.J. (1992) Host range, geographical distribution and probable ecological status of *Catathelasma imperiale* in North America. Mycologia 84: 472–475.

Izzo, A.D., Meyer, M., Trappe, J.M., North, M. and Bruns, T.D. (2005) Hypogeous ectomycorrhizal fungal species on roots and in small mammal diet in a mixed-conifer forest. *Forest Science* 51: 243-254.

Jumpponen, A., Claridge, A.W., Trappe, J.M., Lebel, T. and Claridge, D.L. (2004) Ecological relationships among hypogeous fungi and trees: inferences from association analysis integrated with habitat modeling. *Mycologia* 96: 510-525.

Kennedy, P.G., and Hill, L.T. (2010) A molecular and phylogenetic analysis of the structure and specificity of *Alnus rubra* ectomycorrhizal assemblages. *Fungal Ecology* 3: 195-204.

Kirk, P.M., Cannon, P.F., David, J.C. and Stalpers, J.A. (eds.) (2008) *Ainsworth and Bisby's Dictionary of the Fungi*. 10th Edition. Wallingford, UK: CABI Publishing.

Kohzu, A., Yoshioka, T., Ando, T., Takahashi, M., Koba, K. and Wada, E. (1999) Natural ^{13}C and ^{15}N abundance of field-collected fungi and their ecological implications. *New Phytologist* 144: 323–330.

Kovács, G.M., Trappe, J.M., Alsheikh, A.M., Bóka, K. and Elliott, T. (2008) *Imaia*, a new truffle genus to accommodate *Terfezia gigantea*. *Mycologia* 100: 930-939.

Læssøe, T. and Hansen, K. (2007) Truffle trouble: what happened to the Tuberales? *Mycological Research* 111: 1075-1099.

Larsson, K.-H., Larsson, E. and Kõljalg, U. (2004) High phylogenetic diversity among corticioid homobasidiomycetes. *Mycological Research* 108: 983–1002.

Matheny, P.B., Curtis, J.M., Hofstetter, V., Aime, M.C., Moncalvo, J.-M., Ge, Z.-W., Yang, Z.-L., Slot, J.C., Ammirati, J.F., Baroni, T.J., Bougher, N.L., Hughes, K.W., Lodge, D.J., Kerrigan, R.W., Seidl, M.T., Aanen, D.K., DeNitis, M., Daniele, G.M., Desjardin, D.E., Kropp, B.R., Norvell, L.L., Parker, A., Vellinga, E.C., Vilgalys, R. and Hibbett, D.S. (2006) Major clades of Agaricales: a multi-locus phylogenetic overview. *Mycologia* 98: 984–997.

McGee, P.A. (1996) The Australian zygomycetous mycorrhizal fungi: the genus *Densospora* gen. nov. *Australian Systematic Botany* 9: 329-336.

Moncalvo, J.M., Vilgalys, R., Redhead, S.A., Johnson, J.E., James, T.Y., Aime, M.C., Hofstetter, V., Verduin, S.J.W., Larsson, E., Baroni, T.J., Thorn, R.G., Jacobsson, S., Clémençon, H. and Miller Jr., O.K. (2002) One hundred and seventeen clades of euagarics. *Molecular Phylogenetics and Evolution* 23: 357-400.

Morris, M.H., Smith, M.E., Rizzo, D.M., Rejmánek, M. and Bledsoe, C.S. (2008a) Contrasting ectomycorrhizal fungal communities on the roots of co-occurring oaks (*Quercus* spp.) in a California woodland. *New Phytologist* 178: 167-176.

Morris, M.H., Pérez-Pérez, M.A., Smith, M.E. and Bledsoe, C.S. (2008b) Multiple species of ectomycorrhizal fungi are frequently detected on individual oak root tips in a tropical cloud forest. *Mycorrhiza* 18: 375–383.

Moser, A.M., Frank, J.L., D'Allura, J.A. and Southworth, D. (2009) Ectomycorrhizal communities of *Quercus garryana* are similar on serpentine and nonserpentine soils. *Plant and Soil* 315: 185–194.

Moyersoen, B. (2006) *Pakaraimaea dipterocarpacaea* is ectomycorrhizal, indicating an ancient Gondwanaland origin for the ectomycorrhizal habit in Dipterocarpaceae. *New Phytologist* 172: 753-762,

Münzenberger, B., Bubner, B., Wöllecke, J., Sieber, T.N., Bauer, R., Fladung, M. and Hüttl, R.F. (2009) The ectomycorrhizal morphotype *Pinirhiza sclerotia* is formed by *Acephala macrosclerotiorum* sp. nov., a close relative of *Phialocephala fortinii*. *Mycorrhiza* 19: 481–492.

Norman, J.E. and Egger, K.N. (1999) Molecular phylogenetic analysis of *Peziza* and related genera. *Mycologia* 91: 820-829.

Onguene, N.A. (2000) Diversity and dynamics of mycorrhizal associations in tropical rain forests with different disturbance regimes in South Cameroon. *PhD thesis,* Wageningen University.

Onguene, N.A. and Kuyper, T.W. (2001) Mycorrhizal associations in the rain forest of South Cameroon. *Forest Ecology and Management* 140: 277-287.

Orihara, T., Sawada, F., Ikeda, S., Yamato, M., Tanaka, C., Shimomura, N., Hashiya, M. and Iwase, K. (2010) Taxonomic reconsideration of a sequestrate fungus, *Octaviania columellifera*, with the proposal of a new genus, *Heliogaster*, and its phylogenetic relationships in the Boletales. *Mycologia* 102: 108–121.

Palfner, G. and Agerer, R. (1998) *Leucangium carthusianum* (Tul.) Paol. + *Pseudotsuga menziesii* (Mirb.) Franco. *Descriptions of Ectomycorrhizae 3: 37–42.*

Peintner, U., Bougher, N.L., Castellano, M.-A., Moncalvo, J.-M., Moser, M.M., Trappe, J.M. and Vilgalys, R. (2001) Multiple origins of sequestrate fungi related to *Cortinarius* (Cortinariaceae). *American Journal of Botany* 88: 2168–2179.

Perry, B.A., Hansen, K. and Pfister, D.H. (2007) A phylogenetic overview of the family Pyronemataceae (Ascomycota, Pezizales). *Mycological Research* 111: 549-571.

Porter, T.M., Schadt, C.W., Rizvi, L., Martin, A.P., Schmidt, S.K., Scott-Denton, L., Vilgalys, R. and Moncalvo, J.M. (2008) Widespread occurrence and phylogenetic placement of a soil clone group adds a prominent new branch to the fungal tree of life. *Molecular Phylogenetics and Evolution* 46: 635-644.

Rinaldi, A.C., Comandini, O. and Kuyper, T.W. (2008) Ectomycorrhizal fungal diversity: separating the wheat from the chaff. *Fungal Diversity* 33: 1-45.

Rosling, A., Landeweert, R., Lindahl, B.D., Larsson, K.-H., Kuyper, T.W., Taylor, A.F.S. and Finlay, R.D. (2003) Vertical distribution of ectomycorrhizal fungal taxa in a podzol soil profile. *New Phytologist* 159: 775-783.

Ryberg, M., Larsson, E. and Molau, U. (2009) Ectomycorrhizal diversity in *Dryas octopetala* and *Salix reticulata* in an Alpine cliff ecosystem. *Arctic, Antarctic, and Alpine Research* 41: 506-514.

Selosse, M.A. and Roy, M. (2009) Green plants that feed on fungi: facts and questions about mixotrophy. *Trends in Plant Sciences* 14: 64-70.

Sinsabaugh, R.L. (2010) Phenol oxidase, peroxidase and organic matter dynamics of soil. *Soil Biology & Biochemistry* 42: 391-404.

Smith, S.E. and Read, D.J. (2008) *Mycorrhizal symbiosis*. 3rd Edition. San Diego, USA: Academic Press.

Smith, M.E., Trappe, J.M. and Rizzo, D.M. (2006) *Genea*, *Genabea* and *Gilkeya* gen. nov.: ascomata and ectomycorrhiza formation in a *Quercus* woodland. *Mycologia* 98: 699-716.

Summerbell, R.C. (2005) From Lamarckian fertilizers to fungal castles: recapturing the pre-1985 literature on endophytic and saprotrophic fungi associated with ectomycorrhizal root systems. *Studies in Mycology* 53: 191-256.

Talbot, J.M., Allison, S.D. and Treseder, K.K. (2008) Decomposers in disguise: mycorrhizal fungi as regulators of soil C dynamics in ecosystems under global change. *Functional Ecology* 22: 955-963.

Tedersoo, L., Hansen, K., Perry, B.A. and Kjøller, R. (2006) Molecular and morphological diversity of pezizalean ectomycorrhiza. *New Phytologist* 170: 581-596.

Tedersoo, L., Suvi, T., Beaver, K. and Kõljalg, U. (2007) Ectomycorrhizal fungi of the Seychelles: diversity patterns and host shifts from the native *Vateriopsis seychellarum* (Dipterocarpaceae) and *Intsia bijuga* (Caesalpiniaceae) to the introduced *Eucalyptus robusta* (Myrtaceae), but not *Pinus caribea* (Pinaceae). *New Phytologist* 175: 321–333.

Tedersoo, L., Pärtel, K., Jairus, T., Gates, G., Põldmaa, K. and Tamm, H. (2009) Ascomycetes associated with ectomycorrhizas: molecular diversity and ecology with particular reference to the Helotiales. *Environmental Microbiology* 11: 3166–3178.

Tedersoo, L., May, T.W. and Smith, M.E. (2010) Ectomycorrhizal lifestyle in fungi: global diversity, distribution, and evolution of phylogenetic lineages. *Mycorrhiza* 20: 217-263.

Trappe, J.M. (1969) Mycorrhiza-forming Ascomycetes. In: *Proceedings of the First North American Conference on Mycorrhizae*. Misc. Publication 1189 U.S. Department of Agriculture, *Forest Service,* 19-37.

Trappe, J.M., Kovács, G.M. and Claridge, A.W. (2010) Comparative taxonomy of desert truffles of the Australian outback and the African Kalahari. *Mycological Progress* 9: 131–143.

Trowbridge, J. and Jumpponen, A. (2004) Fungal colonization of shrub willow roots at the forefront of a receding glacier. *Mycorrhiza* 14: 283-293.

Vrålstad, T., Myhre, E. and Schumacher, T. (2002) Molecular diversity and phylogenetic affinities of symbiotic root-associated ascomycetes of the Helotiales in burnt and metal polluted habitats. *New Phytologist* 155: 131–148.

Wang, C.J.K. and Wilcox, H.E. (1985) New species of ectendomycorrhizal and pseudomycorrhizal fungi: *Phialophora finlandia, Chloridium paucisporum*, and *Phialocephala fortinii. Mycologia* 77: 951-958.

Warcup, J.H. (1985) Ectomycorrhiza formation by *Glomus tubiforme. New Phytologist* 99: 267-272.

Warcup, J.H. (1988) Mycorrhizal associations of isolates of *Sebacina vermifera. New Phytologist* 110: 227-231.

Warcup, J.H. (1990) Occurrence of ectomycorrhizal and saprophytic discomycetes after a wild fire in an eucalypt forest. *Mycological Research* 94: 1065-1069.

Warcup, J.H. (1991) The *Rhizoctonia* endophytes of *Rhizanthella* (Orchidaceae). *Mycological Research* 95: 656–659.

Wei, J., Peršoh, D. and Agerer, R. (2010) Four ectomycorrhizae of Pyronemataceae (Pezizomycetes) on Chinese Pine (*Pinus tabulaeformis*): morpho-anatomical and molecular-phylogenetic analyses. *Mycological Progress* 9: 267-280.

Weiß, M., Selosse, M.-A., Rexer, K.-H., Urban, A. and Oberwinkler, F. (2004) Sebacinales: a hitherto overlooked cosm of heterobasidiomycetes with a broad mycorrhizal potential. *Mycological Research* 108: 1003–1010.

Wilcox, H.E. and Wang, C.J.K. (1987) Ectomycorrhizal and ectendomycorrhizal associations of *Phialophora finlandia* with *Pinus resinosa, Picea rubens*, and *Betula alleghanensis. Canadian Journal of Forest Research* 17: 976-990.

Yagame, T., Yamato, M., Suzuki, A. and Iwase, K. (2008) Ceratobasidiaceae mycorrhizal fungi isolated from nonphotosynthetic orchid *Chamaegastrodia sikokiana. Mycorrhiza* 18: 97–101.

Yamada, A., Ogura, T. and Ohmasa, M. (2001) Cultivation of mushrooms of edible ectomycorrhizal fungi associated with *Pinus densiflora* by in vitro mycorrhizal synthesis. II. Morphology of mycorrhizas in open-pot soil. *Mycorrhiza* 11: 67-81.

Zeller, B., Bréchet, C., Maurice, J.-P. and Le Tacon, F. (2008) Saprotrophic versus symbiotic strategy during truffle ascocarp development under holm oak. A response based on ^{13}C and ^{15}N natural abundance. *Annals of Forest Science* 65: 607-616.

In: Mycorrhiza: Occurrence in Natural and Restored...
Editor: Marcela Pagano

ISBN: 978-1-61209-226-3
© 2012 Nova Science Publishers, Inc.

Chapter 10

MYCORRHIZAL DIVERSITY IN NATIVE AND EXOTIC WILLOWS (*SALIX HUMBOLDTIANA AND S. ALBA*) IN ARGENTINA

Mónica A. Lugo[1], Alejandra G. Becerra[2], Eduardo R. Nouhra[2] and Ana C. Ochoa[3]

[1]Instituto Multidisciplinario de Investigaciones Biológicas, UNSL,
[2]Instituto Multidisciplinario de Biología Vegetal, UNC,
C 495; 5000, Córdoba, Argentina
[3]UNSL, Av. Ejército de los Andes 1148, CEP 5700, San Luis, Argentina

ABSTRACT

Mycorrhizal morphology and diversity vary not only within each particular host family of plants, but also with soil characteristics, nutrients availability, spatial-temporal conditions, micro-habitats, and host's age. Ectomycorrhizae (ECM) and arbuscular mycorrhizae (AM) provide nutritional benefits to their hosts, besides their effects in soil aggregation, soil pollutants sequestration and host's interconnection by hyphal network that allows nutrients transport, seedlings establishment and conservation of forest ecosystems. Furthermore, certain hosts species such as *Salix* spp. in the Salicaceae are involved in dual associations with ECM and Glomeromycota fungal symbionts. *Salix*, presents a wide distribution in South America, native *Salix humboldtiana* populations as well us the introduced species are mostly located in riparian ecosystems, or in temporarily flooded areas. Riparian zones have an important role regulating the movement of material and water between soil-river systems. Mycorrhizal diversity and colonization was studied in riparian populations of native *S. humboldtiana* Willd. and of exotic *S. alba* L. in semiarid riparian environments. Differences in ECM morphotypes and mycorrhizal colonization were found in both species. Effects of soil physical-chemical features on ECM diversity are discussed and ECM morphotypes are illustrated.

Keywords: Argentina, mycorrhizal symbiosis, Riparian habitat, *Salix*, semi-arid ecosystems

INTRODUCTION

Mycorrhizal colonization of plants roots produces permanent or temporary changes, depending on the symbionts that define each particular mycorrhizal type (ECM, AM, ectendo- arbutoid-, ericoid-, monotropoid and orchid-mycorrhizas, and dark septate endophyte (DSE) in a broad sense), each one characterized by its root-fungi-anatomy and morphology (Brundrett 2004, Peterson et al. 2004).

Arbuscular mycorrhizal fungi (AMF) are functionally important in riparian areas, but little is known about the role of AMF within these areas. It's also unknown how they are affected by water fluctuations and other associated factors such as changes in the plant community and edaphic conditions. AMF are non host specific (Smith and Read 2008), meaning that several species of fungi can associate with many different plant species; however, some host plants show a better fungal growth and sporulation for some AMF species than others (Bever 2002, van der Heijden et al. 1998, Streitwolf-Engel et al. 2001). As suggested above, changes in the composition of the plant community and soil conditions caused by flow regulation, could also affect the composition of the AM fungal community (Bever et al. 1996, van der Heijden et al. 1998), given that fungal species differ in their tolerances for soil moisture (Miller 2000, Miller and Bever 1999), pH (Abbot and Robson 1977, Porter et al 1987), or nutrient levels (Johnson et al. 1991, He et al. 2002).

Salicaceae is worldwide distributed, most species occurring in the Northern Hemisphere. The genus *Salix* comprises *ca.* 450 species mainly distributed in China with about 270 species, the former Soviet Union with *ca.* 120 species, North America with 103 species and 65 species occurring in Europe. *Salix* also occurs in Japan, Africa, the Middle East, India, and Central and South America, and it has been introduced in Australasia and Oceania (Argus 1997, 1999).

Only two species are native from South America, *S. humboldtiana* Willd. ("sauce criollo", "sauce colorado" or red willow) that is widely distributed in Argentina (Parodi 1978) and *S. martiana* Willd. distributed in Amazonia and other regions of Brazil. *Salix* species are common along water courses, lakes or ponds and has strong preference for temporarily flooded riparian zones; therefore, their roots survive wholly or partially submerged. In central Argentina native *S. humboldtiana* form small populations or co-exists with introduced species such as *Salix alba* L. ("white willow"). Both, native and exotic willows are used for their medicinal properties and their timber in the production of firewood and poles (Demaio et al. 2002).

In the Northern Hemisphere, Salicaceae was reported to form dual ECM-AM (Smith and Read 2008) or triple ECM-AM-DSE (dark-septate fungi) associations (Trowbridge and Jumpponen 2004). Among the reported *Salix* species forming this type of association, *S. viminalis* L., *S. babylonica* Kunth, *S. nigra* Marshall, *S. gooddinguii* Ball, *S. alba* L., *S. dasyclados* Wimm, *S. repens* L., *S. herbacea* L., *S. barrattiana* Hook., *S. caprea* L., *S. commutata* Bebb. and *S. planifolia* Pursh., among others, mainly occurs in temperate and boreal forests (Lodge 1989, Khan 1993, Dhillon 1994, Graf and Brunner 1996, van der Heijden and Vosatka 1999, Kernagham 2001, van der Heijden 2001, Baum et al. 2002, Püttsepp et al. 2004, Trowbridge and Jumpponen 2004, Watling 2005, Beauchamp et al. 2006, Parádi and Baar 2006, Hrynkiewicz et al. 2008).

In Argentina, only few studies have reported the mycorrhizal status (ECM-AM-DSE) of *Salix humboldtiana* (Silva 2004, Becerra et al. 2009). It is worth mentioning that *S. humboldtiana* besides the fact of being a native species occurring in South America, occupies an enormous gradient of habitats in which diverse biotic and abiotic factors might intensely affect not only the diversity of fungi associated but also their symbiotic contribution. These features would facilitate *Salix* populations establishment in extreme habitat conditions.

Salix species can settle as a pioneer community colonizing disturbed sites, accelerating the recovery of damaged ecosystems and re-establishment of natural ecological complexity (Kuzovkina and Quigley 2005). Thus, it can be expected that mycorrhizal infection in riparian areas will prove an important aspect of their ecology and of the restoration of forest as occurred in *Chamaesiparis* wetland trees (Cantelmo and Ehrenfeld 1999). Furthermore, *Salix* species had been successfully employed in phytoremedation programs where their biomass production overlapped waste handling in disturbed sites (Mirck et al. 2005), and in rhizosphere remediation due to the capacity of *Salix* species of self-seeding in polluted areas while its ECM symbiosis remains functional and cleaning-up the soil from persistent organic pollutants (Meharg and Cairney 2000).

The aim of this chapter is to characterize the mycorrhizal diversity in *S. humboldtiana* and the introduced *S. alba* and analyze the information on some environmental factors affecting the ECM diversity (soil gravimetric water, pH) in riparian habitats of Central Argentina.

ENVIRONMENTAL FACTORS AFFECTING ECM DIVERSITY

Considering community diversity composed by the species richness and relative abundance (Magurran 1988), two main groups of environmental factors can affect ECM fungal community diversity: natural habitat features such as soil moisture, nutrients levels, pH, temperature, soil structure, microsites characteristics, wildfires, etc., and anthropogenic perturbations like soil disturbances, heavy metal contamination, acidification, fertilization, CO_2 levels, among others. ECM fungal community diversity in terms of morphotypes diversity is the result of morpho-physiological interactions between ECM fungal species and their host's roots. Thereby, environmental factors also have influence on morphotypes diversity and host physiology. In this chapter, we only considered natural environmental factors that might be affecting *Salix* populations in sites without anthropogenic disturbances.

Soil Moisture

The lost of soil moisture generally is related with a reduction of the fungal community diversity. However, ECM fungi showed a controversial behavior about reduced soil moisture and water excess in wet soils. For example *Cenococcum geophilum* Moug. & Fr. seems to increase the amount of colonized root tips under soil drought conditions as well as in wetland environments; other ECM fungi such as *Tuber* prefer mesic sites. ECM fungi have different and specific soil moisture requirements; this preferential response to soil water content by specific ECM morphotypes was considered as the ecological support to individual fungal species variation of relative "niche width" (Erland and Taylor 2002 and reference therein).

Soil pH

Soil pH is a well known complex factor because of its relation to many soil properties that includes nutrients availability; particularly N and P, mostly involved in mycorrhizal functioning. Soil pH also interacts with other soil factors and changes widely between soil microsites. It is also a key factor in fungal ecology affecting ionic forms of CO_2 within the soil; it also modifies the availability and ionic forms of nutrients and heavy metal ions solubility. Among this Al^{3+}, Cu^{2+} and Fe^{3+} are highly available at high hydrogen concentrations, that as well as Cd^+, Zn^{2+} and heavy metals, can be toxic to fungi (Cooke and Whipps 1993).

Although most fungal species prefer acidic conditions to grow, they can generally tolerate a wide pH range; even, some Basidiomycetes species will develop above pH 7.5, such is the case of ECM fungi (Hung and Trappe 1983). Hyphae have a very effective fungal buffering system that includes an external concentration of hydrogen ions that affects fungal metabolism and indirect effects on growth. Hydrogen ions contribute to soil regulation of soluble salts, facilitating the availability of necessary ions for the fungus; furthermore, pH affects plasmalema permeability and enzyme activity. Therefore, changes in fungal extracellular pH that are out of optimal value can reduce extracellular enzyme catalysis (Neville and Webster 1995).

As mentioned earlier fungal species differed in their optimal pH level for growth and ECM fungi are not exceptions. pH can alter the potential for ECM colonization, influencing ECM enzymatic and competition capabilities, species and morphotypes richness, sporocarps production and ECM community composition (Erland and Taylor 2002).

Soil Nutrients Levels

Ectomycorrizal diversity effects on plant productivity and nutrient capture by host plant is a major benefit when nutrient conditions are poor. In terrestrial ecosystems and particularly in arid environments, plant growth is limited by deficient nutrition of N and P. ECM provide these elements to their host from organic sources such as protein and amino acids; instead, non mycorrhizal plants mainly access to inorganic N supplies. However in some boreal forest, plants are capable of amino acid uptake without ECM assistance (Nordin et al. 2001, Wiemken and Boller 2002). Wiemken and Boller (2002) pointed out that the structure and function of ECM is highly dependant of N sources in soil, and in situations with low N concentration, ECM fungal fruit bodies production had a clustered distribution, suggesting a nutrient patchiness effect (Wiemken and Boller 2006).

In forest soils, only a portion of the organic P is available for ECM fungi (Häussling and Marschner 1989). The hyphae of ECM fungi stored phosphates probably within vacuoles, which are later mobilized to the host plant. In general low levels of available soil nutrients, particularly N and P, affect the internal nutritional status of the host leading to increase ECM infection (Alexander 1983). Bruns (1995) suggested that the high ECM fungal diversity in

Northern Hemisphere forests could be the result of resource partitioning, disturbance and interactions with other soil microorganisms.

ECM fungi can display a wide range from strictly symbiotic to hemi-saprobes (Högberg et al. 1999), based on their high capability to produce a diverse range of extra-cellular and cell wall-bound hydrolytic enzymes to break down soil organic molecules that contain N and P such as amino acids, protein and chitin (Leake and Read 1990, Leake et al. 2002, Smith and Read 2008). Furthermore, ECM fungi are able to produce extra-cellular enzymes for cellulose and lignin breakdown (Courty et al. 2005, 2006, 2007). Nevertheless, this theoretical postulated saprophytic role of ECM fungi (Tarkka et al. 2005; Weiken and Boller 2002) in the acquisition of carbon from soil, based on the increase of enzymatic activity during periods of low photosynthate supply from tree hosts, is actually questioned. Recently, Baldrian (2009) argued that still there is not enough evidence to confirm the saprophytic functioning of ECM fungi.

Soil Structure

Forests soils are a well known heterogeneous environment. The spatial heterogeneity of forest soil affects the vertical distribution of habitats for microbial and fungal communities and the physicochemical soil properties along the gradient.

In temperate and boreal forests the upper soil layer with major litter content houses saprophytic fungi, whereas ECM fungi can colonize deeper soil horizons with low availability of carbon compounds and positive energetic value (Baldrian 2009). Vertical distribution of ECM fungi has been reported previously (Dickie et al. 2002, Neville et al. 2002, Rosling et al. 2003), for example, in a French oak forest it was found that the ECM richness was higher in the undisturbed A1 horizon and in the A2 sieved soil, than undisturbed A2 horizon and in the dead woody debris niches, indicating how the ECM community structure can strongly vary depending on the nature of soil horizon or habitat patches (Buée et al. 2007).

New results considering fine-scale spatial structure of ECM (Amend et al. 2009), distances between host plant and seedlings (Teste et al. 2009) and plot-level effects (Izzo et al. 2005), provided a high support to the earlier Bruns (1995) proposal about the maintenance of the high ECM fungal diversity in Northern Hemisphere forests by the resource partitioning, disturbance and interactions with other microorganisms.

Microsites Effect

ECM fungi are relatively host selective, while AM tend to be generalists. Ectomycorrhizas have been mainly studied in forest systems due to its preference to tree hosts, where a high number of fungal species usually associate to a low number of plant species (Read 1991, Allen et al. 1995). In this type of ecosystems, habitat conditions change with tree age and soil's spatial heterogeneity, usually showing a mosaic of diverse physical-chemical and biological properties along short distances. Microorganisms living within soils microsites, such as mycorrhizal fungi are highly influenced by environmental characteristics that modify their development and functioning (Bledsoe 1992). In mature Douglas-fir old-growth strands, ECM diversity and distribution were related to soil characteristics and microhabitats; ECM density and richness were higher in forest floor over the mineral soil than in mineral soil or logs (Goodman and Trofymow 1998). In addition, ECM systems present a spatially clumped distribution widespread in the forest soil, facilitated by wide variation range of root tips available to be colonized by fungal symbionts. Examples of clumped distribution are those occurring with ECM mats and ECM fungal clones (Dahlberg and Stenlid 1990, 1994, Griffths et al. 1996). Also, environmental factors as soil depth have shown effects on dual ECM-AM in Salicaceae. In *Populus tremuloides* Michx. (Salicaceae) populations, the overall ECM colonization reached the 86%, of the roots, while AM colonization only reached a 6%, with a negative correlation between ECM and AM fungal colonization among all depths with "ECM colonization more abundant in the shallow organic soils and AM colonization more abundant in deeper mineral soils" (Neville et al. 2002). This also suggests that ECM and AM are preferentially partitioned at different soil depths.

ECM FUNCTIONING AMONG AQUATIC ENVIRONMENTS

The presence of mycorrhizas in aquatic plants is controversial. Few years ago, scientists argued that the low concentration of oxygen in waterlogged or swamped soils, were a determining factor that inhibited the formation of mycorrhizal associations. ECM were reported for trees species of *Casuarina cunninghamiana* Miq., *Melaleuca quinquenervia* (Cav.) Blake and *Salix babylonica* Kunth when growth in drained and aerated soils but not in swamps, water or sediments (Khan 1993) and dual associations were confirmed in these trees in aquatic and terrestrial environments (Lodge 1989). Absence of AM where reported in hydrophytes and members of the families Urticaceae, Casuarinaceae, Nyctaginaceae, Portulacaceae, Caryophyllaceae, Zygophyllaceae, Tamarinaceae, Euphorbiaceae and Palmaceae growing in sandy swamp soils (Khan 1974). However, recent studies in terrestrial and aquatic plants with whole or part of the root systems submerge, determined the presence of mycorrhizas (Stenlund and Charvat 1994).

There are no studies that examine the relationship between the concentration of soil water and ECM colonization and diversity, neither the presence of mycorrhizas in submerged roots of *Salix* spp.

SALIX ALBA L. ECM DIVERSITY ALONG AN ARID TO RIPARIAN GRADIENT IN CENTRAL ARGENTINA

Salix alba is an introduced species in South America, it presents a wide distribution in Argentina, mostly growing in riparian ecosystems or in areas temporarily flooded. The riparian zone is the interface area between land and a river or stream. It has an important role regulating the movement of material and water between the ecotonal soil-river systems.

ECM diversity and colonization was studied during autumn-winter in a riparian population of *S. alba* in Potrero de los Funes, San Luis (Central Argentine), a semi-arid environment. Potrero de los Funes is located at 952 m of altitude at 33° 13' 0S and 66° 13' 60W. Average summer temperature is 21-23 °C and 8-9 °C, in autumn-winter. Rains occur mainly in summer. Potrero River is a permanent mountain river, characterized by an area of

2.41 km², a total length of 2987 m, a rundown density of 4.45 km/ km² and a flood flow of 31.6 m³/s. This river is located on a recent geological unit with alluvial sediment (silt, sand and gravel) that partially constitutes the river courses. Riparian soil is sandy, with gravel and rocks (Ojeda et al. 2007). During late autumn of 2006 three transects along Potrero riverside were established and soil samples were collected across riparian zone every 1.5 m from water to land (0, 1.5, 3, 4.5 and 6 m). At each point of distance within the transect, *S. alba* root samples were dug out. At the laboratory the gravimetric soil moisture, soil pH, ECM colonization, ECM relative frequency and ECM diversity and richness, were measured. Duplicate samples of soil and roots were passed through sieves of different diameter to separate roots from soil and debris. After sieving samples were weighed to obtain the initial weight (Wi); subsequently placed in a forced air oven at 65-70 °C, during 72 hours until constant weight to estimate the final dry weight (Wf) considered as the last constant measurement.

A fraction of the roots were clarified and stained (Grace and Stribley 1991) to record AM associations, whereas another root fraction was used to characterize the ECM morphotypes (Agerer 1991, Baum et al. 2002) and estimate their colonization (Gehring and Whitham 1994, Giovannetti and Mosse 1980). ECM diversity was estimated based on the relative frequencies of the ECM morphotypes, using Simpson index (Helm et al. 1999).

Salix alba roots presented AM, ECM and DSE structures in their roots. The DSE structures were also present in the submerged roots. Previously, single ECM or AM or dual AM-ECM associations were reported for this species (Harley and Harley 1987, Lodge 1989, Khan 1993, Parádi and Baar 2006, Sumorok et al. 2008). The tripartite AM-ECM-DSE association had been previously reported only for shrub willows such as *Salix commutata* and *S. planifolia* (Trowbridge and Jumpponen 2004).

Soil water percentage was uniformly low along the distances gradient, except for the highest value at the extreme of the gradient where the roots were submerged. pH did also not differed along the transect (Table 1). We did expect gravimetric soil moisture and pH to diminish while distance from riverside increased but these correlations were not found.

Along the gradient 12 ECM morphotypes were found associated to *Salix alba* roots (Figures 1, 2). In general, the ECM morphotypes richness associated to *Salix* species is low. In native locations, *S. alba*, among Pilica River (Poland) riparian areas, presented 15 ECM morphotypes (Sumorok et al. 2008) and in the The Netherlands up to 20 morphotypes (Parádi and Baar 2006). In this sense, the morphotypes richness of *S. alba* seems to be similar in both, Northern and Southern Hemispheres locations. Another interesting aspect observed is that the introduced *S. alba* in Argentina, presented a relatively higher morphotypes richness than the native *S. humboldtiana* in Central Argentina (Becerra et al. 2009).

Table 1. Soil gravimetric water percentage (%H₂O) and pH along Potrero de los Funes riparian zone. Data are mean value ± SE

Distance from river (m)	H₂O (%)	pH
0	100 ± 0	6.5-7
1.5	9.69 ± 5.79	6.5-7
3	6.73 ± 2.48	6.5-7
4.4	5.70 ± 2.23	6.5-7
6	6.37 ± 4.23	6.5-7

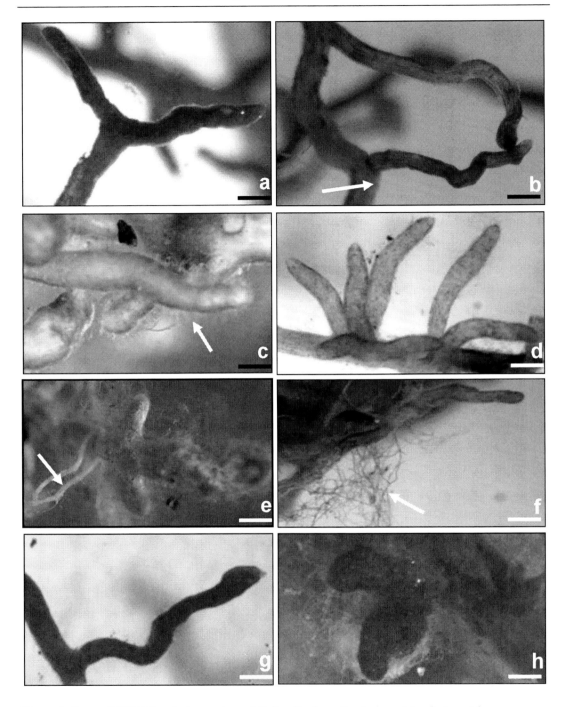

Figure 1. General ECM Morphotypes view of *Salix alba* from San Luis. a. Morphotype 1 b. Morphotype 2. c. Morphotype 3, cistidiae surrounding the root tip (→). d. Morphotype 4. e. Morphotype 5 rizomorph (→). f. Morphotype 6 emmanating hyphae (→). g. Morphotype 8. h. Morphotype 9 (hyphae that appear in the photograph are loose hyphae that do not have connection to the mantle). *Bars* a, b, e: 225 μm; c, f: 800 μm; d: 300μm; g: 400 μm; h: 145μm.

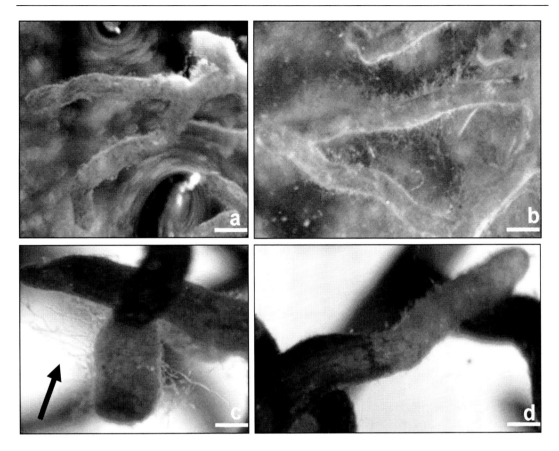

Figure 2. General ECM Morphotypes view of *Salix alba* from San Luis. a. Morphotype 10. b. Morphotype 11. c. Morphotype 12 points out emanating hyphae (→). d. Root tip dark bottom and light top. *Bars* a: 300µm; b: 400 µm; c: 145µm; d: 225µm.

No ECM fungal sporocarps were found at the *Salix alba* riparian sites. The lack of fruiting bodies also occurs in the other *S. alba* locations, for example, in Poland only *Inocybe* and *Hebeloma* sporocarps were found at Pilica River (Sumorok et al. 2008); however, ca. 40 ECM fungi were found under *Salix* spp., *Betula pendula* and *Pinus sylvestris* in dry habitats of this floodplain. In other regions, only *Tricholoma cingulatum* (Almfelt) Jacobasch has been reported fructifying with *S. caprea* (Hrynkiewicz et al. 2008) and *Inocybe glabripes* Ricken, with *S. viminalis* in Germany (Baum et al. 2002). Surprisingly, in non riparian habitats, the ECM fungal diversity supported by sporocarps records is much higher, van der Heijden et al. (1999), registered up to 78 fungal species associated to *Salix repens* in a natural sand dunes system, and 53 ECM sporocarps taxa associated to *Salix herbacea* in alpine habitats (Graf and Brunner 1996). However, the riparian zones of Central Argentina, associated to both *S. alba* and *S. humboldtiana* despite their exotic or native status, sporocarps were uncommon and only one species of *Inocybe* was recorded under *S. humboldtiana* (unpublished data). In the semiarid sites associated to *Salix alba*, sporocarps of saprophytic fungi were also absent but several species were found under *S. humboldtiana* riparian areas (Becerra et al.2009).

Salix alba presented a low AM colonization (15%, data not shown), whereas overall ECM mean frequency of colonization was higher and ranged from 45% to 100% (Figure 3).

None of the variables studied were significantly correlated by the percentage of gravimetric water, and the ECM morphotypes richness and diversity were negatively correlated with the gravimetric water percentage of soil (Spearman ρ=-0.11, p= 0.5; Spearman ρ=-0.02, p=0.92, respectively).

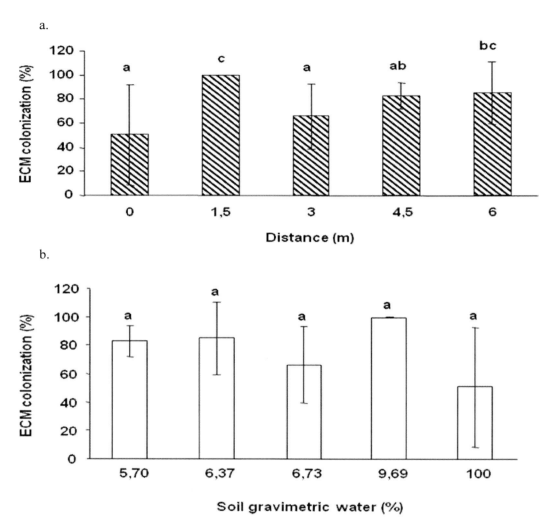

Figure 3. *Salix alba* ECM colonization along Potrero de los Funes riparian zone. a. ECM colonization versus distances from riverside. b. ECM colonization versus soil gravimetric water percentage. Data showed are mean value ± SE. Complete data set were analyzed by Kruskal Wallis non-parametric test. Different letters represent the results of *a posteriori* test of multiple comparisons.

High frequency of ECM means colonization in roots of *S. alba* (Figure 3) was similarly observed in riparian habitats from Poland (Sumorok et al. 2008). However, *S. alba* ECM colonization was lower in flooded areas of Waal River than in dry riparian edges like those of Potrero in Argentina and Pilica in Poland.

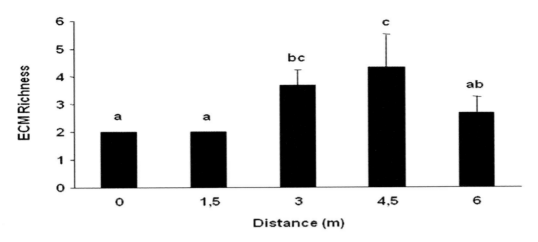

Figure 4. *Salix alba* ECM richness versus distances from riverside. Data showed are mean value ± SE. Complete data set were analyzed by Kruskal Wallis non-parametric test. Different letters represent the results of *a posteriori* test of multiple comparisons.

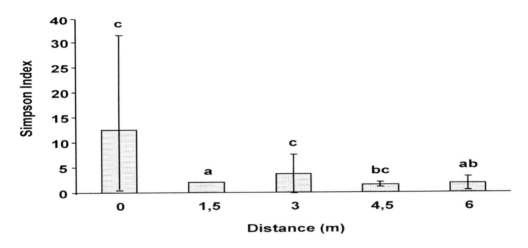

Figure 5. *Salix alba* ECM diversity *versus* distances from riverside. Data showed are Simpson index mean value ± SE. Complete data set were analyzed by Kruskal Wallis non-parametric test. Different letters represent the results of *a posteriori* test of multiple comparisons.

Within the riparian zone, ECM root colonization (Figure 3), richness (Figure 4) and diversity (Figure 5) varied significantly with increasing distance from the river edge (Kruskal Wallis test H= 21.34, p<0.001; H=29.85, p<0.0001; H=21.04, p<0.001, respectively). Furthermore, ECM richness was significantly correlated to distance from shore (Spearman ρ=0.44, p<=0.01). Only morphotypes M1 and M2 were associated with flooded roots, representing the 17% of total ECM morphotypes richness. The remaining 83% were associated to land roots and within these morphotypes, 50% were recorded in only one distance from the river edge. Here, the distance with the highest richness was 4.5 m, in which 9 different morphotypes were recorded; followed by 3 and 6 m distances, each with 6 morphotypes; 1.5 m. distance with 3 morphotypes, and the submerged roots showed the lowest richness with only 2 morphotypes.

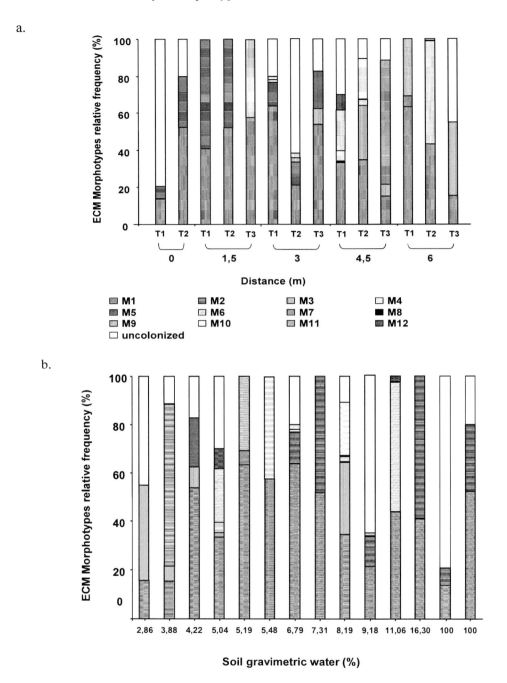

Figure 6. *Salix alba* ECM relative morphotypes frequency along riparian zone. a. ECM relative morphotypes frequency *versus* distance from riverside. Data showed are values from each transect sampled per site. Complete data set were analyzed by Kruskal Wallis non-parametric test. Significant differences were not found by *a posteriori* test of multiple comparisons. References: M: morphotype, followed by successive designation numbers.

ECM morphotypes relative frequency did not change significantly with the distances from the shore (Figure 6a) neither with the gravimetric water percentage (Figure 6b). However, all ECM morphotypes (M1 to M12) presented four differential patterns of distribution or guilds with water soil content and distance along the riparian zone: i) morphotypes distributed along all gravimetric water percentages and distances (M1); ii) morphotypes presented in most distances and in submerged roots (M2); iii) morphotypes presented in most distances, except submerged roots (M3, M4, M6) and iv) morphotypes distributed in only one or two distances or gravimetric water percentages excluding the submerged roots (M5, M7, M8, M9, M10, M11, M12). No ECM morphotypes were found exclusively in submerged roots.

Although, some morphotypes showed weakly significant correlations with soil gravimetric water percentage and distance to the river edge. Thus, only M2 frequency was positively correlated to gravimetric water percentage in soil (Spearman $\rho=0.61$, $p<0.05$) but it was negatively correlated with distance from river edge (Spearman $\rho= -0.62$, $p<0.05$). M6, M7, M9 and M12 frequencies were positively correlated with distance from river edge (Spearman $\rho=0.53$, $p<0.05$; $\rho=0.59$, $p<0.05$; $\rho=0.59$, $p<0.05$ and $\rho=0.57$, $p<0.05$, respectively).

Thus, morphotypes M1 (guild *i*) showed a wide distribution range, from dry soil to the submerged roots. Mophotype M2 (guild *ii*) was found almost in all gravimetric water percentages but shows a preference for high gravimetric water percentage in order to colonize *S. alba* roots; however, soil condition changes along the increased distance to the Potrero River affecting M2 negatively; for instance we could predict that M2 could be better adapted to wet and flooded habitats. M3, M4 and M6 (guild *iii*), could be considered as an intermediate guild, forming ECM evenly in dry soil conditions rather than in the water flow. M5, M7, M8, M9, M10, M11, M12 (guild *iv*) which were never found in the submerged roots, showed restricted distribution along the distance gradient, and could be terrestrial specialists that are absent in flooded environments as indicated in other studies (Khan 1993).

Among Potrero riverside gradient *S. alba* ECM colonization did not vary with soil gravimetric water percentage, although ECM richness and diversity were correlated in opposite senses. These results together with the four mophotypes guilds observed within the *S. alba* populations transects, suggest that gravimetric water percentage in the soil could not affect ECM colonization but could influence the species composition of the ECM fungal community. Thus, in Potrero riparian environment an increase in drought could "cause a shift in plant/fungus communities" as has been proposed by Tarkka et al. (2005).

SALIX HUMBOLDTIANA ECM DIVERSITY IN CENTRAL ARGENTINA

Salix humboldtiana Willd. (Sauce criollo) is the only native species of *Salix* and is one of these so-called dual mycorrhizal plan species. Their populations are distributed along streams, rivers and lakeshores and it occurs in habitats in which soils are heavily influenced by water level fluctuations (Ragonese 1987). As a dual mycorrhizal plant, *S. humboldtiana* may have a selective advantage in these highly dynamic ecosystems.

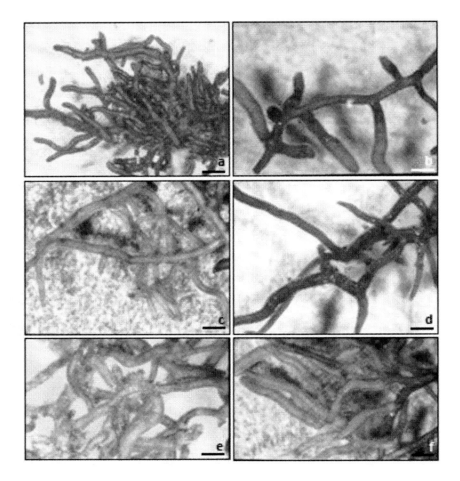

Figure 7. General view patterns of ECM of *Salix humboldtiana* from Central Argentina. a. Morfotype 1, *Inocybe* sp. b. Morfotype 2 *Tomentella* sp. c. Morphotype 3. d. Morphotype 4. e Morphotype 5. f. Morphotype 6. *Bars* a-f: 0.5 mm.

In Argentina, two field sites located in Córdoba province (Central Argentina) were studied (Punilla department and Colón department). Both soils presents a neutral soil pH, but differed in texture and in nutrient content. Soils from Colón Department site had higher contents of organic matter and total N, a higher electrical conductivity and higher levels in P than soils from Punilla department site. *S. humboldtiana* populations (height 6-15 m, age 20-30 years) are located along riverbanks. In both sites, the areas remained water-saturated during the rainy seasons (summer–autumn), having the trees partially submerged the root system during this time. *S. humboldtiana* root samples were taken during autumn, sampling monthly from April to June. Five square plots (10 m x 10 m) were established randomly within a homogenous area (100 x 50 m). The percentage of root tips colonized by ECM fungi was determined as described by Gehring and Whitham (1994). Seven ECM morphotypes were found to be associated with *Salix humboldtiana* (Figures 7, 8). All of them appeared mostly turgid and active, with thin mantle layers and without rhizomorphs. For detailed description of the ECM morphotypes, see Becerra et al. (2009). In few cases, root tips appeared senescent and mycorrhizae were dark colored, probably indicating their inactive stage. Two morphotypes matched ECM taxa when blasted to the respective resulting consensus sequence in the NCBI database. The fungal symbionts in both matches are basidiomycetes, belonging to the genus *Tomentella* in the Thelephoraceae and *Inocybe* sp. within the Cortinariaceae. Amplification failures of the remaining morphotypes were probably due to old and inactive mycorrhizae, or to mycorrhizae morphotypes with poorly or discontinuosly developed mantle layers, features that seem common in some species of *Salix* (Jones et al. 1990; Graf and Brunner 1996; Trowbridge and Jumpponen 2004; Püttsepp et al. 2004). Besides, the DNA amplification from *Salix* can be difficult, attributable to the co-extracting inhibitors (possibly salicylic acid) (Herrera Medina et al. 2003).

The ECM colonization of *S. humboldtiana* differed between the two sites (P<0.001) and sampling dates (P<0.001). ECM colonization was 71.13 % (S.E.= 24.7) at the Hayke site, and 86.31 % (S.E.=17.2) at the La Calera site, varying from 33 to 99 % at both sites. The ECM colonization significantly differed among sampling dates (from April to June), being higher in June and May than April. This variation could be mediated by differences in temperature, soil moisture and soil nutrient status, physiological and phenological changes in the host plant, ultimately affecting the symbiosis development (Harvey et al. 1978; Jones et al. 1990; Swaty et al. 1998; van der Heijden et al. 1999; Baum and Makeschin 2000; Baum et al. 2002; Püttsepp et al. 2004). Indeed, important changes on humidity and temperature usually occurred during the fall months (April to June), as it was previously registered by Luti et al. (1979).

Roots submerged in water (Figure 8) present a complete or incomplete thin mantle when they were observed under microscope. Mainly those roots present a simple ramification and brown color, 4.5-7.5 mm long and 0.1-0.3 mm diameter. No rhizomorphs and emanating hyphae were observed. A transitional type mantle between the plectenchymatous and pseudoparenchymatous mantle, in which irregularly shaped hyphae form a coarse net, 2.5-6 µm diameter, hyphae without clamps. These roots do not present AM colonization. Unique mycorrhizal associations were found in submerged roots of *S. alba* and *S. humboldtiana*. Other studies have shown similar mycorrhizal development in aerobic environments associated to *Alnus glutinosa* L. Gaertner in wet habitats (Baar et al. 2000).

To observe AM colonization, roots were clarified and stained (Grace and Stribley 1991), and the structures described in detail (Becerra et al. 2009), and photographed (Figure 8). AM colonization of *S. humboldtiana* differed between sampling dates (P<0.001), showing a significative interaction effect between sites and sampling dates (P<0.01). AM colonization at the Hayke site was 4.05 % (S.E.= 4.49), and 3.12 % (S.E.= 2.98) at the La Calera site, the colonization ranged between 0 to 17 % for both sites. AM colonization significantly differed among sampling dates being higher in April at Hayke and May at La Calera site. *S. humboldtiana* showed low AM colonization that differed between sampling dates. Our colonization rates resemble those observed by Trowbridge and Jumpponen (2004) on *S. commutata* and *S. planifolia* and van der Heijden and Vosatka (1999) on *S. repens*. Although we observed low levels of infection, AM fungi might still provide benefits to *S. humboldtiana*, as van der Heijden and Vosatka (1999) suggested.

Figure 8. a. General view pattern of morphotype 7 in *Salix humboldtiana* from Central Argentina. b. Coils of arbuscular mycorrhizal colonization. c. Roots submerged in water. d. Incomplete thin mantle of submerged roots. *Bars* a: 0.5 mm; b: 10 μm; c: 1cm.

In both Argentinian sites, *S. humboldtiana* are colonized by ECM- AM-DSE. The occurrence of ECM and AM colonization in *S. humboldtiana* concurred with the observations of Lodge (1989), Khan (1993), van der Heijden and Kuyper (2003) and Hashimoto and Higuchi (2003).

CONCLUSION

In Central Argentina, *S. alba* and *S. humboldtiana* presented multiple fungal symbiosis (AM, DSE and ECM), which might be indicating alternative o synergistic fungal association that would be more efficient for *Salix* spp. within these semiarid ecosystems. Another interesting observation is the scarcity of fungal sporocarps in populations of both *Salix* spp.

The environmental conditions could greatly influence the host species plasticity to form alternative associations, particularly in riparian ecosystems. The ECM frequency, richness and diversity were not affected by soil gravimetric water percentage in *S. alba* but were influenced by distance to the water edge. The absence of moisture effect on ECM could be due to the very low soil gravimetric water percentage and the scarce differences of water values between each distance along the gradient, except the river point. The distance effects on ECM could be related to negligible and subtle changes in microhabitats along the gradient.

ECM morphotypes richness was higher in the exotic *S. alba* than in native *S. humboldtiana* (Silva 2004, Becerra et al. 2009), but similar to *S. alba* populations in European ecosystems. Therefore, dry-riparian condition could benefit some ECM fungal species well adapted to drought which would colonize successfully the roots of *S. alba*.

The four guilds of *S. alba* ECM morphotypes along the Potrero River riparian area could mirror differential responses of the fungal component in this community by microhabitat effects, that probably selects fungi capable of persisting under particular environmental conditions (Jumpponen et al., 1999; Trappe and Luoma, 1992; Trowbridge and Jumpponen, 2004). Thus, in this arid riparian system, *S. alba* ECM association could be more influenced by differential microhabitat features than by water content in soil because ECM fungal species would be resistant to dry environmental conditions.

Salix species pioneer capabilities, phytoremediation and rhizoremediation properties in addition to *S. alba* and *S. humboldtiana* high ECM colonization and multiple symbiosis are valuable treats to consider their potential use in restoration of arid systems. Further research could lead to understand the variation in colonization level among sites and microsites from stream to land in the arid riparian systems.

ACKNOWLEDGMENTS

Mónica A. Lugo, Alejandra G. Becerra and Eduardo R. Nouhra are Researchers from Consejo Nacional de Investigaciones Científicas y Tecnológicas (CONICET), Argentina. This work was supported by Secyt (69/08) and PROICO 2-0203 (FQByF-UNSL).

REFERENCES

Abbot, LK; Robson, AD. The distribution and abundance of vesicular arbuscular endophytes in some western Australian soils. *Aust. J. Bot.*, 1977, 25:515-522.

Agerer, R. Characterization of ectomycorrhiza. In: Norris JR, Read DJ, Varma AK, editors. *Techniques for the study of mycorrhizae. Methods in microbiology*, vol. 23. Academic Press, London, 1991, 25-73.

Alexander, IJ. The significance of ectomycorrhizas in the nitrogen cycle. In: Lee LA, Mc Neill S, Rorison IH, editors. *Nitrogen as an Ecological Factor*. Oxford: Blackwell, 1983, pp 69-94.

Allen, EB; Allen, MF; Helm, DJ; Trappe, JM, Molina, R, Rincón, E. Patterns and regulation of mycorrhizal plant and fungal diversity. *Plant Soil*, 1995, 170, 47-62.

Argus, GW. Infrageneric classification of *Salix* (Salicaceae) in the New World. *Syst. Bot. Monogr.* 52. USA, 1997.

Argus, GW. Classification of *Salix* in the New World. *Botanical Electronic News*, ISSN 1188-603X, 1999, 227.

Amend, A; Keeley, S; Garbelotti, M. Forest age correlates with fine-scale structure of Matsutake mycorrhizas. *Mycol Res.*, 2009, 113, 541-551.

Baar, J; van Groenendael, JM; Roelofs, JGM. Are ectomycorrhizal fungi associated with *Alnus* of importance for forest development in wet environments? *Plant Biol.*, 2000, 2, 505–511.

Baldrian, P. Ectomycorrhizal fungi and their enzymes in soils: is there enough evidence for their role as facultative soil saprotrophs? *Oecologia,* 2009, 161, 657-660.

Baum, C; Makeschin, F. Effects of nitrogen and phosphorus fertilization on mycorrhizal formation of two poplar clones (*Populus trichocapa* and *P. tremula* x *tremuloides*). *J Plant Nutr Soil Sci*, 2000, 163:491-497.

Baum, C; Weih, M; Verwijst, T, Makeschin, F. The effects of nitrogen fertilization and soil properties on mycorrhizal formation of *Salix viminalis*. *Forest Ecol. Managem.*, 2002, 160, 35-43.

Beauchamp, VB; Stromberg, JC; Stutz, JC. Arbuscular mycorrhizal fungi associated with *Populus–Salix* stands in a semiarid riparian ecosystem. *New Phytol.*, 2006, 170; 369-380.

Becerra, AG; Nouhra, ER; Silva, MP; McKay, D. Ectomycorrhizae, arbuscular mycorrhizae, and dark-septate fungi on *Salix humboldtiana* in two riparian populations from central Argentina. *Mycoscience*, 2009, 50, 343–352.

Bever, JD. Host-speficity of AM fungal population growth rates can generate feedback on plant growth. *Plant Soil*, 2002, 244:281-290.

Bever, JD; Morton JB; Antonovics, J; Schultz, PA. Host-dependent sporulation and species diversity of arbuscular mycorrhizal fungi in a mown grassland. *J Ecol.*, 1996, 84:71-82.

Bledsoe, CS. Physiological ecology of ectomycorrhizae: implications for field application. In: Allen MF, editor. *Mycorrhizal Functioning. An integrative plant fungal process.* Chapman & Hall, NY, London, 1992, 424-437.

Brundrett, M. Diversity and classification of mycorrhizal associations. *Biol. Rev.*, 2004, 79, 473-495.

Bruns, T. Thoughts on the processes that maintain local species diversity of ectomycorrhizal fungi. *Plant Soil*, 1995, 170, 63-73.

Buée, M; Courty, PE; Mignot, D; Garbaye, J. Soil niche effect on species diversity and catabolic activities in an ectomycorrhizal fungal community. *Soil Biol Biochem.*, 2007, 39, 1947-1955.

Cantelmo Jr., AJ; Ehrenfeld, JG. Effects of microtopography on mycorrhizal infection in Atlantic white cedar (*Chamaecyparis thyoides* (L.) Mills.). *Mycorrhiza*, 1999, 175-180.

Cooke, RC; Whipps, JM. *Ecophysiology of fungi.* Blacwell Scientific Publications, Oxford, London, 1993, 1-337.

Courty, P-E; Pritsch, K; Schloter, M; Hartmann, A; Garbaye, J. Activity profiling of ectomycorrhiza communities in two forest soils using multiple enzymatic tests. *New Phytol.*, 2005, 167, 309-319.

Courty, P-E; Pouysegur, R; Buée, M; Garbaye, J. Laccase and phosphatase activities of the dominant ectomycorrhizal types in a lowland oak forest. *Soil Biol. Biochem.*, 2006, 38, 1219-1222.

Courty, P-E; Bréda, N; Garbaye, J. Relation between oak tree phenology and the secretion of organic matter degrading enzymes by *Lactarius quietus* ectomycorrhizas before and during bud break. *Soil Biol. Biochem.*, 2007, 39, 1655–1663.

Dahlberg, A; Stenlid, J. Population structure and dynamics in *Suillus bovinus* as indicated by spatial distribution of fungal clones. *New Phytol.*, 1990, 115, 487-493.

Dahlberg, A; Stenlid, J. Size, distribution and biomass of genets in populations of *Suillus hovinus* (L.: Fr.) Roussel revealed by somatic incompatibility. *New Phytol.*, 1994, 228, 225-234.

Demaio, P; Karlin, UO; Medina, M. *Árboles nativos del centro de Argentina*. L.O.L.A. Buenos Aires, Argentina, 2002.

Dhillon, S. Ectomycorrhizae, arbuscular mycorrhizae, and *Rhizoctonia* sp. of alpine and boreal *Salix* spp. in Norway. *Arctic Alpine Res.*, 1994, 26, 304-307.

Dickie, IA; Xu, B; Koide, RT. Vertical niche differentiation of ectomycorrhiza hyphae in soil as shown T-RFLP analysis. *New Phytol.*, 2002, 156, 527-535.

Erland, S; Taylor, AFS. Diversity of ecto-mycorrhizal fungal communities in relation to the abiotic environment. In: van der Heijden MGA and Sanders IR, editors. *Mycorrhizal Ecology*. Springer, Ecological Studies Vol. 157, Berlín, Tokio, 2002, 163-200.

Gehring, CA; Whitham, TC. Comparisons of ectomycorrhizae on pinyon pines (*Pinus edulis*, Pinaceae) across extreme of soil type and herbivory. *Am. J. Bot.*, 1994, 81, 1509-1516.

Giovannetti, M; Mosse, B. An evaluation of techniques for measuring vesicular arbuscular mycorrhizal infection in roots. *New Phytol.*, 1980, 84, 489-500.

Goodman, DM; Trofymow, JA. Distribution of ectomycorrhizas in micro-habitats in mature and old-growth stands of Douglas-fir on Southeastern Vancouver Island. *Soil Biol. Biochem.*, 1998, 30, 2127-2138.

Grace, C; Stribley, DP. A safer procedure for routine staining of vesicular-arbuscular mycorrhizal fungi. *Mycol. Res.*, 1991, 95, 1160-1162.

Graf, F; Brunner, I. Natural and synthesized ectomycorrhizas of the alpine duarf willow *Salix herbacea*. *Mycorrhiza*, 1996, 6, 227-235.

Griffths, RP; Bradshaw, GA; Marks, B; Licakaemper, GW. Spatial distribution of ectomycorrhizal mats in coniferous forest of the Pacific Northwest, U.S.A. *Plant Soil*, 1996, 180, 147-158.

Harley, JL; Harley, EL. A check-list of mycorrhiza in the British flora. *New Phytol.*, 1987, 105, 1-102.

Harvey, AE; Jurgensen, MF; Larsen, MJ. Seasonal distribution of ectomycorrhizae in a mature Douglas-fir/larch forest soil in western Montana. *For. Sci.*, 1978, 24:203–208.

Hashimoto, Y; Higuchi, R. Ectomycorrhizal and arbuscular mycorrhizal colonization of two species of floodplain willows. *Mycoscience*, 2003, 44, 339-343.

Häussling, M; Marschner, H. Organic and inorganic soil phosphates and acid phosphates activity in the rhizosphere of 80-year-old Norway spruce (*Picea abies* (L.) Karst.) trees. *Biol. Fertil. Soils*, 1989, 8: 128-133.

He, XL; Mouratov, S; Steinberger, Y. Temporal and spatial dynamics of vesicular-arbuscular mycorrhizal fungi under the canopy of *Zygophyllum dumosum* Boiss. in the Negev Desert. *J Arid Environm.*, 2002, 52:379-387.

Helm, DJ; Allen, EB; Trappe, JM. Plant growth and ectomycorrhiza formation by transplants on deglaciated land near Exit Glacier, Alaska. *Mycorrhiza*, 1999, 8, 297-304.

Herrera Medina, MJ; Gagnon, H; Piché, Y; Ocampo, JA; García Garrido, JM; Vierheilig, H. Root colonization by arbuscular mycorrhizal fungi is affected by the salicylic acid content of the plant. *Plant Sci.*, 2003, 164:993-998.

Högberg, P; Plamboeck, AH; Taylor, AFS; Fransson, PMA. Natural 13C abundance reveals trophic status of fungi and host-origin of carbon in mycorrhizal fungi in mixed forests. *Proc. Nat. Acad. Sci. USA*, 1999, 96, 8534-8539.

Hrynkiewicz, K; Haug, I; Baum, C. Ectomycorrhizal community structure under willows at former ore mining sites. *European J. Soil Biol.*, 2008, 44, 37-44.

Hung, LL; Trappe, JM. Growth variation between and within species of ectomycorrhizal fungi in response to pH *in vitro*. *Mycologia*, 1983, 75, 234-241.

Johnson, NC; Zak, DR; Tilman, D; Pfleger FL. Dynamics of vesicular-arbuscular mycorrhizae during old field succession. *Oecologia*, 1991, 86:349-358.

Jones, MD; Durall, DM; Tinker, PB. Phosphorus relationships and productions of balsam poplars, aspen and willows on former arable land in the Federal Republic of Germany. III. Soil ecological effects. *For. Eco.l Manag.*, 1990, 121:85-99.

Jumpponen, A; Trappe, J; Cázares, E. Ectomycorrhizal fungi in Lyan Lake Basin: a comparation between primary and secondary successional sites. *Mycologia*, 1999, 91, 575-582.

Kernagham, G. Ectomycorrhizal fungi at tree line in the Canadian Rockies II. Identification of ectomycorrhizae by anatomy and PCR. *Mycorrhiza*, 2001, 10, 217-229.

Khan, AG. The ocurrence of mycorrhizas in halophytes, hydrophytes and xerophytes, and of *Endogone* spores in adjacent soils. *J. Gen. Microbiol.*, 1974, 8I, 7-14.

Khan, AG. Occurrence and importance of mycorrhizae in aquatic trees of New South Wales, Australia. *Mycorrhiza*, 1993, 3, 31-38.

Kuzovkina, Y; Quigley, M. Willows beyond wetlands: uses of *Salix* L. species for environmental projects. *Water Air Soil Poll.*, 2005, 162, 183-204.

Leake, JR; Read, DJ. Chitin as a nitrogen source for mycorrhizal fungi. *Mycol. Res.*, 1990, 94, 993-995.

Leake, JR; Donnelly, DP; Boddy, L. Interactions between ectomycorrhizal and saprotrophic fungi. In: van der Heijden MGA and Sanders IR, editors. *Mycorrhizal Ecology*. Springer Verlag, Germany, 2002, 346-372.

Lodge, DJ. The influence of soil moisture and flooding on VA-endo and ectomycorrhizae in *Populus* and *Salix*. *Plant Soil*, 1989, 117, 243-253.

Luti, R; Bertrán de Solís, M; Galera, F; Ferreira, N; Nores, M; Herrera, M; Barrera, JC. *Vegetación*. In: Boldt (ed) Geografía Física de la Provincia de Córdoba. Argentina, 1979, pp 1-464.

Magurran, AE. *Ecological diversity and its measuremen*t. Croom Helm, UK, 1988.

Meharg, AA ; Cairney, JWG. Ectomycorrhizas-extending the capabilities of rhizosphere remediation? *Soil Biol. Biochem.*, 2000, 32, 1475-1484.

Miller, SP. Arbuscular mycorrhizal colonization of semi-aquatic grasses along a wide hydrologic gradient. *New Phytol*, 2000, 145:145-155.

Miller, SP; Bever, JD. Distribution of arbuscular mycorrhizal fungi in stands of the wetland grass *Panicum hemitomon* along a wide hydrologic gradient. *Oecologia*, 1999, 119:586-592.

Mirck, J; Isebrands, JG; Verwijst, T; Ledin, S. Development of short-rotation willow coppice systems for environmental purposes in Sweden. *Biomass Bioenergy*, 2005, 28, 219-228.

Neville, JD; Tessier, JL; Morrison, I; Scarratt, J; Canning, B; Klironomos, JN. Soil depth distribution of ecto- and arbuscular mycorrhizal fungi associated with *Populus tremuloides* within a 3-year-old boreal forest clear-cut. *Appl. Soil Ecol.*, 2002, 19, 209-216.

Neville, JD; Webster, J. *Fungal Ecology*. Chapman & Hall, UK, 1995.

Nordin, A; Högberg, P; Näsholm, T. Soil nitrogen form and plant nitrogen uptake along a boreal forest productivity gradient. *Oecologia*, 2001, 129, 1125-1132.

Ojeda, GE; Lacreu, HL; Sosa, GR. *Atlas de recursos Geoambientales. Un aporte a la formación de los ciudadanos.* Municipio de Potrero de los Funes-UNSL, ISBN: 978-987-23360-5-9. Argentina, 2007.

Parádi, I; Baar, J. Mycorrhizal fungal diversity in willow forests of different age along the river Waal, the Netherlands. *Forest Ecol. Managem.*, 2006, 237, 366-372.

Parodi, L. *Enciclopedia argentina de agricultura y jardinería* (Tomo I). Tercera Edición. Editorial ACME. Argentina, 1978.

Peterson, RL; Massicotte, HB; Melville, LH. *Mycorrhizas*: *Anatomy and cell biology*. NRC Research press, Canada, 2004.

Porter, WM; Robson, AD; Abbott, LK. Field survey of the distribution of vesicular arbuscular mycorrhizal fungi in relation to soil pH. *J Appl. Ecol.*, 1987, 24:659-662.

Püttsepp, Ü; Rosling, A; Taylor, AFS. Ectomycorrhizal fungal communities associetiated with *Salix viminalis* L. and *S. dasyclados* Wimm. clones in a short- rotation forestry plantation. *Forest Ecol Managem.*, 2004; 196, 413-424.

Ragonese, A. Familia Salicaceae. *Flora Ilustrada de Entre Ríos III*. Colección Científica del INTA. Tomo VI, Buenos Aires, 1987, pp 6–14

Read, DJ. Mycorrhizas in ecosystems. *Experientia*, 1991, 47, 376-391.

Rosling, A; landerweert, T; Larsson, KH; Kuyper, TW; Taylor, AFS; Finaly, ED. Vertical distribution of ectomycorrhizal fungal taxa in a podzol soil profile. *New Phytol.*, 2003, 159, 775-783.

Silva, M. Micorrizas de *Salix humboldtiana* Willd. Trabajo final para optar al título de Biólogo. Universidad Ncional de Córdoba, Argentina, 2004, pp.1-46.

Smith, SE; Read, DJ. *Mycorrhizal Symbiosis*. Academic Press, San Diego, London, New York, Boston, Sydney, Tokio, Toronto, 2008.

Stenlund, DL; Charvat, ID. Vesicular arbuscular mycorrhizae in floating wetland mat communities dominated by *Typha. Mycorrhiza*, 1994, 4, 131-137.

Streitwolf-Engel, R; van der Heijden, MGA; Wiemken, A; Sanders, IR. The ecological significance of arbuscular mycorrhizal fungal effects on clonal reproduction in plants. *Ecology*, 2001, 82:2846-2859.

Sumorok, B; Kosiński, K; Michalska-Hejduk, D; Kiedrzyńska, E. Distribution of ectomycorrhizal fungi in periodically inundated plant communities on the Pilica River foodplain. *Ecohydrol. Hydrobiol.*, 2008, 8, 401-410.

Swaty, RL; Gehring, CA; Van Ert, M; Theimer, TC; Keim, P; Whitman, TG. Temporal variation in temperature and rainfall differentially affects ectomycorrhizal colonization at two contrasting sites. *New Phytol.*, 1998, 139:733-739.

Tarkka, M; Nehls, U; Hampp, R. Chapter: Physiology of Ectomycorrhizae (ECM). In: *Progress in Botany*. Vol. 66. Springer Verlag, Germany, 2005, 247-276.

Trappe, JM; Luoma, DL. The ties that bind: fungi in ecosystems. In: Carrol GC and Wicklow DT, editors. *The fungal community: its organization and role in the ecosystem.* New York: Dekker; 1992; 17-27.

Trowbridge, J; Jumpponen, A. Fungal colonization of shrub willow roots at the forefront of a receding Glacier. *Mycorrhiza*, 2004, 14, 283-293.

van der Heijden, EW. Differential benefits of arbuscular mycorrhizal and ectomycorrhizal infection of *Salix repens. Mycorrhiza*, 2001, 10, 185-193.

van der Heijden, EW; Kuyper, TW. Ecological strategies of ectomycorrhizal fungi of *Salix repens*: root manipulation versus root replacement. *Oikos*, 2003, 103:668-680.

van der Heijden, EW; Vosatka, M. Mycorrhizal associations of *Salix repens* L. communities in succession of dune ecosystems. II. Mycorrhizal dynamics and interactions of ectomycorrhizal and arbuscular mycorrhizal fungi. *Can. J. Bot.*, 1999, 77, 1833-1841.

van der Heijden, MGA; Boller, T; Wiemken, A; Sanders, IR. Different arbuscular mycorrhizal fungal species are potential determinants of plant community structure. *Ecology*, 1998, 79:2082-2091.

van der Heijden, EW; de Vries, FD; Kuyper, ThW. Mycorrhizal associations of *Salix repens* L. communities in succession of dune ecosystems. I. Above-ground and below-ground views of ectomycorrhizal fungi in relation to soil chemistry. *Can. J. Bot.*, 1999, 77, 1821-1832.

Watling, R Fungal associates of *Salix repens* in northern oceanic Britain and their conservation significance. *Mycol. Res.*, 2005, 109, 1418-1424.

Wiemken, V; Boller, T. Delayed succession from alpine grassland to savannah with upright pine: Limitation by ectomycorrhiza for   mation? *Forest Ecol. Managem.*, 2006, 237, 492-502.

Wiemken, V; Boller, T. Ectomycorrhizas: gene expression, metabolism and the wood-wide web. *Curr. Opin. Plant Biol.*, 2002, 5, 355-361.

In: Mycorrhiza: Occurrence in Natural and Restored…
Editor: Marcela Pagano

ISBN: 978-1-61209-226-3
© 2012 Nova Science Publishers, Inc.

Chapter 11

TREE SPECIES COMPOSITION AND DIVERSITY IN BRAZILIAN FRESHWATER FLOODPLAINS

Florian Wittmann
Max Planck Institute for Chemistry, Biogeochemistry Dept.,
Johann J. Becherweg 27, 55128 Mainz-Germany

ABSTRACT

Neotropical freshwater floodplain forests are important landscape units because they consist of both flood-resistant species and of immigrants from the adjacent uplands, thereby concentrating large part of regional floristic biodiversity. They substantially contribute to the food webs of the terrestrial and the aquatic fauna, and fulfil a variety of ecologically and economically important functions. The inundation of roots and aboveground organs reduces oxygen availability to trees and is widely considered as a potential stress factor. Inundation is thus a powerful factor selecting the occurrence and distribution of tree species, which in turn influences species distribution and richness of most Neotropical floodplain forests.

Floodplain forests occur in all tropical biomes of Brazil. Most of its tree species are strongly zoned along the flooding gradient and associated habitat disturbance, thus reflecting the degree of adaptations developed to the unfavourable site conditions. Floodplains mostly are characterized by subsets of species of the adjacent uplands. Few flood-tolerant tree species are widely distributed, with occurrence in nearly all Brazilian biomes. However, these species are rarely restricted to flooded habitats. This evidences that these species not only tolerate flooding – but that they are generally tolerant to a wide range of stressful conditions.

Most floodplain forests are highly diverse, but lack endemic tree species – with exception of Amazonian large-river floodplains. The lack of endemism within flooded forests of the Cerrado and Atlantic forest domain can be traced to both, intense species exchange between floodplains and adjacent uplands and high connectivity by river systems that act as migration routes over huge geographic distance. The occurrence of severe droughts and wet periods during Tertiary and Quaternary climate change repeatedly interrupted evolutional processes of the flood-adapted tree floras. By contrast, elevated degrees of endemism in Amazonian large-river floodplains indicate comparatively s environmental conditions over large part of the Amazon basin since at least the early Paleocene.

1. INTRODUCTION

An excess of water is generally considered to be detrimental to plant health and growth (Schueler & Holland 2000). The exposure to surface water saturation is outside the normal range of tolerable environmental conditions and thus a powerful stressor to many plant species (Crawford 1992, Larcher 2001, Visser et al. 2003). However, many Neotropical landscapes contain ecosystems which are seasonally or episodically flooded by freshwater. Most of these ecosystems are characterized by the occurrence of an exceptional species-rich flora, with many well-adapted and flood-tolerant tree species.

In Brazil, many floodplain forests influenced by freshwater disappeared through intense deforestation especially in areas with high population densities and intense agricultural activities. Large-scale agro-industries especially affected flooded forests of the Cerrado belt (Junk 2002), and in the Atlantic forest domain. Remaining forests often are endangered through an inadequate use of its natural resources, modified flood regimes through dam and water reservoir constructions mostly related to hydropower, and eutrophication and other water pollution which led to habitat fragmentation or even the extinction of the native flora at local and regional scales. In addition, flooded forests react promptly to changes in the hydrological regime and thus are thought to belong to the most sensitive communities affected by climate change (Moore et al. 2007, Wittmann et al. 2010).

Wetlands in general and tropical wetlands in particular are hot spots for the development and maintenance of biodiversity (Gopal et al. 2000). Flooded forests are particularly important because they consist of both flood resistant species and of immigrants from the adjacent uplands, thereby comprising most of the regional floristic biodiversity (Pither & Kellman 2002). This might be especially important in regions where seasonal water shortage is an important determinant of species diversity, such as in large parts of the Brazilian Cerrado and Caatinga (Rizzini 1997). In addition, flooded forests fulfil a variety of important ecological and economical functions, such as the regulation of natural flow and flood regimes, the protection of riverbanks against erosion, local climate regulation and water-quality protection, and they offer food and shelter for the local aquatic and terrestrial fauna.

Despite its ecological and economical importance, Neotropical wetlands have been barely investigated (Naranjo 1995, Esteves 1998), and specific knowledge on tree species composition, distribution, and diversity of freshwater inundation forests as well as its dependence on flood regimes and associated environmental parameters is scarce. The present chapter aims at reviewing the existing literature about the flood-adapted tree flora in Brazil. Species lists from inventories are compared to detect floristic similarities and tree species distribution patterns of generalists in flooded forests of different ecosystems and biomes. Gaps of knowledge are pointed out, and floristic patterns are discussed at landscape-wide and evolutional scales.

2. FLOODING AND ITS CONSEQUENCE FOR TREE VEGETATION

Flooding represents a complex system of stressors that impose several often concurrent challenges to plant functioning (Jackson & Colmer 2005). Trees have evolved to survive in a terrestrial environment (Parolin 2009). Inundated soils turn conditions to hypoxic or anoxic within a few hours as a result of oxygen consumption by respiring roots and microorganisms, and insufficient diffusion of oxygen through water and submerged tissues (Crawford 1989, 1992, Armstrong et al. 1994, Visser et al. 2003). Oxygen depletion is accompanied by increased levels of CO_2, anaerobic decomposition of organic matter, increased solubility of mineral substances, and reduction of the soil redox potential (Joly & Crawford 1982; Kozlowski & Pallardy 2002), which is followed by accumulation of many potentially toxic compounds caused by alterations in the composition of the soil micro-flora (Parolin et al. 2010). In some floodplains, e.g. those of rivers originating from highly erosive areas such as the Andes, sedimentation rates can be extreme and the deposits may additionally deteriorate soil aeration. The often high productivity of floating macrophytes in floodplains results in elevated decomposition rates which further decrease the oxygen level (Armstrong et al. 1994). Moreover, when flooding results in complete submergence of trees, as for example in small individuals and seedlings, shoots are also deprived of sunlight (Jackson & Colmer 2005, Parolin 2009) especially when floodwaters are poorly transparent.

Beside these constraints imposed by flooding with its associated parameters, numerous tree species are highly productive in flood-prone areas (Jackson & Colmer 2005), and several Neotropical floodplains are exceptional rich in tree species (Wittmann et al. 2010). Trees subjected to flooding often compensate anaerobic soil conditions by developing biochemical, physiological and morpho-anatomical adaptations to avoid oxygen deficiency. Many trees that tolerate long-term flooding reduce its metabolism during waterlogging, which results in decreased photosynthetic rates, and reduced wood and shoot growth (Worbes 1986, Fernandez et al. 1999, Schöngart et al. 2002). Further adaptations of trees to flooding include partial or complete leaf shedding with the onset of flooding, the formation of adventitious roots and/or specialized roots such as pneumatophores, increased root biomass during flooding, the formation of hypertrophic lenticels on roots and stems, the formation of root aerenchyma, high resprouting abilities, the induction of activity of fermentative enzymes under anaerobic conditions, as well as the production of elevated levels of anti-oxidant compounds (i.e., Schlüter et al. 1993, Worbes 1997, Larcher 2001, De Simone et al. 2002, Parolin et al. 2002, Wittmann & Parolin 2005, Oliveira Wittmann 2007, Ferreira et al. 2007, Parolin 2009). Moreover, specific adaptations such as timing of seed dispersal (Van Splunder et al. 1995), seed dispersal by river waters or aquatic dispersers, and enhanced germination rates and velocities of seeds in dependence of its contact with river waters are especially common in species subject to a regular *flood-pulse* (*sensu* Junk et al. 1989) (i.e., Andrade et al. 1999, Parolin 2001, Oliveira Wittmann et al. 2007). Many tree species may also dispose of acclimations such as the *physical escape* from a submerged environment (Parolin 2002, Voesenek et al. 2003), i.e. they only establish on elevations where flood height and duration is tolerable.

The degree of adaptations and acclimations developed generally lead to a well-defined zonation of tree species along the flooding gradient. Contrasting heights and periods of flooding thus directly influence on the composition and richness of tree species. This has been investigated in several tropical large-river floodplains which are subjected to seasonal floods, such as those along the Mekong and Okavango Rivers (i.e., resumed in Junk et al. 2006 and Parolin & Wittmann 2010), and, in the Neotropis, along the Orinoco (Godoy et al. 1999, Rosales et al. 2001), Amazon (Junk 1989, Ferreira 1997, Wittmann et al. 2002, 2004), and Paraguay Rivers (Cunha & Junk 2001, Damasceno-Júnior et al. 2005). In tropical South-America, large-river floodplains are characterized by seasonal water-level oscillations that range from approximately 4-5 m (Paraná and Paraguay Rivers) to up to 12 m (Orinoco and Amazon Rivers with its major tributaries), and trees are subject to waterlogging and/or submergence for up to 300 days year[-1]. In these floodplains, the periodically inundations superimpose most other abiotic factors which normally are of crucial importance for tree species establishment, distribution, and growth in a strictly terrestrial environment (Junk et al. 1989).

However, there are numerous forest inventories from seasonally and/or periodically short and low flooded forests which also report on well-defined tree species zonation along the flooding gradients, both from extra-tropical (i.e., Clark & Benforado 1981, Schnitzler et al 2005, Renöfält et al. 2005, Robertson 2006, Budke et al. 2007) and tropical landscapes (i.e., Bongers et al. 1988, Pélissier et al. 2002, Veneklaas et al. 2005, Ramberg et al. 2006, Da Silva & Batalha 2006). These studies suggest that surface water saturation is often a more important determinant on tree species zonation than other abiotic variables, independent if flood regimes are regular or seasonal, and independent of the communities' geographic location. Jackson & Drew (1984) and Malick et al. (2002) stated that even if a short period of waterlogging does not kill vegetation, it has considerable long-term effects on plant growth by reducing the competitive abilities of species. Wherever waterlogging causes local extinctions of intolerant species, it thus may shape the composition of local assemblages (Visser et al. 2003).

Flooding is thus widely considered as stressful to trees, and implies the need for specific adaptations and/or acclimations at some stage in the evolution of a population or species (Otte 2001). However, while some of these adaptations were most probably developed in response to oxygen deficiency as caused by flooding (i.e., pneumatophores, aerenchyma); many of them in fact are effective against a wide range of stressful conditions (Kubitzki 1989, Larcher 2001, Kozlowski & Pallardy 2002). For example, a reduced metabolic activity during climatologically unfavourable periods is of advantage in non-flooded semi-deciduous tropical forests how they i.e. occur next to the Amazon-Cerrado boundary, and at the western extension of the Atlantic forest domain in SE-Brazil (i.e., Rizzini 1997). Many tree species of the Cerrado and also of coastal restingas dispose of high resprouting capacity which is interpret to be an advantageous acclimation against naturally occurring fires (Olson & Platt 1995, Cirne & Scarano 2001) – although it increases stem superfices and thus aeration during flooding (Lüttge 1997, Wittmann & Parolin 2005, Wittmann et al. 2008). Hypertrophic lenticels on tree stems can be found in nearly all Amazonian tree genera (i.e. Ribeiro et al. 1999), even when restricted to non-flooded habitats, and elevated degrees of anti-oxidant compounds are produced to protect against many kinds of stressors, among others such as salinity and drought (Blokhina et al. 2003). Most Neotropical floodplains thus consist of tree species that are not only tolerant to seasonally poor oxygen concentrations, but that are generally tolerant against a wide range of unfavourable environmental conditions (Kubitzki 1989, Parolin et al. 2010). In fact, most Neotropical floodplain and riparian forests are poor in endemic tree species, with most species being characterized by wide distribution ranges, independent if they occur in rainforests or savannas (i.e., Veneklaas et al. 2005, Junk et al. 2006, Wittmann et al. 2006). Intense species exchange between floodplains and adjacent uplands thus are common since the evolution of the modern Neotropical flora, and there is no evidence of the development of specific floodplain floras (Wittmann et al. 2010). However, there are substantial differences of the Neotropical freshwater floodplain flora between ecosystems and biomes, and the combination of taxa of many flooded forests might be characteristically distinct depending on climate, flood and disturbance regimes, paleo-climatic changes, and flora evolution.

3. DEFINITION, EXTENT AND DISTRIBUTION OF TROPICAL FLOODPLAIN FORESTS IN BRAZIL

Brazil is characterized by large-river systems, and most of its tropical wetlands are subjected to relatively regular water-level fluctuations. Flooding can occur by the rise of connected rivers, lakes, and/or by rainwater. Tropical forests influenced by freshwater floods range from evergreen rainforests in the Amazon basin to deciduous forests near the Amazon - Cerrado boundary and in hyperseasonal savannas (*sensu* Eiten 1982), such as the Pantanal and Bananal, riparian communities in the Cerrado and the Atlantic forest, and bad draining depressions in nearly all tropical biomes, such as the *Campinas* and *Campinaranas* (Amazon), *veredas* (Atlantic forest, and Cerrado) and the flooded forests of coastal restingas (Atlantic forest). However, permanent swamps are rare and often part of larger periodical wetlands (Junk 2002). In large river floodplains, flood regimes range from episodic near the headwaters and low-order rivers to periodic along the main streams and larger tributaries, with flood amplitudes of few cm to several meters, and flood durations of trees of few single days to up to 300 days year^{-1}.

Brazil is divided into nine hydrographic basins from which eight are located in tropical latitudes: The Amazon (N), the Oiapoque (NE-Brazil, Amapá), the Araguaia-Tocantins (northern Central Brazil), the Paraguay-lower Paraná (western Brazil), the upper Paraná (southern Central-Brazil), the Pindaré-Parnaíba (NE-Brazil), São Francisco (central to eastern Brazil), and Paraíba do Sul (SE-Brazil) (Ministério dos Transportes 2010). Most of the northern Brazilian Rivers including those of the Amazon basin, the Oiapoque, lower Araguaia and Tocantins, and Paraguay with its associated floodplains still remain in a fairly pristine stage, whereas many rivers of the other basins are characterized by the construction of reservoirs for flood control and hydropower generation (Junk 2002), and further anthropogenic impacts associated with high population densities and intense agricultural activities. The construction of reservoirs modifies natural water-level fluctuations and discharge as well as sediment- and nutrient-input in river-waters, and thus drastically modifies or even destroys flooded vegetation communities.

Episodically or periodically flooded forests occur in nearly all tropical wetlands of Brazil, and they may cover huge areas. Junk (1993), and Junk et al. (in press) estimate that approximately 25-30% of the 7 million square kilometers that make up the Amazon basin fit international criteria for wetland definition. Periodically flooded forests along the Amazon main stem and its major tributaries cover an area of approximately 400.000 km^2 (Melack & Hess 2010). These forests are exceptional diverse in habitats and tree species which - besides the optimal climate conditions for plant growth within the Amazon region - is basically linked to differing sediment and nutrient-loads of flooding waters.

Due to seasonal water-level changes of up to 3 m, the Paraguay River is accompanied by periodically flooded riparian forests. Its upper course meanders through the Pantanal, a wetland of about 170.000 km^2, of which about 150.000 km^2 belong to Brazil. The flooded tree vegetation in the Pantanal can be distinguished in several communities depending on its location along the flooding gradient, and consist of evergreen riparian forests, seasonally flooded woodland savannas, and low tree-and-scrub woodland savannas (Eiten 1982, Ratter et al. 1988, Junk et al. 2006). One further hydro-climatologically forested savanna is that of the Bananal, an interfluvial wetland of approximately 20.000 km^2 fed by the Araguaia and Javaés Rivers. The vegetation of the Bananal is physiognomically similar to the Pantanal, however, due to its hydrologic connectivity to the Amazon main stem, species numbers of flood-tolerant communities in the Bananal are possibly higher (Ratter et al. 2003, Marimon et al. 2006). Besides the Bananal, the Araguaia and Tocantins Rivers are accompanied by riparian forests along its courses of up to 2600 km. However, the headwater region of the Araguaia as well as its major tributaries is already heavily impacted by agricultural activities.

Further tropical floodplains in Brazil are mostly restricted to riparian zones of the Atlantic forest domain and its western extension in SE-Brazil such as the upper Paraná River and its main tributaries. In addition, insular floodplains may fringe the headwater regions or the courses of low-order Rivers. Besides rivers of the hydrographic basins as listed above, these floodplains may also occur along small and isolated rivers without connection to larger basins, such as the Panaguaçu, Pardo, Jequitinonha and Doce Rivers in E and SE-Brazil, all perennial and exorrheic, directly mouthing into the Atlantic Ocean (i.e., Kobiyama 2003, Rolim et al. 2006). However, the lack of detailed vegetation-classifications over huge areas and the variety of local terms found in literature complicate the definition of flooded forest habitats in Brazil, especially when they are comparatively small and of interfluvial origin. Local terms such as *mata de galeria* (gallery forest), *floresta ripária* (riparian forest), *mata ciliar* ("eyelash forest" = riparian forest), *floresta higrófila* (all types of waterlogged forest), *mata de brejo* and *floresta paludosa* (mostly bad draining headwater depressions in the Cerrado and Atlantic forest), as well as *lavrado* and *vereda* (bad draining swamps fed by rainwater with dominance of *Mauritia flexuosa* L.f., mostly savannas N of the Amazon basin and all over the Brazilian Cerrado) (i.e., Hoehne 1936, Hueck 1957, Ducke 1959, Eiten 1972, Rizzini 1997, Oliveira Filho & Ratter 1995) demonstrate the widespread occurrence of seasonally flooded forests by rivers and rainwater in all parts of tropical Brazil.

Independent of the geographic location and biome, remarkably less forest inventories were performed in flooded than in non-flooded habitats, and knowledge about flooded tree communities and its ecology is scarce. According to the extent and economical importance of flooded forests, most specific knowledge about the flooded tree flora in Brazil is available for the Amazon and the Pantanal. On the one hand, this reflects the huge extend of flooded forests in these regions in contrast to the insular occurrence of flooded forest habitats in large parts of the Cerrado and the Atlantic forest domain. However, the inventoried area in the Amazon and the Pantanal is small when compared to the huge extension of these areas, and mostly concentrated to ecosystems or habitats of high economic importance (i.e., highly productive white-water forests in the Amazon basin), while areas of lower economic importance were scarcely inventoried.

In the following, some examples of important flooded forest types in different Brazilian biomes are presented. Depending on the geographic location of the flooded forests, its flora is characterized by distinct hydro-climatic conditions, and consequently varying flood amplitudes. However, the comparison of the tree flora of different Brazilian flooded forests provides important information about similar or diverging species distribution patterns as well as on similar or diverging rules of their development.

4. FLOODED FORESTS OF THE AMAZON BASIN

4.1. Environment

The Amazon forest covers an area of approximately 6 million km^2 and accounts for most of the world's remaining rainforests. With approximately 30.000 species of angiosperms, the Amazonian flora belongs to the most diverse worldwide. Gentry (1982) estimated the degree of endemism at the species-level to account for approximately 76%.

The climate over most of the Amazon basin is hot and humid and characterized by a weak thermal periodicity, with mean monthly temperatures ranging from 25.9 – 27.6 C°. Precipitation over most of the Amazon basin is clearly periodic, with a rainy season from October/November to April/May and a dryer season from May/June to September/October. Total annual rainfall averages approximately 3000 mm in the western part of Brazilian Amazonia, and decreases eastwards to approximately 2100 mm in the region of Manaus, and 1800 mm near the mouth of the Amazon (Ribeiro & Adis 1984, Sombroek 2001). Accordingly, the number of months with precipitation of > 100 mm increases from 0 in the western Brazilian Amazon basin to the 1.5 in central Amazonia, and up to 3 in eastern Amazonia (Sombroek 2001). During these months, evaporation can exceed precipitation, which may results – also in large-river floodplains - in periods with droughts of diverging severity (Parolin et al. 2010).

The high amount of precipitation, its seasonality and the generally low slope inclinations in the Amazonian lowland lead to seasonally flooded areas along the main Amazonian river systems. The major Amazonian rivers are subject to seasonal water-level oscillations of up to 10 m in height (Figure 1). Due to the flat landscape over most of the Amazon basin, there are extensive floodplains which cover an area of up to 400,000 km^2 (Junk 1993, Melack & Hess 2010). They are covered to a large extent by forests. According to the differing origin of the Amazonian river waters, the Amazonian floodplains can roughly be differentiated into nutrient-rich white-water floodplains, and nutrient-poor black-water or clear-water floodplains (Sioli 1954; Prance 1979). White-water (várzea) floodplains occur along the channels that drain the westerly located Andes and/or the Andean foothills, such as those of the Ucayali-Solimões-Amazon, Juruá, Japurá, Purús, and Madeira Rivers, and they cover an area of approximately 300,000 km^2. These rivers are rich in suspension load, which is deposited along the river margins as well as in the channel bars, levees, lakes, and backwater depressions along the river courses. Due to the high input of sediment, white-water rivers form highly dynamic systems of constantly migrating river channels, thus creating a variety of floodplain and forest microhabitats (Salo et al. 1986, Kalliola et al. 1991, Peixoto et al. 2009). Várzea floodplains are exceptionally rich in nutrients, as reflected by the elevated net primary production (NPP) of their vegetation cover (Junk & Piedade 1993, Schöngart et al. 2005). Due to the origin of most of its sediments, the várzea can be considered as a geochemical extension of the Andes (Fittkau et al. 1975), although mineral composition is altered due to the intensive tropical weathering (Irion et al. 1997).

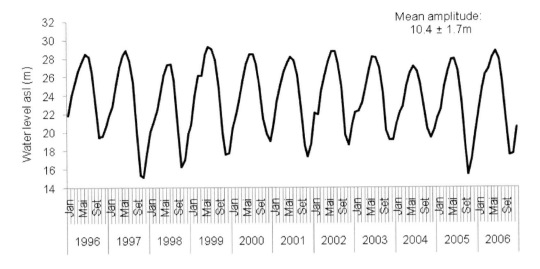

Figure 1. Mean water-level of the Negro River, near its confluence with the Solimões River in the harbor of Manaus, during 1996-2006. Data originate from the Manaus Harbor Authority.

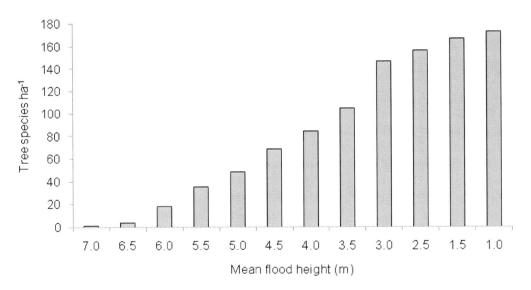

Figure 2. Tree species richness ≥ 10 cm diameter at breast height - dbh of central Amazonian várzea forests along the flooding gradient. Highest variations in species richness occur between forest plots with mean flood heights of 3.5–3.0 m. Data are from forest inventories of the Mamirauá Sustainable Development Reserve, near Tefé, and at the lower Solimões River, near Manaus (Wittmann et al. 2010), where 83 plots of 625 m2 were nested along the mean inundation height (daily water-level records in the harbor of Manaus from 1903 to 2003) of 2,631 individual trees.

Igapó forests occur along rivers that drain the Paleozoic and/or Precambric shields of Guyana (N) and Central Brazil (S) as well as the Cretaceous sediments of Amazonia, e.g., the Negro, Tapajós, and Xingú Rivers. These rivers carry low loads of suspended matter and solutes, resulting in a paucity of nutrients. Compared to the hydro-geomorphologic dynamic várzea, igapó forests form relatively stable floodplain habitats (Wittmann et al. 2010). Periodically flooded igapó floodplains cover an area of approximately 100,000 km^2 (Junk 1989).

The formation of várzea and igapó is closely related to Pleistocenic sea-level fluctuations and thus a direct result of global climate change (Irion et al. 1997). Periods of sea-levels of up to 120 m below its present level (i.e., Würm cold period) promoted incisions of the middle and lower Amazon and its tributaries, while periods of higher sea-levels resulted in a blockage of the river outflow and enlarged river beds (i.e. interglacial periods). Detailed information on the geochemistry of várzea and igapó is described in Sioli (1984) and Junk (1997, 2000).

Seasonal variations in the river levels subject trees to long periods of continuous flooding, with changes of the water levels that can reach up to 10 cm per day (Junk 1989). Mean flood amplitude between lowest and highest water levels of the main Amazonian rivers is highest in the central part of the basin where it averages up to 10 m (Figure 1). Flood amplitudes generally decline in direction to the headwater regions as well as in direction of the lower river courses. However, the *flood pulse* (Junk et al. 1989) is monomodal and therefore predictable, resulting in well-defined high-water (aquatic phase) and low-water (terrestrial phase) period during the year (Figure 1). Tree lines establish at mean maximum flood-levels ranging from 7.5 – 9 m, which corresponds to flood durations of 230-270 days year[-1] (Wittmann et al. 2010). In nutrient-rich várzea, lower elevations are vegetated by semi-aquatic and aquatic herbaceous plants, whereas these are scarce in nutrient-poor igapó (Piedade et al. 1991).

4.2. Tree Vegetation of Seasonally Floodplains along Large Rivers

Amazonian freshwater floodplains are covered by the most species-rich floodplain forests worldwide. More than 1,000 flood-tolerant tree species were recorded to occur in the várzea (Wittmann et al. 2006). From igapó, the comparatively low number of inventories does still not allow for reliable estimations of total species richness. However, comparisons from both local and basin-wide scales indicate lower species richness in igapó than in the várzea (Prance 1979; Ferreira et al. 2005; Haugaasen & Peres 2006, Wittmann et al. 2010). Besides comparatively long floristic evolution since at least the Paleocene (Kubitzki 1989), it is thought that the exceptional tree species richness in Amazonian várzea can be traced back to the high beta-diversity of the alluvial landscape (Salo et al 1986) in combination with a moderate disturbance regime imposed on plant assemblages by the annual floods (Peixoto et al. 2009). The distribution of the different floodplain forest types is thus determined by adaptations of tree species to different levels and periods of flooding, and most habitats and species are strongly zoned along the flooding gradient (Junk 1989; Ferreira 1997; Wittmann et al. 2002, 2006). However, due to the different chemical composition and sediment and nutrient input by the flood waters, both disturbance regimes and the tree flora of the várzea and igapó differ substantially in species composition and diversity (Prance 1979; Kubitzki

1989). Comparisons of forest inventories at local and basin-wide scales suggest overall floristic similarity between both ecosystems to be < 20% (Wittmann et al. 2010). Main reason for the diverging flora is the contrasting nutrient level, which seems to act as a distribution barrier for many white-water species migrating to the igapó and vice versa. In addition, alluvial dynamism is distinct, the white-water floodplains being highly dynamic systems where constant processes of sedimentation and erosion create a high variety of micro-habitats thereby increasing b-diversity (Wittmann et al. 2004).

Tree species composition and diversity along the flooding gradient was investigated by several authors, in both Amazonian várzea and igapó (Junk 1989, Ferreira 1997, Cattanio et al. 2002, Wittmann et al. 2002, 2004, 2006). However, much more floristic inventories were performed in the várzea than in the igapó, and consequently more knowledge on its flora is available at both local and regional scales. Várzea forests can be separated according to two main habitats: a) *low-várzea forests*, influenced by mean inundations with heights between 3.0 and 7.5 m (corresponding to an mean inundation period of 50–230 days year^{-1}), and b) *high-várzea forests*, influenced by mean inundations with heights of less than 3.0 m (mean < 50 days year^{-1}) (Wittmann et al. 2002). From a local study in central Amazonian várzea comprising an area of more than 5 ha, only approximately 17.5% of all inventoried tree species occurred in both low várzea and high várzea, demonstrating the striking difference between the two habitats (Wittmann et al. 2002, 2004). Variations in tree species richness were highest between sites subjected to mean inundations between 3.0 and 2.5 m, thus justifying the habitat differentiation between low várzea and high várzea at these elevations (Figure 2). The continuous increase in species richness with decreasing inundation height and length, and the pronounced species zonation of trees in várzea forests along the flooding gradient were later investigated on a continental-wide scale: Wittmann et al. (2006) made use of data from 44 floristic inventories totalling an inventoried area of 62.34 ha scattered over ten regions along the Amazonian várzea river system, including sites in Brazil, Peru, Bolivia, Colombia, and Ecuador. These authors found that mean species richness in low-várzea forests averaged 56.9 species ha^{-1}, with 100.8 species ha^{-1} in high-várzea forests ($\geq$ 10 cm dbh).

The pronounced zonation of várzea tree species along the fooding gradient leads to characteristic species associations and forest types. Only a few tree species occur along the entire flooding gradient; instead, most of them are restricted to very small topographic amplitude (Wittmann et al. 2002). Várzea forests establish in well-defined successional seres, and patterns of forest structure such as stem densities, basal areas, mean and maximum tree heights, sizes of individual tree crowns, and the crown area coverage show a strong relationship to the mean stand age of forests, which in turn is traced to the position of forest stands along the flooding gradient (Wittmann et al. 2002; 2010). Pioneer and early-secondary forests thus normally are higher flooded than later successional stages. Exceptions may occur when river or river-channels migrate and forests are subjected to extreme hydro-geomorphic disturbance (Wittmann et al. 2010).

Tree species of Amazonian large-river floodplains developed different levels of acclimations and adaptations to cope with the seasonally hypoxic site conditions. Trees may disperse to higher flooded sites than the parent trees and establish during the terrestrial phase but either are intolerant to the peculiar site conditions or quickly loose competition to better adapted species (Parolin & Wittmann 2010). In fact, many Amazonian floodplain tree species that tolerate high and prolonged inundations near the physiologically induced tree lines show adaptations against a wide range of potentially stressful conditions: they tolerate high sedimentation rates when located near the river channels of white-water rivers, or scarcely aerated soils when located in backwater swamps, and they often are tolerant to full sunlight and drought during the terrestrial phases when low river water levels coincide with seasonally low precipitation. Trees establishing at highly flooded sites are therefore light-demanding pioneer species which also dispose of the capacity of vegetative reproduction (Worbes et al. 1992, Parolin et al. 2002, Wittmann & Parolin 2005). They grow fast and exhibit relative short life cycles, as in the white-water pioneers *Salix martiana* and *Cecropia latiloba* (Worbes et al. 1992; Parolin et al 2002). With their establishment, these pioneer trees modify the local site conditions to that way that they itself are no longer able to establish (Wittmann et al. 2010), but facilitate the establishment of species of later successional stages.

Species-poor low-várzea forests floristically resemble each other throughout the Amazon basin, even when separated by long geographic distances. Species-rich high-várzea forests may exhibit floristic distinctness, but share approximately 30% of all tree species with the adjacent uplands (Wittmann et al. 2006). In fact, tree species richness and alpha-diversity of várzea forests are significantly correlated to flood height and length, and to the forest stand age (Wittmann et al. 2010): Maximum species richness ($\geq$ 10 cm dbh) recorded in high-várzea forests of Amazonia amount to 84 species ha^{-1} in the eastern parts of the basin, to 142 species ha^{-1} in Central Amazonia, and to 157 species ha^{-1} in the southern part of western Amazonia.

Endemism is highest in highly flooded low-várzea forests, and was estimated to account for approximately 39% in the 186 most common Central Amazonian várzea tree species (Wittmann et al. 2010). However, recent data suggest that this number is overestimated. Investigating the occurrence of 658 várzea tree species across the South-American continent including both flooded and non-flooded habitats and ecosystems, Wittmann (unpubl. data) suggests that only 71 tree species (approximately 11 %) have restricted occurrence in Amazonian várzea, whereas up to 89% of all investigated várzea tree species occur in one or more Amazonian and/or extra-Amazonian flooded and/or non-flooded ecosystems.

Much less phytogeographic knowledge is available for the Amazonian igapó than for the várzea floodplains (resumed in Wittmann et al. 2010). The low number of floristic inventories in the igapó can be traced to its comparatively low economic importance in comparison to the nutrient-rich várzea which offers a lot of timber products and other natural resources to the local population. In addition, the igapó varies significantly in water types and nutrient-levels depending on the origin of rivers and substrates, and forests may vary considerably in tree species composition and diversity, reinforcing the need of higher resolute habitat and vegetation classifications than those proposed by Sioli (1954) and Prance (1979). A modern and higher-resolute classification of Amazonian wetlands was recently elaborated by Junk et al. (in press). For example, the numerous ria-lakes and black-water tributaries bordering the middle Solimões/Amazon River in western Brazilian Amazonia are black-water lakes (igapó *sensu* Sioli 1954), but most of them drain paleo-alluvial substrates deposited by abandoned channels of white-water rivers. The flora of these igapó lakes is much more related to the várzea than the flora of the Negro River, which is characterized by sediment- and nutrient-poor, 'real' blackwaters.

The flora of Amazonian várzea is distinct and contrasts remarkably from that of the surrounding uplands and also from that of the igapó. Fabaceae are the most important várzea tree family, followed by the Malvaceae, Euphorbiaceae, Moraceae, Palmae, and Salicaceae (Table 1).

Florian Wittmann

Table 1. Family distribution of the 10 most important families in Amazonian white-water (várzea) forests along vertical and longitudinal gradients

	Family	Vertical distribution				
		LV	HV	EA	CA	WA
1	Fabaceae	12.46	8.45	7.62	12.23	8.79
2	Malvaceae	12.45	8.34	30.54	14.09	6.88
3	Euphorbiaceae	7.35	8.47	4.37	8.41	7.34
4	Moraceae	5.67	11.42	5.74	5.51	11.8
5	Arecaceae	5.09	12.33	5.58	0.71	11.42
6	Salicaceae	6.73	1.99	11.94	6.62	4.14
7	Urticaceae	5.23	2.53	4.63	5.69	3.21
8	Annonaceae	3.79	5.49	1.4	3.98	4.23
9	Brassicaceae	5.06	0.17	3.98	6.21	0.13
10	Sapotaceae	3.34	3.99	1.62	3.06	3.69
Total		**67.17**	**63.18**	**77.42**	**66.51**	**61.63**
No. of families		**63**	**64**	**35**	**48**	**64**

Numbers represent percentages of total importance values (TIV, Curtish & McIntosh 1951). LV = low várzea, HV = high várzea; EA = eastern Amazonia, CA = central Amazonia, WA = western Amazonia. Data originate from several floristic inventories resumed in Wittmann et al. (2006)

Table 2. Overall importance of the 15 most-important tree species in a) low várzea and b) high várzea. rIVI = relative importance value index (Curtish & McIntosh 1951), OIV = overall importance value (= IVI + relative frequency in 44 inventories scattered over the Amazon basin, see Wittmann et al. 2006)

a)

	Family	Species	rIVI	OIV
1	Malvaceae	*Pseudobombax munguba* (Mart. & Zucc.) Dugand	8.48	9.95
2	Salicaceae	*Laetia corymbulosa* Spruce ex Benth.	5.33	6.51
3	Brassicaceae	*Crataeva benthamii* Eichler	5.13	6.10
4	Urticaceae	*Cecropia latiloba* Miq.	4.51	5.70
5	Fabaceae	*Pterocarpus amazonum* (Mart. ex Benth.) Amshoff	3.23	4.49
6	Verbenaceae	*Vitex cymosa* Bert. ex Spreng.	3.76	4.32
7	Lauraceae	*Nectandra amazonum* Nees	3.11	4.16
8	Euphorbiaceae	*Piranhea trifoliata* Baill.	2.73	3.71
9	Malvaceae	*Luehea cymulosa* Spruce ex Benth.	2.90	3.60
10	Arecaceae	*Euterpe oleracea* Mart.	2.67	3.23
11	Moraceae	*Ficus trigona* L. f.	2.40	2.82
12	Polygonaceae	*Triplaris surinamensis* Cham.	1.46	2.16
13	Bignoniaceae	*Tabebuia barbata* (E. Mey.) Sandwith	1.16	2.14
14	Fabaceae	*Macrolobium acaciifolium* (Benth.) Benth.	1.16	2.07
15	Aquifoliaceae	*Ilex inundata* Poepp. ex Reissek	1.67	1.96
Σ			49.70	62.92
Σ 16-617			50.30	137.08
Total			**100**	**200**

b)

	Family	Species	rIVI	OIV
1	Moraceae	*Brosimum lactescens* (S. Moore) C.C. Berg	2.56	3.36
2	Malvaceae	*Theobroma cacao* L.	1.83	2.48
3	Euphorbiaceae	*Hura crepitans* L.	1.65	2.23
4	Meliaceae	*Trichilia septentrionalis* C. DC.	1.82	2.11
5	Anacardiaceae	*Spondias lutea* L.	1.59	2.10
6	Moraceae	*Maquira coriacea* (H. Karst.) C.C. Berg	1.53	2.04
7	Arecaceae	*Astrocaryum chonta* Mart.	1.64	2.00
8	Boraginaceae	*Cordia nodosa* Lam.	1.20	1.93
9	Arecaceae	*Iriartea deltoidea* Ruiz & Pav.	1.46	1.90
10	Euphorbiaceae	*Drypetes amazonica* Steyerm.	1.49	1.85
11	Arecaceae	*Euterpe oleracea* Mart.	1.04	1.69
12	Euphorbiaceae	*Alchornea triplinervia* (Spreng.) Müll. Arg.	1.49	1.64
13	Malvaceae	*Ceiba samauma* (Mart.) K. Schum.	1.24	1.60
14	Dichapetalaceae	*Tapura acreana* (Ule) Rizzini	1.36	1.58
15	Myristicaceae	*Otoba parvifolia* (Markgr.) A.H. Gentry	1.40	1.55
Σ			23.30	30.06
Σ 16-686			76.70	169.94
Total			**100**	**200**

The family importance, however, depends strongly on the location of the forest along the flood-level gradient, the successional stage, and the geographic location of the inventories. Independent of vertical or geographic gradients, the 10 most important tree families account for about 60–80% of all várzea tree species; however, the distribution of várzea tree families differs considerably between low-várzea and high-várzea forests (Table 1).

At the basin-wide scale, (918 tree species in 62.34 ha), the 15 most important low-várzea species account for 31.5% of the overall importance [OI = sum of the importance value index (IVI; Curtis & McIntosh 1951) and the relative frequency (rF) in 44 inventories scattered over the Amazon basin (Wittmann et al. 2006)], whereas the 15 most important high-várzea species account for only 15% of the OI (Table 2). Floristic resemblance between low-várzea forests and high-várzea forests is 35% in eastern Amazonian várzea (data from Cattanio et al. 2002), but only 17.5% in central Amazonian várzea (Wittmann et al. 2002). For the western part of the basin, no data allowing for a comparison between low várzea and high várzea are available.

4.3. Further Flooded Forest Types in the Amazon Basin

Besides várzea and igapó along the large river system, the Amazon basin consists of several wetland types with mostly insufficient description of vegetation. The river system drains the excess of rainwater so slowly that many interfluvial areas become waterlogged or shallowly flooded during the rainy season (Junk 1993, Junk et al. in press). The thousands of small rivers (locally named *igarapés*) that cross the Amazon landscape mostly react to local rainfall, therefore being characterized by an unpredictable, polymodal flood-pulse (*flashfloods* after local rainfall events *sensu* Junk 1993). The riparian zones of these rivers are covered by forests which are floristically and physiognomically distinct from both,

Amazonian uplands and seasonal floodplains. In central Amazonia, they are named *baixios* (bottomlands).

Trees of Amazonian *baixios* are adapted to cope with episodically inundations that range from few hours after heavy rains to up to several days or even weeks during the rainy seasons. Besides moderate adaptation to flooding and associated constraints, they are tolerant to mostly extremely weathered, acidic sandy soils of low nutrient status (i.e., Ribeiro et al. 1999, Castilho et al. 2006). Most trees of the *baixios* have pronounced superficial root systems, and growth heights are commonly lower in comparison to adjacent upland forests (Costa et al. 2009). In central Amazonia, the *baixios* show generally higher floristic similarity to upland forests than to seasonally flooded forests, and are mostly dominated by palms of the genera *Oenocarpus, Mauritia, Mauritiella,* and *Euterpe*. Abundant dicots belong to the genera *Virola, Iryanthera, Bellucia, Protium,* and *Goupia*.

One further vegetation type in the Amazon basin called *Campina* is often related to hydromorphic soils. *Campinas* establish upon sandy, extremely leached and nutrient-poor (oligotrophic) soils with bad drainage due to the formation of an impermeable C-horizon. Vegetation varies from open grasslands to scrub- and shrub-dominated grasslands until closed-canopy forests in transition to upland forest, named *Campinaranas*. Its occurrence is often fragmented and insular, but very frequent in the central part of the basin. Its total coverage was estimated to account for 7% (> 400,000 km^2) of the area of the Amazon basin (Prance & Daly 1989).

The woody vegetation of the *Campinas* is often sclerophyllous and xeromorphic (Jordan 1985, Medina et al. 1990), generally species-poor in comparison to the adjacent uplands and composed of many endemic tree species (i.e., Steyermark 1986, Gentry 1988, Huber 1988, Prance 1996). One distinct characteristic of the woody vegetation is the formation of an expressive fine-root layer above the soil surface, which locally may become up to 1 m thick (Klinge & Herrera 1978, Silveira 2003). It is interpreted to be an adaptation to both, extreme scarcity in nutrients and against the lack of oxygen during flooding. In fact, the *Campinas* share many tree species with seasonally flooded igapó forests upon white-sand soils along the upper Negro River (Prance & Schubart 1978), where superficial fine-root layer are common (own obs.). In addition, the *Campina* was also described to share many tree species with montane forests of the Guyana Shield (Bongers et al. 1985, Steyermark 1986).

Due to its heterogeneous species composition and vegetation structure in combination with its huge extension over the Amazon basin, the *Campina* is still insufficiently described. In central and the southern part of Amnazonia, common tree species of flooded tree communities of the *Campina* belong to the genera *Macrolobium, Dimorphandra, Aldina, Hevea, Mauritiella, Bactris, and Protium* (i.e., Silveira 2003, Vicentini 2004).

5. FLOODED FORESTS OF THE CERRADO

5.1. Environment

The Brazilian Cerrado is one of the world's centers of biodiversity (Myers et al. 2000). However, biodiversity is highly endangered through expansion of modern agriculture, already having lost nearly 50% of its original 2 million km^2 area (Prance 1996, Daly & Mitchell

2000). The Cerrado is very heterogeneous in vegetation coverage, but most biodiversity changes are thought to be traced to edaphic (i.e. mesotrophic *versus* dystrophic), and climatic factors (Ratter & Dargie 1992, Castro & Martins 1999). For example, temperature and soil hydric deficiency generally increase in a SE-NE direction and two climatic barriers influencing species distribution patterns cut across the Cerrado: Occurrence of frost to the south of 20° S, and of severe droughts to the north and east of 15° S, 45° W (Castro & Martins 1999). In addition, altitude and topography influence species distribution patterns at local scales (Oliveira-Filho et al. 1994, Ratter et al. 2003). Depending on the density of woody species, vegetation structure varies from open grasslands to close-canopy forests (Silva & Bates 2002, Ratter et al. 2003). Approximately 35% of all tree species occurring in the Cerrado were estimated to be endemic to this biome (Pennington et al. 2006, Ratter et al. 2006), whereas the overwhelming part of species is associated either within humid and dry forests of the Atlantic forest domain (i.e., Rizzini 1997, Méio et al. 2003), or the Amazon (i.e., Gonçalves 2004, Marimon et al. 2006).

The climate in the Cerrado is hot with a pronounced dry season commonly lasting from April/May to September/October and a rainy season lasting from September/October to March/April. Annual rainfall ranges from approximately 1000 mm to 1700 mm, with annual evaporation rates generally exceeding 900 mm (Rizzini 1997). Mean monthly temperatures range from 21°C (July) to 27.5°C (December).

The Cerrado consists of three different ecosystems where flooded forests occur: 1) hyperseasonal savannas such as the Pantanal and the Bananal, those linked to the seasonal inundation of large rivers such as the Paraguay and Araguaia, 2) riparian forests along rivers of different size, those being seasonally flooded when rivers are comparatively large, and episodically flooded in headwater regions or when rivers are relatively small, and 3) *veredas*, those establishing in interfluvial depressions and usually fed by rainwater. According to Rizzini (1997), most of the bad-draining floodplains fed by rainwater in Central and eastern Brazil including those located in the Caatinga and Cerrado are floristic variations of the Atlantic forest. The hygrophyle associations are successional seres where the silting up of the flooded soils and the accumulation of humus lead to a climax pluvial forest. The dominating species of the *veredas* is *Mauritia flexuosa*, a widespread palm species that occurs all over tropical Brazil. *M. flexuosa* is well-adapted to permanent swamps with oxygen-poor conditions at the root-level through the formation of mostly small pneumatophores. Thus, the species preferentially occurs in areas where flood amplitudes are low. Due to its insular occurrence over different Brazilian biomes, and because they lack a specific tree flora, the *veredas* will not be specifically addressed in this chapter.

5.2. Tree Vegetation of Hyperseasonal Savannas: The Pantanal

The Pantanal is a seasonal wetland that is located in the depression of the upper Paraguay River, and covers an area of approximately 160.000 km^2. It belongs to the large-river floodplains and thus is subject to a predictable, monomodal flood-pulse (Junk et al. 1989). It extends between the old crystalline shield of Central Brazil and its transition zone to the foothills of the geologically young Andes. The main period of subsidence resulting in the wetland depression is very likely related to the last compression pulse of the Andes during the upper Pliocene-lower Pleistocene about 2.5 million years ago. The depression is surrounded by different geological formations which form the catchment area of the upper Paraguay River and its tributaries (Ussami et al. 1999). Eastwards most common are sandstones of different age, limestones and some minor granitic outcrops, the western border is built by Precambrian massifs (Junk et al. 2006).

Due to its heterogeneous geomorphology formed by spreadings, depressions, relict dunes, inselbergs, and differently-aged alluvial deposits, the Pantanal is composed of a patchwork of both, seasonally inundated and non-inundated habitats (Zeilhofer & Schessl 1999). The seasonally inundated vegetation consists of open grasslands and different forest types.

During the period of 1996-2006, mean flood amplitude between highest and lowest water levels of the Paraguay River averaged approximately 3.0 ± 0.8 m (Figure 3). Few centimeters change in elevation has dramatic importance for the environmental conditions in floodplain habitats of the Pantanal because they influence height and duration of inundation and drought stress (Junk 1993). In fact, less than 0.5 m of difference in topography was found to influence flood duration on trees varying from 0 to up to 160 days year^{-1} in a floristic inventory of a riparian forest of the lower Miranda River, southern Pantanal (Wittmann et al. 2008, Figure 4). Because of the slight inclination of the terrain of 2-3 cm km^{-1} in N to S and 5-25 cm km^{-1} in E to W floodwaters require about 3-4 months to pass the Pantanal (Alvarenga et al. 1984). Highest parts in the Pantanal floodplain reach about two meters above the mean flood level and are permanently dry or flooded for very short periods only during extreme flood events (Junk et al. 2006). The elevations are called *capão* when they are of rounded shape and *cordilheira* when they are of linear shape. They are covered by deciduous and/or semi-deciduous forest.

The Pantanal is highly diverse in flooded and non-flooded habitats, and different vegetation classifications were presented mainly in dependence of height and period of seasonal inundations and physical and chemical soil properties (i.e., Cunha 1980, Prance & Schaller 1982, Pott 1988, Ratter et al. 1988, Boock et al. 1989, IBGE 1992, Guarim-Neto et al. 1994, Ponce & Cunha 1993, Zeilhofer & Schessel 1995, Abdon et al. 1998, Silva et al. 2000). Seasonal semi-deciduous forests and woodlands predominate on well-drained sites, while lower, periodically inundated floodplains are dominated by open savanna formations interspersed with evergreen forests (Prance & Schaller 1982).

There are comparatively few studies on the floristic composition of seasonally inundated forests within the Pantanal (but see Pott & Pott 1994, Cunha & Junk 1999, 2001, Nunes et al. 2004, Damasceno Junior et al. 2005, Arieira & Cunha 2006, Wittmann et al. 2008). Periodically inundated forests within the Pantanal are often mono-dominant, or composed of a few co-dominant species (Pott & Pott 1994). Extreme climatic events with high inundations and severe droughts, but also spatial-temporary changes in sediment load of the Paraguay River and its tributaries crossing the Pantanal are thought to cause disturbance which leads to the mono-dominance of flooded tree communities (i.e. Nascimento & Cunha 1989, Silva et al. 2000, Colischonn et al. 2001). The often by only few cm to up to 1.5 m height flooded forest types are locally named by the dominating species' vernacular name, e.g. *abobral* (*Erythrina fusca* Lour.), *paratudal* (*Tabebuia aurea* (Silva Manso) Benth. & Hook f. ex S. Moore), *carandazal* (*Copernicia alba* Morong), *canjiqueiral* (*Byrsonima orbignyana* A.Juss), *cambarazal* (*Vochysia divergens* Pohl), *carvoeiro* (*Callisthene fasciculata*), *piuval* (*Tabebuia heptaphylla*), *pimenteiral* (*Licania parvifolia*), and others (i.e., Nascimento & Cunha 1989, Arieira & Cunha 2006, Pott & Pott 2009).

Pott & Pott (1994) listed 756 woody plant species in the Pantanal (scrubs, shrubs, trees, and lianas). Most species are drought-resistant savanna species. Cunha & Junk (2001) attributed less than 50% of all woody species occurring in the Pantanal as flood-tolerant. Estimations of species with tolerance to prolonged floods range from 5% (Cunha & Junk 1999) to 20% (Cunha & Junk 2001) of all woody species in the Pantanal. On the other hand, more than 50% of all tree species are restricted to non-flooded habitats.

Riparian forests, which accompany the main rivers and the secondary river-channels, were described as the most species-rich inundation forests within the Pantanal (Pott & Pott 1994, Damasceno-Junior et al. 2005, Wittmann et al. 2008). Mono-dominant tree communities how they occur in shallowly flooded plains of the Pantanal are scarce along river channels, where high abundances normally are shared by several, co-dominant tree species. Comparatively high species richness in riparian forests is thus most probably traced to high habitat diversity, as riparian zones consist of the entire gradient from seasonal long-term floods (low positions) to occasionally short-term floods (high positions) at comparatively small scales.

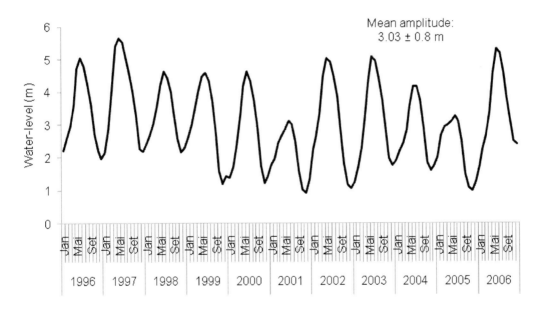

Figure 3. Mean flooding amplitude of the Paraguay River at Ladário gauge during 1996-2006. Data originate from DNAEE – Departamento Nacional de Águas e Energia Elétrica, Brazil.

Maximum flood heights on trees of > 3 m were recorded by Damasceno-Júnior et al. (2005) in riparian forests of the Paraguay River near the city of Corumbá. Due to the general flat topography of flooded lowlands, these elevations are flooded of up to 220 days year^{-1} (> 7 months) and as thus close to maximum flood periods described by Wittmann et al. (2002) in Amazonian floodplains. However, tree species of flooded Amazonian forests maybe subject to much higher water columns that increase probability of full submergence, whereas highest flooded trees of the Pantanal maintain parts of aboveground organs above highest water-levels.

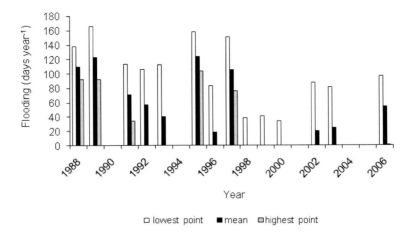

Figure 4. Mean flooding period of trees ≥ 10 cm diameter at breast height - dbh in 16 study plots of riparian forest of the lower Miranda River, during 1996-2006, estimated from daily flood-level data of the Ladário gauge, Paraguay River (DNAEE – Departamento Nacional de Águas e Energia Elétrica, Brazil). The topographically lowest point of the study area is seasonally flooded by a water column with mean heights of approximately 65 cm, the topographically highest point by a water column of mean heights of approximately 18 cm. Modified from Wittmann et al. (2008). Most tree species occurring in the Pantanal have wide distribution ranges in both humid Neotropical forests and semi-humid to dry savannas and there are no endemic tree species (Pott & Pott 1994, 2009). Characteristic morpho-anatomical and/or physiological adaptations of tree species to flooding, such as they occur in many species of Amazonian floodplain forests are unknown within tree species of the Pantanal. Increasing diversity with decreasing impact of flooding is strongly linked to the presence of rare species that are represented by a single or few individuals. These species are tolerant of seasonal inundations, but not restricted to highly inundated habitats (Wittmann et al. 2008).

Table 3. Overall importance of the 15 most-important tree species in flooded forests of the Pantanal

	Family	Species	rIVI	OIV
1	Fabaceae	*Inga vera* Willd.	12.32	14.6
2	Vochysiaceae	*Vochysia divergens* Pohl	10.57	11.94
3	Vochysiaceae	*Callisthene fasciculata* Mart.	8.96	10.55
4	Rubiaceae	*Duroia duckei* Huber	7.68	9.51
5	Urticaceae	*Cecropia pachystachya* Trécul	5.38	8.12
6	Arecaceae	*Attalea phalerata* Mart. ex Spreng.	6.33	7.24
7	Euphorbiaceae	*Alchornea discolor* Poepp. & Endl.	3.73	6.01
8	Melastomataceae	*Mouriri guianensis* Aubl.	1.59	4.79
9	Dilleniaceae	*Curatella americana* L.	3.87	5.24
10	Lauraceae	*Ocotea longifolia* H.B.K.	2.94	4.76
11	Chrysobalanaceae	*Licania parvifolia* Huber	2.73	4.55
12	Hippocrateaceae	*Peritassa dulcis* (Benth.) Miers	1.93	3.76
13	Malpighiaceae	*Byrsonima orbignyana* A.Juss.	2.93	3.84
14	Bignoniaceae	*Tabebuia heptaphylla* (Vell.) Toledo	2.22	3.59
15	Fabaceae	*Dypterix alata* Vogel	1.67	3.04
Σ			74.85	101.54
Σ 16-123			25.15	98.46
Total			100	200

rIVI = relative importance value index (Curtish & McIntosh 1951), OIV = overall importance value (= IVI + relative frequency in 9 inventories scattered over the Pantanal. Data originate from Corsini & Guarim Neto 2000, Fonseca et al. 2004, Salis et al. 2004, Damsceno Júnior et al. 2005, Arieira & Cunha 2006, and Wittmann et al. 2008)

Several authors indicated that tree species richness in the Pantanal generally decreases with increasing height and duration flooding (Pott & Pott 1994, Cunha & Junk 1999, 2001, Damasceno-Júnior et al. 2005, Arieira & Cunha 2006, Wittmann et al. 2008). Species richness in seasonally flooded forests is much lower than that of non-flooded savannas. These findings are well within the range of those of Da Silva & Batalha (2006) in a hyperseasonal savanna of the Emas National Park (SW Goiás State, central Brazil), which found species richness in flooded sites to be nearly exactly half of that found in non-flooded Cerrado; and also well within the range of species richness patterns reported in the Pantanal of the Mortes-Ararguaia Rivers in Mato Grosso (Marimon & Lima 2001), and the hyperseasonal savannas of Colombia (Blydenstein 1967, Veneklaas et al. 2005).

Table 3 lists the 15 most common flood-tolerant tree species of the Pantanal, according to floristic inventories of several authors. In total, 123 flood-tolerant tree species were recorded, belonging to 90 genera and 31 families. The 10 most important families accounted for approximately 56% of total overall importance. Fabaceae was the most important family, its species summarizing more than 15% of the OIVr, followed by Vochysiaceae, Rubiaceae, Arecaceae, and Euphorbiaceae (Figure 5).

5.3. Tree Vegetation of Riparian forests of the Cerrado

Riparian forests belong to the most species-rich habitats of the Cerrado biome (Felfili 1995), as they represent forest refuges within landscapes dominated by open savannas (Meave et al. 1991). Comparatively high diversity in riparian forests is explained by ground water availability which contrasts the soil water deficiency over most of the savanna landscape. Seasonal drought, but also fire frequency is comparatively low in riparian forests (Kellmann & Meave 1997). Most tree species are evergreen, and reproductive phenology is distributed throughout the year (Gouveia & Felfili 1998).

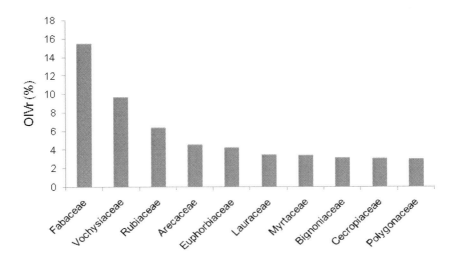

Figure 5. Relative overall importance (OIVr) of the 10 most important tree families of flood-tolerant tree species in the Pantanal, according to the inventories of Corsini & Guarim Neto (2000), Fonseca et al. (2004), Salis et al. (2004), Damsceno Júnior et al. (2005), Arieira & Cunha (2006), and Wittmann et al. (2008).

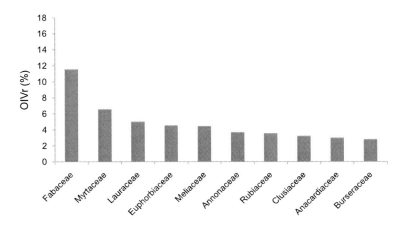

Figure 6. Relative overall importance (OIVr) of the 10 most important tree families of flood-tolerant tree species in riparian forests of the Brazilian Cerrado, according to the inventories of Toniato et al. (1998), Vilela et al. (1999), Van den Berg & Oliveira-Filho (2000), Bertani et al. (2001), Souza et al. (2003), Silva Júnior (2004), Battilani et al. (2005), Guarino & Walter (2005), Rocha et al. (2005), Teixeira & Assis (2005), Santos & Vieira (2006), Teixeira & Rodrigues (2006), Brito et al. (2008), and Camargo et al. (2008).

Table 4. Overall importance of the 15 most-important tree species in episodically to periodically flooded riparian forests of the Brazilian Cerrado.

	Family	Species	rIVI	OIV
1	Clusiaceae	*Calophyllum brasiliense* Cambess	8.72	6.82
2	Fabaceae	*Copaifera langsdorffii* Desv.	6.63	5.77
3	Magnoliaceae	*Talauma ovata* A. St.-Hil.	5.87	4.72
4	Anacardiaceae	*Tapirira guianensis* Aubl.	4.18	3.66
5	Salicaceae	*Casearia sylvestris* Sw.	3.61	3.16
6	Arecaceae	*Euterpe edulis* Mart.	2.92	2.10
7	Burseraceae	*Protium spruceanum* (Benth.) Engl.	2.60	2.08
8	Annonaceae	*Xylopia emarginata* Mart.	1.63	0.99
9	Burseraceae	*Protium heptaphyllum* (Aubl.) Marchand	1.10	0.91
10	Ulmaceae	*Celtis iguana* (Jacq.) Sarg.	1.47	0.81
11	Lauraceae	*Cryptocarya aschersoniana* Mez	0.92	0.78
12	Urticaceae	*Cecropia pachystachya* Trécul	0.58	0.75
13	Myrtaceae	*Eugenia florida* DC.	0.71	0.74
14	Anacardiaceae	*Tapirira obtusa* (Benth.) D.J. Mitch.	0.85	0.73
15	Sapindaceae	*Cupania vernalis* Camb.	0.63	0.70
Σ			42.42	34.72
Σ 16-318			57.58	165.28
Total			100	200

rIVI = relative importance value index (Curtish & McIntosh 1951), OIV = overall importance value (= IVI + relative frequency in 19 inventories scattered over the Brazilian Cerrado). Data originate from Toniato et al. (1998), Vilela et al. (1999), Van den Berg & Oliveira-Filho (2000), Bertani et al. (2001), Souza et al. (2003), Silva Júnior (2004), Battilani et al. (2005), Guarino & Walter (2005), Rocha et al. (2005), Teixeira & Assis (2005), Santos & Vieira (2006), Teixeira & Rodrigues (2006), Brito et al. (2008), and Camargo et al. (2008)

Although many floristic inventories in riparian forests across the Brazilian Cerrado are available, only few of them specifically addressed waterlogged tree communities. This is very likely traced to the fact that several riparian forests only inundate aboveground organs of trees (i.e., become visible) for comparatively short periods after rainfall events. However, riparian zones extend until the open water bodies of rivers (Kobiyama 2003), thus many trees establish roots near or even below mean ground-water levels. Waterlogging of roots thus is considered a limiting factor in riparian forest communities, and both floristic composition and forest structure are mainly dependent on soil topography and water saturation (i.e., Oliveira-Filho et al. 1994, Ivanauskas et al. 1997, Rodrigues & Nave 2000, Souza et al. 2004).

Similar to the Pantanal, tree vegetation of riparian forests of the Cerrado is a subset of the woody Cerrado flora, and there are no endemic tree species. However, species richness seems to be higher in riparian forests than in seasonally flooded forests of the Pantanal. For example, Rodrigues & Nave (2000) analyzed 43 inventories of extra-Amazonian riparian forests, and found general low floristic similarity between different rivers and forest types. Souza (2004) found similar results comparing nine floristic inventories in riparian forests of the upper Paraná River, where more than 50% of all inventoried woody species were exclusively of one location. From a tributary of the Tiete River in the State of São Paulo, Metzger et al. (1997) reported on 42% of rare species that only occurred in one of four investigated riparian forest types. Comparatively high species richness in riparian forests possibly reflects the high b-diversity of the Cerrado landscape, where water and alluvial soils differ in sediment load and nutrients according to the geology and geomorphology of the rivers' catchment areas. Moreover, the flora of riparian forests may especially differ near the borders to other biomes, such as the Atlantic forest in the E and S, and the Amazon in the N (i.e., Rizzini 1997, Ratter et al. 2003, Marimon et al. 2006), and many inventories near the border to the Atlantic forest are difficult to assign either to the Cerrado or the Atlantic forest domain.

The species list derived from floristic inventories in 19 riparian forests of the Brazilian States of São Paulo (4), Minas Gerais (10), Distrito Federal (2), Goiás (1), Mato Grosso do Sul (1), and Tocantins (1) indicated 318 flood-tolerant woody species (Table 4). The 15 most important tree species in these inventories summarize approximately 17% of the OIVr (Table 4). Most important families were Fabaceae (11.5%), followed by Myrtaceae (6.5%), Lauraceae (5%), Euphorbiaceae (4.5%), and Meliaceae (4.4%) (Figure 6).

6. THE ATLANTIC FOREST DOMAIN

6.1. Environment

The Atlantic forest originally covered an area of approximately 1.5 million km^2 from which only approximately 12% remained due to intense habitat destruction and fragmentation through colonization and agriculture (Ribeiro et al. 2009). It extends with variable widths parallel to the Brazilian coast from the tropical NE (Brazilian State of Rio Grande do Norte) until the subtropical state of Santa Catharina in S-Brazil. Especially in its southern part there is still a controversial discussion if the vegetation of the interior of the Brazilian States of São Paulo, Paraná, and Santa Catharina are extensions of the Atlantic forest, or if they belong either to the Cerrado (São Paulo and NW Paraná) or Pinheiral (Central Paraná and interior of Santa Catharina) (i.e., Rizzini 1997, Prado 2000, Oliveira-Filho et al. 2006, Pennington et al. 2006, Fiaschi & Pirani 2009), basically covered by seasonally dry forest in the N and *Araucaria* forest in the S. In the present review, the Atlantic forest is defined as coastal domain without this large western extension.

The climate is basically that of the Cerrado with the difference that there is only an inexpressive meteorologically dry season, with the number of dry months varying from 0 (i.e. near the sea-level in SE Brazil) to up to four (i.e., Caxambu, Minas Gerais). Annual precipitation ranges from 1300 – 2200 mm, and mean monthly temperatures from 14.4°C (July) to 21.0°C (January), with minima of up to -6°C near its southern border (Rizzini 1997).

The vegetation of the Atlantic forest domain is very heterogeneous, and includes mangroves and coastal restingas (shrub vegetation upon sandy soils) near the sea-level to evergreen tropical forests, semi-deciduous montane forests, and high altitude grasslands (>2000 m) (Oliveira-Filho & Fontes 2000, Scarano 2002, Nogueira-Ferraz et al. 2004, Safford 2007). With approximately 20.000 vascular plant species (Myers et al. 2000) and degrees of endemism between 40-45% (Thomas et al. 1998), diversity and endemism in the Atlantic forest are among the highest in the world (Martini et al. 2007). The remaining fragments of vegetation are therefore considered one of the world's priorities for biodiversity conservation (Myers et al. 2000).

Flooded forests form a complex and mostly fragmented landscape component of the Atlantic forest domain, and enclose riparian forests along rivers and creeks, interfluvial montane fens, bogs and hygrophile forests (*matas de brejo*), and bad draining depressions in coastal restingas (i.e., Araújo 1992, Sugiyama 1998, Assis et al. 2004). While there are several floristic inventories in riparian forests especially in SE-Brazil (Brazilian States of Paraná, São Paulo, Rio de Janeiro, and Espiritu Santo), much less knowledge is available about the non-riparian flooded tree communities.

6.2. Tree Vegetation of Flooded Forests

Riparian forests of the Atlantic forest domain occur along the Paraiba do Sul, Pardo, Jequitinonha and Doce Rivers in E and SE-Brazil, and the thousands of small creeks that drain the interior and the coastal mountains (*Serra do Mar, Serra da Mantiqueira*). Riparian forests thus are in direct neighbourhood of semi-deciduous (at higher altitudes, mostly in the western part) or evergreen (lowlands, mostly in the eastern part) forests. Species richness is generally higher than species richness of riparian forests of the Cerrado, which can be traced to the absence of a clear seasonal climatic water shortage and a comparatively high beta-diversity.

Although most of riparian forests are heavily affected (i.e., fragmented) by human activities, they considerably increase habitat and species diversity within the Atlantic forest domain (i.e., Toniato et al. 1998, Sánchez et al. 1999, Rolim et al. 2006). There are numerous forest inventories in riparian forests all over the Atlantic biome, with highest concentrations in its southern part (i.e., Sánchez et al. 1999, Bertani et al. 2001, Sztutman & Rodrigues 2002, Marques et al. 2003, Meira Neto et al. 2003, Rolim & Chiarello 2004, Rolim et al. 2006). However, inventories indicating the importance of single tree species per area are scarce, and comparisons with flooded forests of other biomes restricted to the absence and/or occurrence of tree species. Common flood-tolerant tree species in riparian forests of the Atlantic forest domain belong to the genera *Lithraea* spp. and *Astronium* spp. (Anacardiaceae), *Tabebuia* and *Jacaranda* (Bignoniaceae), *Cordia* (Boraginaceae), *Andira, Platymiscium* and *Inga* (Fabaceae), *Ocotea* (Lauraceae), *Maytenus* (Celastraceae), *Sloanea* (Elaeocarpaceae), *Alchornea* and *Sapium* (Euphorbiaceae), *Clidemia* (Melastomataceae), *Calyptranthes* and *Eugenia* (Myrtaceae), and *Annona* and *Rollinia* (Annonaceae). Most cited tree species in riparian forests of coastal areas in the states of São Paulo (Sztutman & Rodrigues 2002) and Rio de Janeiro (Carvalho et al. 2006) were *Calophyllum brasiliense* Camb., *Lithraea brasiliensis* Marchand, *Tabebuia cassinoides* (Lam.) DC., *Symphonia globulifera* L.f., *Cordia sellowiana* Cham., and *Astronium fraxinifolium* Schott.

A further, special flooded forest type may establishes in interfluvial, montane areas where soil water saturation is fed by river- or rainwater (including condensation in cloud forests) leading to montane bogs and fens, and, as consequence of the natural silting up during vegetation succession, in hygrophile evergreen forests locally named *florestas paludosas*, or *matas de brejo* (i.e., Leitão-Filho 1982, Torres et al. 1994, Ivanauskas et al. 1997, Toniato et al. 1998, Carvalho et al. 2006). Theses swamp forests are considered to be a mediterranean extension of the Atlantic forest, independent if they occur in its northern or southern part (Rizzini 1997). Soil water saturation and flooding ranges from episodic to permanent and the floristic composition varies significantly due to fragment size, geology, topography, microclimate, and the surrounding vegetation of evergreen or semi-deciduous forests (i.e., Pagano & Leitão Filho 1987, Rodrigues et al. 1989, Toniato et al. 1998). According to Leitão Filho (1982), tree species richness and structure of the *matas de brejo* is distinct from riparian forests, the latter being more diverse and of higher trees. Maximum tree heights in the *matas de brejo* are of approximately 12 m, a further mostly shrubby sub-canopy stratum establishes at 1-3 m. These forests mostly lack epiphytes in the southern part of the Atlantic forest domain, whereas these are present in the northern part (Rizzini 1997). Floristically, most dominant tree species in the *matas de brejo* also dominate many riparian communities. Besides *Calophyllum brasiliensis*, common tree species belong to the genera *Protium* (Burseraceae), *Styrax* (Styracaceae), *Talauma* (Magnoliaceae), and *Guarea* (Meliaceae), common palms to the genera *Syragus* and *Geonoma*.

A third, but very scarcely studied seasonally flooded forest type in the Atlantic forest domain is located in coastal restingas. In restinga forests, sand deposited by the sea occasionally generates a wave-like topography with topographic differences of few meters. While the soils on the ridges are well-drained, the water table in the depressions is close to the surface and eventually rises above the soil surface during the rainy season, forming canals of 20-50 cm depth draining to the sea (Marques et al. 2009). Substrate nutrient availability in these depressions is very low, the soils being characterized by comparatively low pH (4.0), often with high concentrations of Aluminum (Marques et al. 2009).

In contrast to non-flooded restinga forests which are characterized by a dense understory, the sub-canopy of flooded restinga forests is sparsely developed. Trees of flooded restinga forests often are higher than those growing upon the ridges (Marques et al. 2009). The flooded forest has fewer individuals, is more species-rich than the non-flooded forest, and has higher growth and mortality rates. This reflects the edaphically induced water shortage and scarcity of nutrients as limiting factors for plant growth off the ground-water levels in coastal restingas. Britez et al. (1997) stated that the availability of water may affect plant growth by enhancing the mineralization of nutrients in soil organic matter and promoting the transport of water-soluble nutrients to the roots, as well as contributing directly to avoidance of drought-related mortality.

Common tree species in flooded restinga forests are *Calophyllum brasiliensis*, *Tapirira guianensis* Aubl., *Ocotea pulchella* Mart., *Faramea marginata* Cham, *Myrcia racemosa* (Berg) Legrand, *Alchornea triplinervia* (Spreng.) Müll.Arg., and *Schefflera angustissima* Marchand.

7. FLOOD-TOLERANT TREE ASSEMBLAGES IN DIFFERENT BRAZILIAN BIOMES: OLIGARCHIES AND ENDEMISM

Much more publications and species lists can be used for floristic comparison between flooded forests of different biomes, as they can base on the presence/absence of species only, without the need for species-area indices. However, the estimation of the degree of endemism of a given ecosystem and/or biome is difficult and needs extensive analyses and comparisons. Wittmann (unpubl. data) compared the flora of Amazonian várzea forests based on 62 inventories performed by different authors, comprising up to 80 ha of inventoried area and up to 92,000 trees (inventories summarized in Wittmann et al. 2006, 2010). All species were checked in several herbaria (among others the Missouri Botanical Gardens, the New York Botanical Gardens, and the Royal Botanical Gardens Kew), and morphospecies, unidentified species, synonymous species, and species not documented in these herbaria were excluded. As they may not rooted in flooded soils, lianas and vines also were excluded from the list, except when species were described to have multiple habits (e.g., as shrubs or trees). After that, the list contained 1014 flood-tolerant várzea species. In order to reduce the probability of erroneous species determinations, how they may occur especially in rare species, only tree species were selected that occurred in at least two distinct floristic inventories, performed by distinct authors. Singletons thus were excluded from the list. The number of várzea tree species that occurred in at least two inventories thus was reduced to 658.

The occurrence of each of the 658 várzea tree species in South-America including Panama was investigated in electronic databases from several herbaria as mentioned above. In addition, the occurrence of the várzea tree species was checked in up to 350 species lists from floristic inventories across the whole South-American continent. After that, two degrees of endemism were distinguished: 1) tree species are restricted to episodically, seasonally, and/or permanently flood habitats, independent of their geographic distribution, and 2) tree species are restricted to Amazonian white-water forests. The second classification excludes tree species that may also occur in basins with hydrographically extensions to the Amazon basin, such as the Bolivian Llanos of Moxos. It further excludes tree species that may also occur in Amazonian black- and clear-water floodplains, such as along the Negro and Tapajós Rivers, and along the several black-water tributaries of the Orinoco and Amazon basins.

With 490 (74.4%) shared tree species, overall highest species resemblance of Amazonian várzea forests was detected to Amazonian uplands (terra firme), while only 136 tree species (20.6%) were shared with Amazonian black- and clearwater floodplains (igapó). In comparison with extra-Amazonian forests, a total of 116 várzea tree species (17.6%) occurred in Brazilian savannas, either in unflooded Cerrado (78 species, 11.8%), in seasonally flooded hyperseasonal savannas (74 species, 11.2%), or in ocasionally flooded riparian forests (50 species, 7.6%). Examples of important várzea tree species with elevated frequencies in the Brazilian Cerrado are *Himatanthus bracteatus* (A.DC.) Woodson, *Myrciaria floribunda* (H.West ex Willd.) O. Berg, *Zollernia paraensis* Huber, *Crataeva tapia* L. and *Crataeva benthamii* Eichler in the unflooded Cerrado, and *Calophyllum brasiliense* Camb., *Maclura tinctoria* (L.) D. Don ex Steud., *Guarea guidonia* (L.) Sleumer, *Guarea macrophylla* Vahl, *Ficus anthelmintica* Rich. ex DC. and *Vitex cymosa* Bert. ex Spreng. in occasionally flooded riparian forests. Most of the latter tree species also have elevated frequencies in seasonally flooded forests of the Brazilian Pantanal. In addition, *Attalea phalerata* Mart. ex Spreng. and *Alchornea discolor* Poepp. & Endl. are important várzea tree species with elevated frequencies in the Pantanal.

The comparison further indicated that 131 várzea tree species (19.9%) occur in the Atlantic forest domain, either in flooded or non-flooded habitats. There was no significant difference in species occurrence between N-Atlantic (82 species) and S-Atlantic (79 species) forests. Examples of important várzea tree species occurring in the Atlantic forest are *Guarea guidonia*, *Guarea macrophylla*, *Symphonia globulifera* L.f., *Calophyllum brasiliense*, *Inga edulis* Mart., *Brosimum lactescens* (S.Moore) C.C. Berg, and *Spondias lutea* L.

Table 5. Flood-tolerant tree species with occurrence in the Amazon, the Cerrado, and the Atlantic forest domain, ordered by species importance

Species	Amazon				Cerrado			Alantic forest		
	1	2	3	4	5	6	7	8	9	10
Tapirira guianensis Aubl.	X	X	X	X	X	X	X	X	X	X
Calophyllum brasiliense Cambess.	X	X	X	X	X	X	X	X	X	X
Genipa americana L.	X	X	X	X	X	X	X	X	X	X
Protium heptaphyllum (Aubl.) Marchand	X		X	X	X	X	X	X	X	X
Guarea guidonia (L.) Sleumer	X			X	X	X	X	X	X	X
Eugenia florida DC.	X	X		X	X	X	X	X	X	X
Sloanea guianensis (Aubl.) Benth.	X		X	X	X	X	X	X	X	X
Symphonia globulifera L.f.	X	X	X	X	X	X		X	X	X
Simarouba amara Aubl.	X		X	X			X	X		X
Spondias lutea L.	X	X	X	X		X	X	X		X
Hymenaea courbaril L.	X		X		X	X	X	X		X
Inga marginata Willd.	X		X	X		X	X	X	X	X
Guazuma ulmifolia Lam.	X				X	X	X	X	X	X
Guarea kunthiana A.Juss.	X			X	X	X		X	X	X
Brosimum guianense (Aubl.) Huber	X		X	X	X					X
Casearia decandra Jacq.	X				X	X	X	X	X	X
Tabebuia serratifolia (Vahl) G. Nicholson	X			X	X		X	X		X
Guarea macrophylla Vahl	X			X	X	X		X	X	X
Minquartia guianensis Aubl.	X	X	X	X			X	X	X	X
Randia armata (Sw.) DC.	X				X	X	X	X	X	X
Casearia aculeata Jacq.	X	X			X	X				X
Alchornea triplinervia (Spreng.) Müll.Arg.	X			X	X	X	X	X	X	X
Cheiloclinium cognatum (Miers) A.C.Sm.	X			X	X		X			X

With 1 = Amazonian white-water floodplains (várzea), 2 = Amazonian black- and clearwater floodplains (igapó), 3 = *baixios* and *Campinas* of the Amazonian terra firme, 4 = non-flooded terra firme, 5 = Riparian forests of the Cerrado, 6 = Pantanal, 7 = non-flooded Cerrado, 8 = Riparian forest of the Atlantic forest domain, 9 = *Matas de brejo*, 10 = Non-flooded forests of the Atlantic forest domain

Out of all várzea tree species, 130 (19.75%) were restricted to flooded habitats of the South-American continent. None of these species occurred in ocasionally flooded riparian forests of the Cerrado. Most várzea tree species restricted to freshwater habitats occurred in the Orinoco basin (46), followed by Amazonian igapó (31) and the Pantanal (2).

None of the species restricted to South-American freshwater habitats occurred in all investigated wetlands, Orinoco, Amazonian blackwaters and the Pantanal, whereas the Orinoco and Amazonian black- and clear-waters shared 19 tree species. The Pantanal shared one tree species with the Orinoco basin (*Tabernaemontana siphillitica* L.f. Leeuwenb.), and one tree species exclusively with Amazonian várzea (*Pithecellobium multiflorum* Benth.), whereas it shared no species with Amazonian black- and clearwaters. Important várzea tree species that had restricted occurrence in other South-American freshwater floodplains were *Astrocaryum jauari* Mart., *Tabebuia barbata* (E.Mey) Sandwith, *Symmeria paniculata* Benth., *and Piranhea trifoliate* Baill. (all Orinoco and Igapó), as well as *Simaba multiflora* A. Juss., *Crescentia amazonica* Ducke, and *Etaballia dúbia* (Kunth) Rudd. (Orinoco).

In total, 71 tree species (10.8%) only occurred in Amazonian várzea. They belonged preferentially to Fabaceae (14), Annonaceae (8), Arecaceae and Malvaceae (each 5), Lauraceae (4), and Combretaceae, Melastomataceae, Rubiaceae and Vochysiaceae (each 3). Important endemic várzea tree species are *Triplaris surinamensis* Cham., *Terminalia dichotoma* G.Mey., *Discocarpus brasiliensis* Klotzsch, *Laetia corymbulosa* Spuce ex Benth., *Oxandra polyantha* R.E.Fries, *Sorocea duckei* W.C. Burger, *Sloanea porphyrocarpa* Ducke, and *Mora paraensis* Ducke.

These kind of floristic comparisons are very important to understand floristic relationships between different biomes and the evolution of specific landscapes and floras. While the origin of the woody species of the Pantanal was recently resumed in Pott & Pott (2009), similar information is completely lacking for flood-tolerant floras of other parts of the Cerrado and the Atlantic forest domain.

The comparison of floristic inventories in flooded forests within different Brazilian biomes resulted in 23 flood-tolerant tree species that occur in the three Brazilian biomes, Amazon, Cerrado, and Altlantic forest domain (Table 5). However, some of these species not necessarily occur in flooded environments, as e.g. *Minquartia guianensis* Aubl., which was detected to occur in flooded forests of the Amazon and the Atlantic forest, but not in flooded forests of the Cerrado (Table 5). *Tapirira guianensis* Aubl., *Calophyllum brasiliense* Camb. and *Genipa americana* L. are the most widely distributed flood-tolerant tree species of tropical Brazil. Accordingly, some specific information about its ecology and ecophysiology is available, although it not necessarily originates from flooded populations (but see Marques & Joly 2000, Fischer & Santos 2001, Phillips et al. 2001, Lenza & Oliveira 2005, Oliveira & Joly 2010).

CONCLUSIONS

Flooded forests in Brazil are exceptional rich in tree species and considerably increase species diversity of humid forest and savanna biomes by increasing habitat diversity. Except the Amazonian large-river floodplains, flooded forests are scarce in endemic tree species, and composed of species with wide distribution ranges in both tropical forests and savannas.

However, flooding has considerable effects on species distribution of flooded forests by local extinctions of intolerant species leading to a characteristic subset of species in each biome (i.e., Malick et al. 2002, Visser et al. 2003, Veneklaas et al. 2005). Waterlogging assembles phylogenetically unrelated species that have converged on similar habitat use (Da Silva & Batalha 2006). Species of typically rare occurrence in non-flooded environments often dominate flooded habitats with high abundance. This seems to be equally valid in all flooded forests of Brazil, independent if they partially consist or not of endemic tree species. For instance, Central Amazonian várzea forests contain several tree species of high importance that are extremely rare in non-flooded Central Amazonian forests, such as *Hura crepitans* L., *Sterculia apetala* (Jacq.) H. Karst, and *Ocotea cymbarum* Kunth. On the other hand, these species occur with moderate to high abundance in western Amazonian uplands and upland forests of Central America and the Caribbean Islands (Wittmann et al. 2010). Numerous other tree species with high abundance in Amazonian várzea occur in habitats that are geographically (i.e., more distant) and ecologically (i.e., dryer) much more distinct from this ecosystem than non-flooded moist Amazonian upland forests. The same holds true in flooded forests of the Cerrado and the Atlantic forest domain, where species assemblages often relate to those of other biomes rather than to those of the surrounding uplands.

Why these under non-flooded conditions commonly rare species are so successful and abundant in flooded habitats? One explanation is that flood-tolerant tree species often are characterized by abilities and functionl traits which makes them competitive in a flooded environment – although these adaptations not necessarily had to be developed in flooded ecosystems. The origin of a given species and specific life-history traits hereby seem to be of crucial importance: most competitive tree species in flooded environments such as the large-river floodplains of the Amazon and the Pantanal are those considered as pioneers, where characteristics such as fast growth, comparatively short life cycles and high reproduction rates are of advantage in highly dynamic ecosystems (Wittmann et al. 2004). These species probably originated from regions that experienced extreme environmental conditions during past climate change, and species thus became highly competitive to external stress factors. Kubitzki (1989) stated that many flood-tolerant tree species of Amazonian várzea and igapó originate from regions with pronounced edaphically and/or climatically aridity, such as the Caatinga, Cerrado, or savannas of the northern hemisphere. For the Atlantic forest domain, Scarano (2002, 2009) postulated that plant colonization of marginal habitats with extreme environmental conditions is often related to terrestrial nurse plants that originally, in their rainforest habitat, were canopy plants such as epiphytes and hemi-epiphytes, and thus plants well-adapted to water and nutrient restrictions. Once established in stressful habitats such as swamps, these species ameliorate local site conditions (i.e., by silting up), which in turn allows less tolerant species to establish. Wittmann et al. (2008) observed that many abundant species in a seasonally flooded riparian forest of the Pantanal were those with high resprouting tendencies –which enhances oxygenation through the increase in root and stem superfices – although resprouting is generally interpret to be an adaptation against fires how they are frequent in non-flooded habitats of the Cerrado (i.e. Cirne & Scarano 2001).

In addition, well-adapted tree species of flooded forests are competitive especially during the earliest stages of life, which normally are considered to be the most vulnerable stages in the life of a plant (Li et al. 1996, Zagt & Werger 1998, Oliveira Wittmann et al. 2010). The restricted distribution along the flooding gradient combined with highly variable inundation periods affects tree species at the population level, and probably defines tree establishment and distribution. There is some evidence that many tree species of flooded forests in both episodically and periodically flooded environments depend for establishment on consecutive periods with exceptional low inundations (Wittmann & Junk 2003, Schnitzler et al. 2005, Marinho et al. in press), thus reinforcing the *physical escape* from a non-tolerable environment *sensu* Voesenek et al. (2003). Flood variability favours the dominance of a few tree species that regenerate during consecutive years with low water levels, and that are highly competitive especially during the early stages of life (Wittmann et al. 2008).

The absence or presence of endemic tree species in most flooded forests is very likely directly related to the evolution of its flora during climatic conditions in the past. Large-river floodplain forests of the Amazon basin consist partially of endemic tree species because freshwater floodplains are a substantial component of the Amazon landscape probably since the existence of Neotropical rainforest. The formation of the Pebas Lake in western Amazonia (c. 18-11 million years B.P) and repeated marine incursions leading to the extension of estuarine systems (i.e., c. 11-8 million years B.P.) emphasize the dominance of palaeo-floodplains in the equatorial part of the Amazon basin during the mid- to upper Miocene (i.e., Räsanen et al. 1995, Irion et al. 1997, Godoy et al. 1999). Climatic changes during the Tertiary and Quaternary directly affected the sea level and thus the size of floodplains by periodic reductions (i.e., during the glacial maxima) and enlargements (i.e., during the interglacials, formation of the Amazonas Lake) of their areas (Irion et al. 1995, 1997; Oliveira and Mori 1999). The postulated species shift from dry to moist climatic conditions as postulated by the refugia hypothesys combined with a spatial reduction of flooded areas, however, affected floodplain species to a lesser extent than upland species, because the flooded ecosystems persisted, at smaller scales, even during the glacial maxima. The riparian connectivity and the highly adapted dispersal mechanisms of floodplain trees would have reduced species losses at regional scales with the floodplains themselves acting as linear refuges for sensitive terra firme species during periods with postulated dryer climatic conditions (Pires 1984, Wittmann et al 2010). Due to the constant habitat availability even through tertiary and quaternary climatic changes, a comparatively long floristic evolution and thus the development of elevated degrees of endemism is evident in Amazonian floodplain forests.

By contrast, biomes with dryer climates and/or with smaller and/or more fragmented flooded habitats were much more affected by climate changes of Terciary and Quarternary glacials. Flooded forests of the Brazilian Cerrado including its hyperseasonal savannas lack endemic tree species (Pott & Pott 1994, Rizzini 1997, Junk et al. 2006). This reflects the comparatively short period of floristic evolution and climatic instability within these biomes in the past in comparison to other floodplains like in the Amazon. Climatic changes during the Quaternary led to intermittent periods of large-scale flooding and severe droughts and interrupted evolutional processes in the Pantanal (Ab'Saber 1988, Assine & Soares 2004). The instability of the actual climate, which is characterized by multi-annual wet and dry periods with extreme flood and drought events combined with large wild fires, has a strong impact on distribution, community structure and population size of many plant and animal species and hinders species segregation of populations (Junk 1993, Junk et al. 2006).

Although located in a biome with several centers of endemism (i.e., Mori et al. 1981, Thomas et al. 1998) and although still not sufficiently inventoried, flooded forests of the Atlantic forest domain are composed by tree species with wide distribution ranges, and none woody species with restricted occurrence to freshwater habitats is known. The lack of specifically flood-adapted endemists in the Atlantic forest domain is possibly traced to the following reasons: 1) The generally small and fragmented flooded habitats such as the *matas de brejo* easily drought up during past glacial periods, thus interrupting the evolution of a specifically flood-resistant flora, and 2) Water discharge of rivers became reduced during glacial maxima, and the moderately flood-adapted flora became replaced by moist-sensitive species from the surrounding uplands. Additionally, species exchange between the Cerrado and the Atlantic forest domain is probable especially along riparian corridors. Oliveira-Filho & Ratter (2005) considered riparian communities to be one of possible floristic migration routes between the Atlantic rainforest and the Cerrado, and even between the Atlantic forest and the Amazon. These riparian corridors possibly hindered the evolution of a specific flood-adapted flora in the Atlantic forest.

ACKNOWLEDGMENTS

I wish to thank the National Institute for Amazon Research (INPA), the Institute of Sustainable Development Mamirauá and the Federal University of Mato Grosso do Sul for logistic support in the field, and Prof. Dr. Wolfgang J. Junk for valuable comments on the manuscript.

REFERENCES

Abdon, MM; Silva, JSV; Pott, VJ; Pott, A; Silva, MP. Utilizacão de dados analógicos do Landsat TM na discriminação da vegetação de parte da sub-região de Nhecolandia no Pantanal. *Pesq. Agropec. Bras.*,1998, 33, 1799-1830.

Ab'Saber, NA. O Pantanal mato-grossense e a teoria dos refúgios. *Revista Brasileira de Geografia*, 1988, 50, 9-57.

Alvarenga, SM; Brasil, AE; Pinheiro, R; Kux, HJH. Estudo geomorfológico aplicado a Bacia do alto Rio Paraguai e Pantanais Matogrossenses. *Boletim Técnico Projeto RADAM. Série Geomorfologia*, 1984, 187, 89-183.

Andrade, ACS; Ramos, FN; Souza, AF; Loureiro, MB; Bastos, R. Flooding effects in seedlings of *Cytharexyllum myrianthum* Cham. and *Genipa americana* L.:responses of two neotropical lowland tree species. *Revista Brasileira de Botânica*, 1999, 22(2), 281-285.

Araújo, DSD. Vegetation types of sandy coastal plains of tropical Brazil: a first approximation. In: Seeliger, U. editor. *Coastal plant communities of Latin America*. New York: Academic Press, 1992, 337-381.

Arieira, J; Cunha CN. Fitosociologia de uma floresta inundável monodominante de Vochysia divergens Pohl (Vochysiaceae) no Pantanal Norte, MT, Brasil. *Acta Botanica Brasilica*, 2006, 20(3), 569-580.

Armstrong, W; Brändle, R; Jackson, MB. Mechanisms of flood tolerance in plants. *Acta Botanica Neerlandica*, 1994, 43, 307-358.

Assine, ML; Soares, PC. Quaternary of the Pantanal, west-central Brazil. *Quaternary International*, 2004, 114, 23-34.

Assis, AM; Pereira OJ; Thomaz, LD. Fitosociologia de uma floresta de restinga no Parque Estadual Paulo César Vinha, Setiba, município de Guarapari (ES). *Revista Brasileira de Botânica*, 2004, 27, 349-361.

Battilani, JL; Scremin-Dias, E; Souza, ALT. Fitosociologia de um trecho de mata ciliar do Rio da Prata, Jardim, MS, Brasil. *Acta Botânica Brasilica*, 2005, 19(3), 597-608.

Bertani, DF; Rodrigues, RR; Batista, JFS; Shepherd, GJ. Análise temporal da heterogeneidade floristica e estrutural em uma floresta ribeirinha. *Revista Brasileira de Botânica*, 2001, 24(1), 11-23.

Blokhina, O; Virolainen, E; Fagerstedt, KV. Antioxidants, oxidative damage, and oxygen deprivation stress. *Annals of Botany*, 2003, 90, 179-194.

Blydenstein, J. Tropical savanna vegetation of the Llanos of Columbia. *Ecology,* 1967, 48, 1-15.

Bongers, F; Engelen, D; Klinge H. Phytomass structure of natural plant communities on spodosols in southern Venezuela: the Bana woodland. *Vegetatio*, 1985, 63, 13-34.

Bongers, F; Popma, J; del Castillo, JM; Carabias, J. Structure and floristic composition of the lowland rain forest of Los Tuxtlas, Mexico. *Vegetatio*, 1988, 74, 55-80.

Boock, A; Araújo, M; Pott, A; Pessotti, JE; Silva, M; Pott, VJ; Souza, O. *Estratégia de ocupacão e uso das pastagens nativas do Pantanal do Nabileque em Mato Grosso do Sul.* EMBRAPA/CPAP/CNPGC and Fazenda Bodoquena S.A., Corumbá-MS, 1989.

Britez, RM; Santos-Filho, A; Reissmann, CB; Silva, SM; Athayde, SF; Lima, RX; Quadros, RMB. Nutrientes do solo de duas florestas da Planície Litorânea da Ilha do Mel, Paranaguá, PR. *Revista Brasileira de Ciências do Solo*, 1997, 21, 625-634.

Brito, ER; Martins, SV; Oliveira-Filho, AT; Silva, E; Silva, AF. Estrutura fitosociológica de um fragmento natural de floresta inundável em área de Campo Sujo, Lagoa da Confusão, Tocantins. *Acta Amazônica*, 2008, 38(3), 379-386.

Budke, JC; Jarenkow, JA; Oliveira-Filho, AT. Relationships between tree component structure, topography and soils of a riverside forest, Rio Botucaraí, Southern Brazil. *Plant Ecology*, 2007, 189, 187-200.

Camargos, VL; Silva, AF; Neto, JAAM; Martins, SV. Influencia de fatores edáficos sobre variações floristicas na floresta estacional semidecídua no entorno da Lagoa Carioca, Parque Estadual do Rio Doce, MG, Brasil. *Acta Botânica Brasilica*, 2008, 22(1), 75-84.

Carvalho, FA; Nascimento, MT; Braga, JMA; Rodrigues, PJF. Estrutura da comunidade arbórea da floresta atlântica de baixada periodicamente inundada na reserva Biológico de Poço das Antas, Rio de Janeiro, Brasil. *Rodriguésia*, 2006, 57(3), 503-518.

Castilho, CV; Magnusson, WE; Araujo, RNO; Luizão, RCC; Luizão, FJ; Lima, AP; Higuchi, N. Variation in aboveground tree live biomass in a central Amazonian forest: effects of soil and topography. *Forest Ecology and Management*, 2006, 234, 85-96.

Cattanio, JH; Anderson, AB; Carvalho, M. Floristic composition and topographic variation in a tidal floodplain forest in the Amazon Estuary. *Revista Brasileira de Botânica*, 2002, 25(4), 419-430.

Cirne, P; Scarano FR. Resprouting and growth dynamics after fire of the clonal shrub *Andira legalis* (Leguminosae) in a sandy coastal plain in south-eastern Brazil. *Journal of Ecology*, 2001, 89, 351-357.

Clark, JR; Benforado, J. *Wetlands of bottomland hardwood forests.* Developments in agricultural managed-forest ecology, Vol. 11. Elsevier, Amsterdam, Oxford, New York, 1981.

Collischonn, W; Tucci, CEM; Clarke, RT. Further evidence of changes in the hydrological regime of the river Paraguay: part of a wider phenomenon of climate change? *Journal of Hydrology* 2001, 245, 218-238.

Costa, FRC; Guillaumet, JL; Lima, AP; Pereira, OS. Gradients within gradients: The mesoscale distribution patterns of palms in a central Amazonian forest. *Journal of Vegetation Science*, 2009, 20, 69-78.

Crawford, RMM. The anaerobic retreat. In: Crawford, RMM editor. *Studies in plant survival. Ecological case histories of plant adaptation to adversity.* Studies in Ecology 11, Blackwell Scientific Publ. Oxford, 1989, pp. 105-129.

Crawford, RMM. Oxygen availability as an ecological limit to plant distribution. *Advances in Ecological Research*, 199, 223, 93-185.

Cunha, NG. Consideracões sobre os solos da sub-região da Nhecolandia, Pantanal Matogrossense. *Circular técnica da EMBRAPA-UEPAE Corumbá*, 1980, 1, 1-45.

Cunha, CN; Junk, WJ. Composicão floristica de capões e cordilheiras: Localizacão das espécies lenhosas quanto ao gradiente de inundação no Pantanal de Poconé, MT Brasil. In: Anais do II simpósio sobre recursos naturais e sócio-economicos do Pantanal – Manejo e Conservação. EMBRAPA-CPAP Corumbá, 1999, pp. 387-406.

Cunha, CN; Junk, WJ. Distribution of woody plant communities along the flood gradient in the Pantanal of Poconé, Mato Grosso, Brazil. *International Journal of Ecology and Environmental Sciences*, 2001, 27, 63-70.

Curtis JT; McIntosh RP. An upland forest continuum in the prairie-forest border region of Wisconsin. *Ecology*, 1951, 32(3), 476-496.

Damasceno Júnior GA; Semir J; Santos FAM.; Leitão Filho HF. Structure, distribution of species, and inundation in a riparian forest of Rio Paraguai, Pantanal, Brazil. *Flora*, 2005, 200, 119-135.

Da Silva, IA; Batalha, MA. Taxonomic distinctness and diversity of a hyperseasonal savanna in central Brazil. *Diversity and Distributions*, 2006, 12, 725-730.

De Simone, O; Haase, K; Müller, E; Junk, WJ; Gonsior, GA; Schmidt, W. Impact of root morphology on metabolism and oxygen distribution in roots and rhizosphere from two Central Amazon floodplain tree species. *Functional Plant Biology*, 2002, 29, 1025-1035.

Ducke, A. Estudos botanicos no Ceará. *Anais da Academia Brasileira de Ciências*, 1959, 31(2), 211-308.

Eiten, G. The cerrado vegetation of Brazil. *Botanical Review*, 1972, 38(2), pp. 341.

Eiten, G. Brazilian "Savannas". In: Huntley, BJ; Walker, BH editors. *Ecology of Tropical Savannas*. Ecological Studies 42, Springer Verlag Berlin, 1982, pp. 27-47.

Esteves, FA. Considerations on the ecology of wetlands, with emphasis on Brazilian floodplain systems. In: Scarano, FR; Franco, AC editors. *Ecophysiological strategies of xerophytic and amphibious plants in the Neotropics*. Oecologia Brasiliensis IV, PPGE-UFRJ, Rio de Janeiro, 1998, pp. 111-135.

Felfili, JM. Diversity, structure and dynamics of a gallery forest in Central Brazil. *Vegetatio,* 1995, 117, 1-15.

Fernandez MD; Pieters, A; Donoso, C; Herrera, C; Tezara, W; Rengifo, E; Herrera, A. Seasonal changes in photosynthesis of trees in the flooded forest of the Mapire River. *Tree Physiology*, 1999, 19, 79-85.

Ferraz, EMN; Araújo, EL; Silva, SI. Floristic similarities between lowland and montane areas of Atlantic coastal forest in northeastern Brazil. *Plant Ecology*, 2004, 174, 59-70.

Ferreira, LV. Effects of the duration of flooding on species richness and floristic composition in three hectares in the Jaú National Park in floodplain forests in central Amazonia. *Biodiversity and Conservation,* 1997, 6, 1353-1363.

Ferreira, CS; Piedade, MTF; Junk, WJ; Parolin, P. Floodplain and upland populationms of Amazonian *Himatanthus sucuuba*: effects of flooding on germination, seedling growth, and mortality. *Environmental and Experimental Botany*, 2007, 60, 477-483.

Fiaschi, P; Pirani, JR. Review of plant biogeographic studies in Brazil. *Journal of Systematics and Evolution*, 2009, 47(5), 477-496.

Fischer, EA; Santos, FAM. Demography, phenology and sex of *Calophyllum brasiliense* (Clusiaceae) trees in the Atlantic forest. *Journal of Tropical Ecology*, 2001, 17, 903-909.

Fittkau, E-J; Irmler, U; Junk, WJ; Reiss, F; Schmidt, GW. Productivity, biomass, and population dynamics in Amazonian water bodies. In: Golley, FB; Medina, E. editors. *Tropical ecological systems*. Springer, New York, 1975, pp. 289-311.

Gentry, AH. Neotropical floristic diversity: phytogeographical connections between Central and South America, Pleistocene climatic fluctuations, or an accident of the Andean orogeny? *Annals of the Missouri Botanical Garden*, 1982, 69, 557-593.

Gentry, AH. Changes in plant community diversity and floristic composition on environmental and geographical gradients. *Annals of the Missouri Botanical Garden*, 1988, 75, 1-34.

Gopal, B; Junk, WJ; Davis, JA. Editors. *Biodiversity in Wetlands: Assessment, Function and Conservation,* Volume I. Backhuys Publ., Leiden, The Netherlands, 2000.

Gouveia, GP; Felfili, JM. Fenologia de comunidades de Cerrado e de mata de galeria no Brasil Central. *Revista Árvore*, 1998, 22, 443-450.

Godoy, JR; Petts, G; Salo J. Riparian flooded forests of the Orinoco and Amazon basins: a comparative review. *Biodiversity and Conservation*, 1999, 8, 551-586.

Guarim-Neto, G; Guarim, VLM; Prance, GT. Structure and floristic composition of the trees of an area of cerrado near Cuiabá, Mato Grosso, Brazil. *Kew Bulletin*, 1994, 49(3), 499-509.

Guarino, ESG; Walter, BMT. Fitosociologia de dois trechos inundáveis de matas de galeria no Distrito Federal, Brasil. *Acta Botânica Brasilica*, 2005, 19(3), 431-442.

Haugaasen, T; Peres, CA. Floristic, edaphic, and structural characteristics of flooded and unflooded forests in the lower Rio Purús region of central Amazonia, Brazil. *Acta Amazônica*, 2006, 36(1), 25-36.

Hoehne, FC. O grande Pantanal de Mato Grosso. *Boletim de Agricultura de São Paulo*, 1936, Série 37a, 443-470.

Huber, O. Guyana highlands versus Guyana lowlands, a reappraisal. *Taxon*, 1988, 37(3), 595-614.

Hueck, K. Sobre a origem dos campos cerrados no Brasil e algumas novas observações no seu limite meridional. *Revista Brasileira de Geografia*, 1957, 19(1), 67-82.

IBGE. *Manual técnico da vegetação brasileira*. Rio de Janeiro, 1992, pp. 92.

Irion, G; Junk, WJ; De Mello, JA. The Large Central Amazonian River Floodplains near Manaus: Geological, climatological, hydrological and geomorphological aspects. In: Junk, WJ editor. *The central Amazon floodplain: Ecology of a pulsating system*. Ecological Studies, Vol. 126, 23-46. Berlin, 1997, pp. 23-46.

Ivanauskas, NM; Rodrigues, RR; Nave, AG. Aspectos ecológicos de um trecho de floresta de brejo em Itatinga, SP: floristica, fitosociologia e seletividade de espécies. *Revista Brasileira de Botânica*, 1997, 20(2), 139-153.

Jackson, MB; Drew, MC. Effects of flooding on growth and metabolism of herbaceous plants. In: Kozlowski, TT. editor. *Flooding and plant growth*. Academic Press, San Francisco, 1984, pp. 47-128.

Jackson MB; Colmer TD. Response and adaptation by plants to flooding stress. *Annals of Botany*, 2005, 96, 501-505.

Joly, CA; Crawford, RMM. Variation in tolerance and metabolic responses to flooding in some tropical trees. *Journal of Experimental Botany*, 1982, 33, 799-809.

Jordan, CF. Soils of the Amazon rainforest. In: Prance, GT; Lovejoy, TE. Editors. *Key environments: Amazonia*. Pergamon Press, Oxford, 1985, pp. 83-105.

Junk, WJ. Flood tolerance and tree distribution in central Amazonian floodplains. In: Holm-Nielsen, LB; Nielsen, IC; Balslev, H. Editors. *Tropical Forests: botanical dynamics, speciation and diversity*. Academic Press, London, 1989, pp. 47-64.

Junk WJ; Bayley PB; Sparks, RE. The flood pulse concept in river-floodplain systems. In: Dodge, D. Editor. Proceedings of the International Large River Symposium. *Canadian Special Publication of Fisheries and Aquatic Science*, 1989, 106, 110-127.

Junk, WJ. Wetlands of tropical South America. In: Wigham, DF; Hejny, S; Dykyjova, D. Editors. *Wetlands of the World IC*. Kluwer, Dordrecht, 1993, pp. 679-739.

Junk, WJ; Piedade, MTF. Herbaceous plants in the floodplain near Manaus: species diversity and adaptations to the flood pulse. *Amazoniana*, 1993, 12, 467-484.

Junk, WJ. Editor. *The Central Amazon floodplain. Ecology of a pulsing system*. Ecological Studies 126, Springer New York, 1997.

Junk, WJ. The central Amazon River floodplain: Concepts for the sustainable use of its resources. In: Junk, WJ; Ohly, JJ; Piedade, MTF; Soares, MGM. Editors. *The Central Amazon Floodplain: Actual Use and Options for a Sustainable Management*. Backhuys Publ. Leiden, 2000, pp. 75-94.

Junk, WJ. Long-term environmental trends and the future of tropical wetlands. *Environmental Conservation*, 2002, 29(4), 414-435.

Junk, WJ; Cunha, CN; Wantzen, KM; Petermann, P; Strüssmann, C; Marques, MI; Adis, J. Biodiversity and its conservation in the Pantanal of Mato Grosso, Brazil. *Aquatic Sciences*, 2006, 68, 278-309.

Junk, WJ; Piedade, MTF; Schöngart, J; Cohn-Haft, M; Adeney, JM; Wittmann, F. (in press) A classification of major Amazonian wetlands. *Brazilian Journal of Botany*.

Kalliola, R; Salo, J; Puhakka, M; Rajasilta, M. New site formation and colonizing vegetation in primary succession on the western Amazon floodplains. *Journal of Ecology* 1991, 79, 877-901.

Kellman, M; Meave, J. Fire in the tropical gallery forests of Belize. *Journal of Biogeography* 1997, 24(1), 23-24.

Klinge, H; Herrera, R. Biomass studies in Amazon caatinga forest in southern Venezuela 1. Standing crop of composite root mass in selected stands. *Journal of Tropical Ecology*, 1978, 19(1), 93-110.

Kobiyama, M. Conceitos de zona ripária e seus aspectos geobiohidrológicos. In: *Anais do I Seminário de Hidrologia Florestal: Zonas ripárias*. Universidade Federal de Santa Catarina, Centro Tecnológico PPGEA Alfredo Wagner-SC, 2003, pp. 1-14.

Kozlowski, TT; Pallardy, SG. Acclimation and adaptive responses of woody plants to environmental stresses. *The Botanical Review*, 2002, 68(2), 270-334.

Kubitzki, K. The ecogeographical differentiation of Amazonian inundation forests. *Plant Systematics and Evolution*, 1989, 163, 285-304.

Larcher, W. *Ökophysiologie der Pflanzen: Leben, Leistung und Streßbewältigung der Pflanzen in ihrer Umwelt*. 6. Aufl. Ulmer Stuttgart, UTB für Wissenschaft, 2001.

Leitão Filho, HF. Aspectos taxonômicos das florestas do estado de São Paulo. *Silvicultura em São Paulo*, 1982, 16a, 197-206.

Lenza, E; Oliveira, PE. Biologia reprodutiva de Tapirira guianensis Aubl. (Anacardiaceae), uma espécie dióica em mata de galeria do Triângulo Mineiro, Brasil. *Revista Brasileira de Botânica*, 2005, 28(1), 179-190.

Lüttge, U. *Physiological ecology of tropical plants*. Springer Verlag Berlin, 1997, pp. 384.

Malick, AI; Colmer, TD; Lambers, H; Setter, TL; Schortemeyer, M. Short-term waterlogging has long-term effects on the growth and physiology of wheat. *New Phytologist*, 2002, 153, 225-236.

Marimon, BS; Lima ES. Caractrerizacão fotofisionomica e levantamento floristico preliminar no Pantanal dos Rios Morte-Araguaia, Cocalinho, Mato Grosso, Brazil. *Acta Botânica Brasilica*, 2001, 15(2), 213-229.

Marimon, BS; Lima, ES; Duarte, TG; Chieregatto, LC; Ratter, JA. Observations on the vegetation of Northeastern Mato Grosso, Brazil. IV. An analysis of the Cerrado-Amazonian forest ecotone. *Edinburgh Journal of Botany* 2006, 62(2/3), 323-341.

Marinho, TAS; Piedade, MTF; Wittmann, F. (in press): Distribution and population structure of four central Amazonian high-várzea timber species. *Wetlands Ecology and Management*.

Marques, MCM; Joly, CA. Germinacão e crescimento de *Calophyllum brasiliense* (Clusiaceae), uma espécie típica de florestas inundadas. *Acta Botânica Brasilica* 2000, 14, 113-120.

Marques, MCM; Silva, SM; Salino, A. Floristica e estrutura do componente arbustivo-arbóreo de uma floresta higrófila da bacia do Rio Jacaré-Pepira, SP-Brasil. *Acta Botânica Brasilica*, 2003, 17, 495-506.

Marques, MCM; Burslem, DFRP; Britez, RM; Silva, SM. Dynamics and diversity of flooded and unflooded forest in a Brazilian Atlantic rainforest: a 16-year study. *Plant Ecology and Diversity*, 2009, 2(1), 57-64.

Martini, AMZ; Fiaschi, P; Amorim, AMA; Paixão, JL. A hot-point within a hot-spot: a high diversity site in Brazilian's Atlantic rainforest. *Biodiversity and Conservation*, 2007, 16, 3111-3128.

Meave, J; Kellman, M; Mac Dougall, D; Rosales, J. Riparian habitats as tropical forest refugia. *Global Ecology and Biogeography Letters*, 1991, 1, 69-76.

Medina, E; García, V; Cuevas, E. Sclerophylly and oligotrophic environments: relationships between leaf structure, mineral nutrient content, and drought resistance in tropical rainforest of the upper Negro region. *Biotropica*, 1990, 22(1), 51-64.

Méio, BB; Freitas, CV; Jatobá, I; Silva, MEF; Ribeiro, JF; Henriques, RPB. Influencia da flora das florestas e Atlântica na vegetação do cerrado *sensu stricto*. *Revista Brasileira de Botânica*, 2003, 26, 437-444.

Meira Neto, JAA; Rego, MM; Coelho, DJS; Ribeiro, FG. Origem, sucessão, e estrutura de uma floresta de galeria periódicamente alagada em Viçosa-MG. *Revista Árvore*, 2003, 27(4), 561-574.

Melack, JM; Hess, LL. Remote sensing of the distribution and extent of wetlands in the Amazon basin. In: In: Junk, WJ; Piedade, MTF; Wittmann, F; Schöngart, J; Parolin, P. Editors. *Ecology and management of Amazonian floodplain forests.* Ecological Series, Springer Verlag Berlin, 2010.

Metzger, JP; Bernacci, LC; Goldenberg, R. Pattern of tree species diversity in riparian forest fragments of different widths. (SE Brazil). *Plant Ecology*, 1997, 133, 135-152.

Ministério dos Transportes. http://www.transportes.gov.br/bit/inhidro.htm. Accessed January 6th.

Moore, N; Arima, E; Walker, R; da Silva, RR. Uncertainty and the changing hydroclimatology of the Amazon. *Geophysical Research Letters*, 2007, 34, L14707.

Mori, SA; Boom, BA; Prance GT. Distribution patterns and conservation of eastern Brazilian coastal forest tree species. *Brittonia*, 1981, 33, 233-245.

Myers, N; Mittermeier, RA; Mittermeier, CG; Fonseca, GAB; Kent, J. Biodiversity hotspots for conservation priorities. *Nature*, 2000, 403, 853-858.

Naranjo, LG. An evaluation of the first inventory of South American wetlands. *Vegetatio*, 1995, 118, 125-129.

Nascimento, NT; Cunha, CN. Estrutura e composição floristica de um cambarazal no Pantanal de Poconé-MT. *Acta Botânica Brasilica*, 1989, 3(1), 3-11.

Nunes, JRS; Favalessa, O; Lula, GAFL; Nunes, PAS; Ferraz, L; Guarim-Neto; Macedo, M. Distribuição de canjiqueira *Byrsonima orbignyana* A. Juss. (Malpighiaceae) em uma área de Pantanal, no município de Santo Antônio de Leverger, Mato Grosso. IV Simpósio sobre recursos naturais e socioeconômicos do Pantanal-SIMPAN, Corumbá, 2004.

Oliveira-Filho, AT; Ratter, JA. A study of the origin of Central Brazilian forests by the analysis of plant species distribution patterns. *Edinburgh Journal of Botany*, 1995, 52, 141-194.

Oliveira-Filho, AT; Almeida, RJ; Mello, JM; Gavilanes, ML. Estrutura fitosociológica e variáveis ambientais em um trecho de mata ciliar dos córregos das Vilas Boas, Reserva Biológica do Poço Bonito, Lavras (MG). *Revista Brasileira de Botânica*, 1994, 17(1), 67-85.

Oliveira-Filho, AT; Fontes, MAL. Patterns of floristic differentiation among Atlantic forests in southeastern Brazil and the influence of climate. *Biotropica*, 2000, 32, 793-810.

Oliveira-Filho, AT; Jarenkow, JÁ; Rodal MJN. Floristic relationships of seasonally dry forests of eastern South America based on tree species distribution patterns. In: Pennington, RT; Lewis, GP; Ratter, JA Editors. Neotropical savannas and seasonally dry forests: Plant diversity, biogeography and conservation. The Systematics Association Special Volume, Series 69. CRC Press, Boca Raton, 2006, pp. 159-192.

Oliveira Wittmann, A. Conteúdo de tococromanóis em espécies arbóreas de várzea da Amazônia Central sob condições controladas. Tese de Doutorado – INPA/UFAM, 2007, Manaus.

Oliveira Wittmann, A; Piedade, MTF; Wittmann, F; Parolin, P. Germination in four low-várzea tree species of Central Amazonia. *Aquatic Botany*, 2007, 86(3), 197-203.

Oliveira, VC; Joly, CA. Flooding tolerance of *Calophyllum brasiliense* (Camb.) Clusiaceae: morphological, physiological, and growth responses. Trees, 2010, 24, 185-193.

Olson MS; Platt, W. Effects of habitat and growing season fires on resprouting of shrubs in longleaf pine savannas. *Vegetatio*, 1995, 119, 101-118.

Otte, ML. What is stress to a wetland plant? *Environmental and Experimental Botany*, 2001, 46, 195-202.

Pagano, SN; Leitão Filho, HF. Composição florística do estrato arbóreo de mata mesófila semidecídua, no município de Rio Claro (Estado de São Paulo). *Revista Brasileira de Botânica*, 1987, 10, 37-47.

Parolin, P. Morphological and physiological adjustments to waterlogging and drought in seedlings of Amazonian floodplain trees. *Oecologia*, 2001, 128, 326-335.

Parolin, P. Submergence tolerance versus escape from submergence: two strategies of seedling estabnlishment in Amazonian floodplains. *Environmental and Experimental Botany*, 2002, 48, 177-186.

Parolin, P; Oliveira, AC; Piedade, MTF; Wittmann, F; Junk, WJ. Pioneer trees in Amazonian floodplains: Key species form monospecific stands in different habitats. *Folia Geobotanica*, 2002, 37, 225-238.

Parolin, P. Submerged in darkness: Adaptations to prolonged submergence by woody species of the Amazonian floodplains. *Annals of Botany*, 2009, 103, 359-376.

Parolin, P; Lucas, C; Piedade, MTF; Wittmann, F. Drought responses of flood-tolerant trees in Amazonian floodplains. *Annals of Botany*, 2010, 105(1), 129-139.

Parolin, P; Wittmann, F. Struggle in the flood – Tree responses to flooding stress in four tropical floodplain systems. *Annals of Botany Plants*, 2010, doi: 10.1093/aobpla/plq003

Peixoto, JMA; Nelson, BW; Wittmann, F. Spatial and temporal dynamics of alluvial geomorphology and vegetation in central Amazonian white-water floodplains by remote-sensing techniques. *Remote Sensing of Environment*, 2009, 113(10), 2258-2266.

Pélissier, R; Dray, S; Sabatier, D. Within-plot relationships between tree species occurrences and hydrological soil constraints: an example in French Guiana investigated through canonical correlation analysis. *Plant Ecology*, 2002, 162: 143-156.

Pennington, RT; Lewis, GP; Ratter, JA. An overview of the plant diversity, biogeography, and conservation of Neotropical savannas and seasonally dry forests. In: Pennington, RT; Lewis, GP; Ratter, JA. Editors. *Neotropical savannas and seasonally dry forests: Plant diversity, biogeography and conservation.* The Systematics Association Special Volume, Series 69. CRC Press, Boca Raton, 2006, pp. 1-29.

Phillips, N; Bond, BJ; Ryan, MG. Gas exchange and hydraulic properties in the crowns of two tree species in a Panamanian moist forest. *Trees – Structure and Function*, 2001, 15, 123-130.

Piedade, MTF; Junk, WJ; Long, SP. The productivity of the C_4 grass *Echinochloa polystachia* on the Amazon Floodplain. *Ecology*, 1991, 72(4), 1456-1463.

Pither, R; Kellman, M. Tree species diversity in small tropical riparian forest fragments in Belize, Central America. *Biodiversity and Conservation*, 2002, 11(9), 1623-1636.

Ponce, VM; Cunha, CN. Vegetated earthmounds in tropical savannas of central Brazil: a synthesis. *Journal of Biogeography*, 1993, 20, 219-225.

Pott, A. *Pastagens no Pantanal*. EMBRAPA CPAP Documentos, 1988, 7, 1-58.

Pott, A; Pott VJ. *Plantas do Pantanal*. Empresa Brasileira de Pesquisa Agropecuária (EMBRAPA), Brasilia, 1994. 320 pp.

Pott, A; Pott VJ. *Vegetação do Pantanal: fitogeografia e dinâmica*. Anais do 2. Simpósio de Geotecnologias no Pantanal, Corumbá. EMBRAPA Informática Agropecuária/INPE, 2009, pp. 1065-1076.

Prado, DE. Seasonally dry forests of tropical South America: from forgotten ecosystems to a new phytogeographical unit. *Edinburgh Journal of Botany*, 2000, 57, 437-461.

Prance, GT; Schubart, FOR. Nota preliminar sobre a origem das campinas abertas de areia branca do baixo Rio Negro. *Acta Amazônica*, 1978, 7(4), 567-569.

Prance, GT. Notes on the vegetation of Amazonia III. The terminology of Amazonian forest types subject to inundation. *Brittonia*, 1979, 3(1), 26-38.

Prance, GT; Schaller GB. Preliminary study of some vegetation types of the Pantanal, Mato Grosso, Brazil. *Brittonia*, 1982, 32, 228-251.

Prance, GT; Daly, D. Brazilian Amazon. In: Campbell, DG; Hammond, HD. Editors. *Floristic inventory of tropical countries*. New York Botanical Garden Press, 1989, pp. 523-533.

Prance, GT. Islands in Amazonia. *Phil Trans. of the Royal Society of London B Biol. Sci.*, 1996, 351, 823-833.

Ramberg, L; Hancock, P; Lindholm, M; Meyer, T; Ringrose, S; Sliva, J; As, JV; VanderPost, JC. Species diversity of the Okavango Delta, Botswana. *Aquatic Sciences*, 2006, 68, 310-337.

Ratter, JA; Pott, A; Pott, VJ; Cunha, CN; Haridasan, M. Observations on woody vegetation types in the Pantanal and at Corumbá. Brazil. *Notes of the Royal Botanical Garden of Edinburgh*, 1988, 45, 503-525.

Ratter, JA; Bridgewater, S; Ribeiro, JF. Analysis of the floristic composition of the Brazilian Cerrado vegetation III: Comparison of the woody vegetation of 376 areas. *Edinburgh Journal of Botany*, 2003, 60(1), 57-109.

Renöfält, BM; Nilsson, C; Jansson, R. Spatial and temporal patterns of species richness in a riparian landscape. *Journal of Biogeography*, 2005, 32, 2025-2037.

Ribeiro, MNG; Adis, J. Local rainfall variability - a potential bias for bioecological studies in the Central Amazon. *Acta Amazônica*, 1984, 14(1/2), 159-174.

Ribeiro, JE; Hopkins, M; Vicentini, A; Sothers, CA; Costa, MA; Brito, JM; Souza, MA; Martins, LH; Lohmann, LG; Assunção, PA; Pereira, E; Silva, CF; Mesquita, MR; Procópio, LC. *Flora da Reserva Ducke, Guia de identificação das plantas vasculares de uma floresta de terra firme na Amazônia Central*. INPA-DFID, Manaus, 1999.

Ribeiro, MC; Metzger, JP; Martensen, AC; Ponzoni, F; Hirota, MM. Brazilian Atlantic forest: How much is left and how is the remaining forest distributed? Implications for conservation. *Biological Conservation* 2009, 142, 1141-1153.

Rizzini, CT. *Tratado de fitogeografia do Brasil. Aspectos ecológicos, sociológicos e florísticos*. Âmbito Cultural Ed. Rio de Janeiro, 1997, pp. 747.

Robertson, KM. Distributions of tree species along point bars of 10 rivers in the south-eastern US coastal plain. *Journal of Biogeography*, 2006, 33, 121-132.

Rocha, CTV; Carvalho, DA; Fontes, MAL; Oliveira-Filho, AT; Van den Berg, E; Marques, JJ. Comunidade arbórea de um continuum entre floresta paludosa e de encosta em Coqueiral, Minas gerais, Brasil. *Revista Brasileira de Botânica*, 2005, 28(2), 203-218.

Rodrigues, RR; Morellato, LPC; Joly, CA; Leitão Filho, HF. Estudo florístico e fitossociológico em um gradiente altitudinal de mata estacional mesófila semidecídua na Serra do Japi, Jundiaí, SP. *Revista Brasileira de Botânica*, 1989, 12, 71-84.

Rodrigues, RR; Nave, AG. Heterogeneidade floristica das matas ciliares. In: Rodrigues, RR; Leitão-Filho, HF. Editors. *Matas ciliares: conservação e recuperação*. Edusp/Fapesp São Paulo, 2000, pp. 45-71.

Rolim, SG; Chiarello, AG. Slow death of Atlantic forest trees in cocoa agroforestry in Southeastern Brazil. *Biodiversity and Conservation*, 2004, 13, 2679-2694.

Rolim, SG; Ivanauskas, NM; Rodrigues, RR; Nascimento, MT; Gomes, JML; Folli, DA; Couto, HDZ. Composicão floristica do estrato arbóreo da floresta estacional semidecidual na planície alluvial do Rio Doce, Linhares, ES, Brasil. *Acta Botânica Brasilica*, 2006, 20(3), 549-561.

Rosales, J; Petts, G; Knab-Vispo, C. Ecological gradients within the riparian forest of the lower Caura River, Venezuela. *Plant Ecology*, 2001, 152, 101-118.

Safford, HD. Brazilian páramos IV. Phytogepgraphy of the campos de altitude. *Journal of Biogeography*, 2007, 34, 1701-1722.

Salo, J; Kalliola, R; Häkkinen, I; Mäkinen, Y; Niemelä, P; Puhakka, M; Coley, PD. River dynamics and the diversity of the Amazon lowland forest. *Nature*, 1986, 322, 254-258.

Sanchez, M; Pedroni, F; Leitão-Filho, HF; Cesar, O. Composição florística de um trecho de floresta ripária na Mata Atlântica em Picinguaba, Ubatuba, SP. *Revista Brasileira de Botânica*, 1999, 22(1), 31-42.

Sarmiento G; Monasterio M. A critical consideration of the environmental conditions associated with the occurrence of savanna ecosystems in tropical America. Tropical Ecological Systems, Springer Berlin, 1975, pp. 223-250.

Scarano, FR; Ribeiro, KT; Moraes, LFD; Lima, HC. Plant establishment on flooded and unflooded patches of a freshwater swamp forest in southeastern Brazil. *Journal of Tropical Ecology*, 1997, 14, 793-803.

Scarano, FR. Structure, function, and floristic relationships of plant communities in stressful habitats marginal to Brazilian Atlantic forest. *Annals of Botany*, 2002, 90, 517-524.

Scarano, FR. Plant communities at the periphery of the Atlantic rainforest: Rare-species and its risks for conservation. *Biological Conservation*, 2009, 142, 1201-1208.

Schlüter, UB; Furch, B; Joly, CA. Physiological and anatomical adaptations by young *Astrocaryum jauari* Mart. (Arecaceae) in periodically inundated biotopes of Central Amazonia. *Biotropica*, 1993, 25, 384-396.

Schnitzler, A; Hale, BW; Alsum, E. Biodiversity of floodplain forests in Europe and esatern North America: A comparative study of the Rhine and Mississippi valleys. *Biodiversity and Conservation*, 2005, 14, 97-117.

Schöngart, J; Piedade, MFT; Ludwigshausen, S; Horna, V; Worbes, M. Phenology and stem-growth periodicity of tree species in Amazonian floodplain forests. *Journal of Tropical Ecology*, 2002, 18, 581-597.

Schöngart, J; Wittmann, F; Piedade, MTF; Junk, WJ; Worbes, M. Wood growth patterns of *Macrolobium acaciifolium* (Benth.) Benth. (Fabaceae) in Amazonian black-water and white-water floodplain forests. *Oecologia*, 2005, 145, 454-461.

Schueler, TR; Holland, HK. Practice of watershed protection. Wetter is not always better: flood tolerance of woody species. *Technical Note 52 from Watershed Protection Techniques*, 2000, 1, 208-210. Ellicott City, MD.

Silva, MP; Mauro, R; Mourão, G, Coutinho, M. Distribution and quantification of vegetation classes by aerial survey in the Brazilian Pantanal. *Revista Brasileira de Botânica*, 2000, 23(2), 143-152.

Silva Júnior, MC. Fitosociologia e estrutura diametrica da mata de galeria do Taquara, na Reserva Ecológica do IBGE, DF. *Revista Árvore*, 2004, 28(3), 419-428.

Silveira, M. *Vegetação e flora das campinaranas do sudoeste Amazônico*. Associação S.O.S. Amazônia, ARPA-IBAMA, 2003, Rio Branco.

Sioli, H. Beiträge zur regionalen Limnologie des Amazonasgebietes. *Archiv für Hydrobiologie*, 1954, 45, 267-283.

Sioli, H. The Amazon and its main affluents: hydrography, morphology of the river courses and river types. In: Sioli, H. Editor. *The Amazon. Limnology and landscape ecology of a mighty tropical river and its basin*. Dordrecht, 1984, 127-165.

Sombroek, W. Spatial and temporal patterns of Amazonian rainfall. Consequences for the planning of agricultural occupation and the protection of primary forests. *Ambio*, 2001, 30(7), 388-396.

Souza, JS; Espírito Santo, FDB; Fontes, MAL; Oliveira-Filho, AT; Botezelli, L. Análise das variações floristicas e estruturais da comunidade arbórea de um fragmento de floresta semidecídua as margens do Rio Capivarí, Lavras-MG. *Revista Árvore*, 2003, 27(2), 185-206.

Souza, MC; Romagnolo, MB; Kita, KK. Riparian vegetation: Ecotones and plant communities. In: Thomaz, SM; Agostinho, AA; Hahn, NS. Editors. *The upper Paraná River and its floodplain: physical aspects, ecology and conservation*. Bachhuys Publishers, Leiden, 2004, pp. 353-367.

Steyermark, JA. Speciation and endemism in the flora of the Venezuelan tepuis. In: Vuilleumier, F; Monasterio, M. Editors. *High altitude tropical biogeography*. Oxford University Press, 1986, pp. 317-373.

Sugiyama, M. Estude de florestas da restinga da Ilha do Cardoso, Cananéia, São Paulo, Brasil. *Boletim do Instituto de Botânica*, 1998, 11, 119-159.

Szutman, M; Rodrigues, RR. O mosaico vegetacional numa área de floresta contínua da planície litorânea, Parque Estadual da Campina do Encantando, Pariquera-Açu, SP. *Revista Brasileira de Botânica*, 2002, 25(2), 161-176.

Teixeira, AP; Assis, MA. Caracterização floristica e fitosociólogica do componente arbustivo-arbóreo de uma floresta paludosa no município de Rio Claro (SP), Brasil. *Revista Brasileira de Botânica*, 2005, 28(3), 467-476.

Teixeira, AP; Rodrigues, RR. Análise floristica e estrutural do componente arbustivo-arbóreo de uma floresta de galeria no município de Cristais Paulista, SP, Brasil. *Acta Botânica Brasilica*, 2006, 20(4), 803-813.

Thomas, WW; Carvalho, AMV; Amorim, AMA; Garrison, J; Arbeláez, AL. Plant endemism in two forests in southern Bahia, Brazil. *Biodiversity and Conservation*, 1998, 7, 311-322.

Toniato, MTZ; Leitão Filho, HF; Rodrigues, RR. Fitosociologia de um remanescente de floresta higrófila (mata de brejo) em Campinas, SP. *Revista Brasileira de Botânica*, 1998, 21(2), 197-210.

Torres, RB; Matthes, LAF; Rodrigues, RR. Floristica e estrutura do componente arbóreo de mata de brejo em Campinas, SP. *Revista Brasileira de Botânica*, 1994, 17, 189-194.

Ussami, N; Shiraiwa, S; Dominguez, JML. Basement reactivation in a Sub-Andean foreland flexural bulge: The Pantanal wetland, SW Brazil. *Tectonics,* 1999, 18, 25-39.

Van den Berg, E; Oliveira-Filho, AT. Composicão floristica e estrutura fitosociológica de uma floresta ripária em Itutinga, MG, e comparação com outras áreas. *Revista Brasileira de Botânica*, 2000, 23(3), 231-253.

Van Splunder, I; Coops, H; Voesenek, LACJ; Blom, CWPM. Establishment of alluvial forest species in floodplains: the role of dispersal timing, germination characteristics and water level fluctuations. *Acta Botanica Neerlandica*, 1995, 44(3), 269-278.

Veneklaas EJ; Fajardo, A; Obregon, S; Lozano, J. Gallery forest types and their environmental correlates in a Colombian Savanna landscape. *Ecography*, 2005, 28, 236-252.

Vicentini, A. A vegetacão ao longo de um gradient edáfico no Parque Nacional do Jaú. In: Borges, SH; Iwanaga S; Durigan, CC; Pinheiro, RR. Editors. *Janelas para a biodiversidade no Parque Nacional do Jaú: uma estratégia para o estudo da biodiversidade na Amazônia.* Fundacão Vitória Amazonica, WWF, IBAMA, Manaus, 2004, pp. 117-143.

Vilela, EA; Oliveira-Filho, AT; Carvalho, DA. Fitosociologia de floresta ripária do baixo Rio Grande, Conquista-MG. *Revista Árvore*, 1999, 23(4), 423-433.

Visser, EJW; Voesenek, LACJ; Vatapetian, BB; Jackson, MB. Flooding and plant growth. *Annals of Botany*, 2003, 91, 107-109.

Voesenek, LACJ; Benschop, JJ; Bou, J; Cox, MCH; Groeneveld, HW; Millenaar FF; Vreeburg, RAM; Peeters, AJM. Interactions between plant hormones regulate submergence-induced shoot elongation in the flooding-tolerant dicot *Rumex palustris*. *Annals of Botany*, 2003, 91, 205-211.

Wittmann F; Anhuf D; Junk WJ. Tree species distribution and community structure of Central Amazonian várzea forests by remote sensing techniques. *Journal of Tropical Ecology*, 2002, 18, 805-820.

Wittmann F; Junk WJ. Sapling communities in Amazonian white-water forests. *Journal of Biogeography*, 2003, 30(10), 1533-1544.

Wittmann, F; Junk, WJ; Piedade, MTF. The várzea forests in Amazonia: flooding and the highly dynamic geomorphology interact with natural forest succession. *Forest Ecology and Management,* 2004, 196, 199-212.

Wittmann, F; Parolin, P. Aboveground roots in Amazonian floodplain trees. *Biotropica*, 2005, 37(4), 609-619.

Wittmann, F; Schöngart, J; Montero, JC; Motzer T; Junk, WJ; Piedade, MTF; Queiroz, HL; Worbes, M. Tree species composition and diversity gradients in white-water forests across the Amazon basin. *Journal of Biogeography*, 2006, 33, 1334-1347.

Wittmann, F; Zorzi, BT; Tizianel, FAT; Urquiza, MVS; Faria, RR; Sousa, NM; Módena, ES; Gamarra, RM; Rosa, ALM. Tree species composition, structure, and aboveground wood biomass of a riparian forest of the lower Miranda River, Southern Pantanal, Brazil. *Folia Geobotanica,* 2008, 43, 397-411.

Wittmann, F; Schöngart, J; Junk, WJ. (2010): Phytogeography, species diversity, community structure and dynamics of Amazonian várzea forests. In: Junk, WJ; Piedade, MTF; Wittmann, F; Schöngart, J; Parolin, P. Editors. *Ecology and management of Amazonian floodplain forests*. Ecological Series, Springer Verlag, Berlin.

Worbes, M. Lebensbedingungen und Holzwachstum in zentral-amazonischen Überschwemmungswäldern. *Scripta Geobotanica,* 1986, 17, Göttingen.

Worbes, M; Klinge, H; Revilla, JD; Martius, C. On the dynamics, floristic subdivision and geographical distribution of várzea forests in Central Amazonia. *Journal of Vegetation Science*, 1992, 3, 553-564.

Worbes M. The forest ecosystem of the floodplains. In: Junk WJ. Editor. *The Central Amazon floodplains. Ecology of a pulsing system.* Springer Verlag Berlin, Heidelberg, New York, 1997, pp. 223-266.

Zeilhofer, P; Schessl, M. Observations on inundation dynamics and soil properties of selected vegetation types in the Pantanal of Poconé. II. SHIFT Workshop, Cuiabá, 1995, pp. 101.

Zeilhofer, P; Schessl, M. Relationship between vegetation and environmental conditions in the northern Pantanal of Mato Grosso, Brazil. *Journal of Biogeography*, 1999, 27, 159-168.

In: Mycorrhiza: Occurrence in Natural and Restored...
Editor: Marcela Pagano

ISBN: 978-1-61209-226-3
© 2012 Nova Science Publishers, Inc.

Chapter 12

ARBUSCULAR MYCORRHIZAE IN AQUATIC PLANTS, INDIA

K. P. Radhika, James D'Souza and B. F. Rodrigues
Department of Botany, Goa University, Goa 403206, India

ABSTRACT

Plant root fungal interactions called mycorrhizae are found in approximately 90% of all vascular plants. Hydrophytes were regarded as nonmycorrhizal until a decade ago as arbuscular mycorrhizal (AM) fungi require oxygen to thrive and it had been assumed AM fungi have little significance in wetland ecosystems. In India, studies are scarce on mycorrhizal association in wetland habitats viz., aquatic and marshy lands, khazan lands, and mangroves ecosystem. The first report of mycorrhizal colonization in aquatic plants was recorded in modified leaves of *Salvinia cucullata.* Recent study indicated that 16 of 20 plant species from aquatic and marshy habitats in Goa exhibited AM fungal root colonization. A study on occurrence of AM fungal association in five subtropical ponds of marshy habitats from Shillong revealed that nonmycorrhizal plants exhibited poor plant growth and biomass. The presence of AM fungal association in 24 plant species of sedges (representing six genera) was confirmed from different vegetation types of Western Ghats. In Goa, occurrence of AM fungal association in 28 plant species belonging to 16 families from khazan land ecosystem indicated adaptability to salt stress conditions in plant growth and survival. Mycorrhizal association in mangroves ecosystem in the Ganges river estuary indicated tolerance of mycorrhizae to various types of physical and chemical stresses in soil. This paper reviews the work carried out on diversity and role of AM fungi in wetland ecosystems from India.

INTRODUCTION

Aquatic plants also called hydrophytic plants or hydrophytes are those which have adapted to living in or on freshwater and saltwater environments. Mycorrhizae form a mutualistic relationship with the roots of most plant species and play an important role in terrestrial ecosystems, where they influence plant community structure and nutrient cycling

(Jackson and Mason 1984). Since AM fungi require oxygen to thrive and as many wetland plants have been described as nonmycorrhizal (Mosse et al. 1981, Anderson et al. 1984, Mejstrik 1984) it has been assumed that AM fungi have little significance in wetland ecosystems. However, recent field studies show that AM fungi exist in wetlands and colonize many hydrophytic plants (Brown and Bledsoe 1996, Cooke and Lefor 1998) *viz.*, roots of submerged macrophytes (Clayton and Bagyaraj 1984, Tanner and Clayton 1985), salt marsh plants (Rozema et al. 1986, Van Duin et al. 1990), plants in oligotropic wetlands (Sondegaard and Laegaard 1977), wetland woody species (Keeley 1980, Lodge 1989), plants in prairie potholes (Wetzel and van der Valk 1996) wetland plants in the Everglades (Azi et al. 1995) and plants in recently rehabilitated wetlands (Turner and Friese 1998). Nine of 49 species examined in aquatic habitats of Denmark were found to be mycorrhizal (Beck-Nielsen and Vindaek, 2001). One of the main functions and ecological roles of AM fungi is enhanced phosphorus nutrition to plants; thus the effect of soil available phosphorus is often assessed in wetland ecosystems (Smith and Read 1997). Reduced plant performance in the wetland ecosystems is generally due to salinity and inundation stress and limitation of nutrients (Valiela and Teal 1979). Plants occurring in wetlands tend to acquire mycorrhizal colonization under dry conditions (Newman and Koske, personal communication), and once the plants are colonized, soil moisture has a negligible effect on the intensity of colonization.

One of the recently accepted theories of arbuscular mycorrhizae suggest that these fungi were influential in colonization of land by aquatic plants, based on fossil evidence from Rhynie chert (Smith and Read 1997). In India, studies are scarce in wetland ecosystems. This review paper discusses the role of mycorrhizal association in different wetland ecosystems of India.

AM FUNGAL ASSOCIATION IN AQUATIC PLANTS

There are only a few reports on the occurrence of AM in natural aquatic conditions. Bagyaraj et al. (1979) first reported mycorrhizal association in modified leaves of *Salvinia cucullata*. Vesicular colonization and characteristic oil droplets were observed but arbuscules were absent. Chaubal et al. (1982) reported the occurrence of mycorrhizal association in five subtropical ponds viz., eutrophic (P1, P2 and P3), running water (P4), oligotrophic lake (P5) and marshy plant community (M). They observed plants growing in one eutrophic (P1), the oligotrophic lake and marshy habitats exhibited vesicular association whereas absence of mycorrhizal association was observed in the other two eutrophic habitats (P2 and P3) and running water. Ragupathy et al. (1990) studied AM fungal association in 22 aquatic and semi aquatic plants of Thanjavur district and recorded mycorrhizal colonization in 10 plant species where percent colonization, size and types of AM fungal spores differed.

Sedges are among the primary colonizers of disturbed areas like mine spoil, volcanic substrates and fallow land which are devoid of mycorrhizal association. Bagyaraj et al. (1979) reported that Cyperaceae members are nonmycorrhizal. Later Muthukumar et al. (1996) reported mycorrhizal association in 24 species of sedges (representing six genera) from different vegetation types. All the sedges recorded mycorrhizal colonization with hyphae and vesicles, but arbuscules were observed only in 42% of the total species. The AM fungal colonization varied considerably between species, ranging from 9 (*Cyperus brevifolius*) to

62% (*Scleria lithosperma*). The number of AM fungal spores in the rhizosphere varied from 5 to 86 g^{-1} soil. *Glomus* spp. (4), *Acaulospora* sp. (1), *Sclerocystis* spp. (1) and *Gigaspora* sp. (1) were identified among the AM fungal spores. Sporocarps of *Sclerocystis sinuosa* and spores of *Glomus* spp. were most abundant in the rhizosphere soils at different sites, including *G. aggregatum, G. geosporum, G. microaggregatum* and *G. mosseae*. They observed that AM colonization in sedges was positively correlated to root diameter and negatively correlated to the root-hair length. Soil edaphic factors had no significant effect on either root colonization or spore density. This is the first report on 22 species of sedges from India other than *Cyperus rotundus* and *C. kyllingia* reported earlier by other workers (Saif et al. 1977, Nadarajah and Nawawi 1988, Koske et al. 1992, Louis 1990). Their study also reported that the low incidence of AM association in the Cyperaceae was based mostly on temperate studies, which may differ substantially from findings in tropical and subtropical conditions and could also be attributed to the habitats in which they occur with few mycorrhizal spores or habitats not conducive for AM formation.

Radhika and Rodrigues (2007) reported AM fungal association in 10 aquatic plant species and 6 marshy plant species from Goa. Vesicular colonization was the most prevalent type and was recorded in 12 plant species (Table 1). In rooted plants with floating leaves, *Marsilea quadrifolia* showed the presence of vesicular colonization whereas no AM fungal root colonization was observed in *Nymphaea stellata*. The two submerged floating plants viz., *Najas minor* and *Ceraptoteris thalictriodes* did not show any AM fungal root colonization. Rooted submerged plants viz., *Blyxa echinosperma, Isoetes coromandelina* (pteridophyte), *Eriacaulon cinereum, Rotala malampuzhensis* and *Rotala densiflora* recorded the presence of AM fungal colonization. The rooted emergent plants *viz.*, *Lymnophila indica* and *Monocoria vaginalis* showed vesicular colonization in the roots. Out of a total of five marshy plant species, five species *viz.*, *Drosera indica, Ludwigia parviflora, Murdania semeteris, Centella asiatica* and *Lindernia ciliata* showed the presence of AM fungal colonization. Except for *C. asiatica*, the remaining four marshy species exhibited vesicular colonization. Arbuscules were found only in three plant species.

The occurrence of mycorrhiza in aquatic conditions indicates its role in nutrient absorption especially in plants growing in nutrionally poor sediments. The association between AM fungi and aquatic plants may mediate co-existence of aquatic plant species and maintain the balance of the hydrophytes community as in terrestrial ecosystems. The primary abiotic factors known to influence the abundance and distribution of AM fungi in aquatic ecosystems are water, nutrient and oxygen availability. Although the roles of AM fungi are still not fully understood in aquatic and marshy environments, the results from this study imply that AM associations are functional in, and are a significant component of aquatic plant communities, mainly in plant competition, succession and diversity, in fens and marshes (Facelli and Facelli 2002). Because of these ecological implications, AM associations and the seasonal dynamics of AM colonization should be considered for the ecological restoration of functional wetlands and should be a significant component of studies assessing aquatic ecosystem dynamics.

Table 1. Arbuscular mycorrhizal fungal colonization and AM fungal species identified in aquatic and marshy plants (Radhika and Rodrigues 2007)

Sr. No.	Plant species	Family	Locality	AM fungal colonization	AM fungi identified
I.	Rooted submerged plants				
	1. *Blyxa echniosperma* (Clarke) Hook. f	Hydrocha-ritaceae	Taleigao	H V	*Glomus claroideum,*
	2. *Eriacualon cinereum* R. Br	Eriocaulaceae	Taleigao	H V	*Scutellospora* sp. *Glomus*
	3. *Isoetes coromandelina* L.	Isoetaceae	Taleigao	H V	*claroideum*
	4. *Rotala malampuzhensis* Nair ex cooke.	Lythraceae	Taleigao	H V	*Glomus claroideum*
	5. *Rotala densiflora* (Roth) Koehne	Lythraceae	Taleigao	H A	*Glomus claroideum*
	6. *Scripus lateriflorus* Gmel.	Cyperaceae	Taleigao	_	*Glomus claroideum Glomus claroideum*
II.	Rooted emergent plants				
	7. *Lymnophila indica* (L.) Druce	Scrophula-riaceae	Santa cruz	H V	*Glomus claroideum*
	8. *Monochoria vaginalis* (Burm. f.) Persi ex Kunth	Pontederiaceae	Taleigao	H V	*Glomus claroideum*
III.	Rooted plants with floating leaves				
	9. *Marsilea quadrifolia* L.	Marsilaceae	Santa cruz	H V	*Glomus* sp.
	10. *Nymphae stellata* Willd.	Nymphaceae	Taleigao	_	_
IV.	Submerged floating plants				
	11. *Najas minor* (Pers.) All Fl.	Najadaceae	Santacruz	_	_
	12. *Ceraptoteris thalictriodes* L.	Ceratophy-llaceae	Taleigao	_	_
V.	Free floating plants				
	13. *Pistia stratiotes* L.	Araceae	Vasco	H V	*Glomus claroideum, Scutellospora verrucosa*
	14. *Salvinia natans* Allioni.	Lamiaceae	Santa cruz	H V	*Glomus claroideum*
VI.	Marshy plants				
	15. *Drosera indica* L.	Droseraceae	Taleigao	H V	*Glomus claroideum*
	16. *Lindernai ciliata* Cols (Penn).	Veronicaceae	Taleigao	H V	*Glomus claroideum*
	17. *Ludwigia parviflora* Roxb.	Onagraceae	Taleigao	H V	*Glomus claroideum*
	18. *Murdannia semeteres* Dalz.	Commeli-naceae Lentibu-lariaceae	Taleigao	H V	*Glomus claroideum*
	19. *Utricularia reticulate* Smith.	Apiaceae	Taleigao	-	*Glomus claroideum*
	20. *Centella asiatica* (L.) Urb.		Taleigao	H A	*Glomus claroideum*

*V, Vesicles; A, Arbuscules; H, Hyphae; (-), absent.

AM Fungal Association in Khazan Lands

'Khazan Lands' are partially modified mangrove and coastal wetland ecosystems of the West Coast of India. As Khazan Lands are sites of extreme conditions (e.g., salinity, flooding, physiological drought, siltation), it affords the opportunity to study the interactions between plants and microbes, particularly AM fungi associated with pioneering vegetation. In this unique stressed environment, there appears to be considerable reduction in AM species diversity. Rodrigues and Anuradha (2009) studied roots of 28 plant taxa belonging to 16 families from Goa region for mycorrhizal association. The colonization ranged from 23 to 81%. A total of 17 AM fungi belonging to three genera (*Acaulospora, Glomus* and *Scutellospora*) were recovered. Among them, *Acaulospora* and *Glomus* with a total of eight species each dominated the Khazan Lands, *Glomus claroideum* the most abundant followed by *Glomus fasciculatum* and *Acaulospora delicata* (Table 2 and 3). The study reveals the role of AM fungi in plant growth promotion and survival under adverse conditions and also forms a baseline for future investigations on the ecology and application of AM fungal species for rehabilitation, reclamation and to improve productivity of Khazan Land and associated biomes.

Arbuscular Mycorrhizal (AM) Fungal Association in Mangroves

Mangroves, the climax formation of hydrohalophytes belonging to several plant families, inhabit tropical and sub-tropical estuaries and marine salt marshes. Mangrove forests are considered as open 'interface' ecosystems connecting upland terrestrial and coastal estuarine ecosystems (Lugo and Snedaker 1974).

AM fungi have occasionally been reported absent in marine salt marsh plants (Mohankumar and Mahadevan 1986). Sengupta and Chaudhuri (1990) reported the occurrence of AM in four species of pioneer salt marsh plants. Sengupta and Chaudhari (1990) studied 53 plant species belonging to 25 families found in three successional stages as true mangroves, mangrove associates and non-littoral species. Mycorrhizal association was observed in 24 mangrove species, 10 mangrove associates and 18 non littoral species. Nine AM fungal species were isolated and identified from the rhizosphere soil.

The presence of seemingly mycorrhizal, dark septate mycelial endophytes in mangrove roots was also recorded (Sengupta and Chaudhuri 1994). Hence the knowledge regarding ecosystem adaptations of mangroves may help in conservation of wetland ecosystems. Histological and plant growth response data have shown that besides AM association some endophytes also have mycorrhizal functions for stress adapted plants of this ecosystems.

Sengupta and Chaudhari (2002) examined 31 species of mangrove and mangrove associates and 23 species of transported flora, belonging to 25 families at four physiographic stages of succession of the mangrove plant community at the terminal part of the Ganges river estuary in India, for AM root association. All the 31 mangrove and associated species belonging to 18 plant families showed the presence of root endophytes.

Table 2. Arbuscular mycorrhizal colonization in roots and spore density in rhizosphere of plant species of Khazan Land (Rodrigues and Anuradha 2009)

Plant species	Family	Colo-nization*			Coloni-zation (%)	Spore density**
		V	A	H		
Acanthus ilicifolius L.	Acanthaceae	-	-	-	00	123
Acrosticum aureum L.	Pteridaceae	-	-	-	00	127
Ammannia multiflora Roxb.	Lythraceae	+	-	+	54	310
Avicennia officinalis L.	Verbenaceae	-	-	-	00	66
Ceratopteris thalictroides (L.) Brongn.	Parkeriaceae	+	-	+	80	370
Cyprus cyperoides (L.) O. Kuntze.	Cyperaceae	+	-	+	39	330
Cyperus rotundus L.	Cyperaceae	+	-	+	33	78
Derris uliginosa Benth.	Papilionaceae	-	-	-	00	156
Echinochloa colonum (L.) Link.	Poaceae	-	-	-	00	64
Eclipta alba (L.) Hassk.	Asteraceae	-	-	-	00	88
Eleocharis geniculata (L.) Roem & Schult.	Cyperaceae	-	-	-	00	96
Excoecaria agallocha L.	Euphorbiaceae	+	-	+	36	100
Fimbristylis aestivalis (Retz.) Vahl.	Cyperaceae	-	-	-	00	87
Fuirena ciliaris (L.) Roxb.	Cyperaceae	+	-	+	37	104
Ipomoea muricata Jacq.	Convolvulaceae	+	-	+	35	98
Isachne globosa (Thunb.) O. Kuntze.	Poaceae	+	-	+	27	210
Ludwigia perennis L.	Onagraceae	+	-	+	45	63
Nymphaea nouchali Burm.	Nymphaeaceae	-	-	-	00	61
Oryza sativa L. var. Jyoti	Poaceae	+	-	+	32	134
Oryza sativa L. var. Asgo	Poaceae	+	+	+	57	110
Oryza sativa L. var. Korgut	Poaceae	+	-	+	42	116
Paspalum scrobiculatum L.	Poaceae	+	-	-	23	53
Phyllanthus niruri L.	Euphorbiaceae	-	-	-	00	78
Salvinia molesta Mitch.	Salviniaceae	-	-	-	00	23
Scirpus lateriflorus Gmel	Cyperaceae	+	-	+	34	60
Sesuvium portulacastrum L.	Aizoaceae	+	+	+	76	93
Sphaeranthus africanus L.	Asteraceae	+	+	+	81	266
Sphenoclea zeylanica Goert. Frud.	Companulaceae	-	-	-	00	59

* V, Vesicles; A, Arbuscules; H, Hyphae.
** Spores/100 g rhizosphere soil (mean, n=3).

The non-littoral species, belonging to 13 families, including some commonly reported non-mycotrophic plant families but excluding those belonging to the Cyperaceae, also showed the presence of AM endophytes in their roots in their respective physiographic situations. Colonization of most of the plant species structurally resembled 'Paris' type AM.

Intracellular hyphae, hyphal coils and intracellular vesicles were the most common structures in the majority of the plant roots.

Table 3. Arbuscular mycorrhizal fungi of rhizosphere soils of selected plants growing in Khazan Lands (Rodrigues and Anuradha 2009)

Plant species	AM fungal species
Acanthus ilicifolius L.	*Glomus claroideum, G. geosporum, G. multicaule*
Acrosticum aureum L.	*Acaulospora delicata, A. spinosa, Glomus claroideum*
Ammannia multiflora Roxb.	*Glomus claroideum G. fasciculatum, G. formosanum*
Avicennia officinalis L.	*Acaulospora dilatata, A. nicolsonii*
Ceratopteris thalictroides (L.) Brongn.	*Glomus aggregatum, G. claroideum, G. geosporum, G. intraradices, Scutellospora weresubiae*
Cyprus cyperoides (L.) O. Kuntze.	*Acaulospora dilatata, A. delicata, A. mellea, A. myriocarpa, A. spinosa, Glomus aggregatum, G. claroideum, G. fasciculatum*
Cyperus rotundus L.	*Glomus claroideum, G. fasciculatum*
Derris uliginosa Benth.	*Acaulospora delicata, A. spinosa, Glomus fasciculatum*
Echinochloa colonum (L.) Link.	*Glomus claroideum, G. geosporum, G. multicaule*
Eclipta alba (L.) Hassk.	*Glomus claroideum, Glomus formosanum*
Eleocharis geniculata (L.) Roem & Schult.	*Acaulospora nicolsonii, Glomus claroideum, G. multicaule*
Excoecaria agallocha L.	*Acaulospora delicata, A. rehmii, A. spinosa, Glomus claroideum*
Fimbristylis aestivalis (Retz.) Vahl.	*Glomus claroideum, G. formosanum G. geosporum, Glomus mosseae*
Fuirena ciliaris (L.) Roxb.	*Glomus claroideum, G. geosporum*
Ipomoea muricata Jacq.	*Acaulopsora dilatata, Glomus geosporum*
Isachne globosa (Thunb.) O. Kuntze.	*Glomus fasciculatum, Acaulospora nicolsonii*
Ludwigia perennis L.	*Acaulospora dilatata, Glomus aggregatum, G. claroideum*
Nymphaea nouchali Burm.	*Acaulospora dilatata, Glomus formosanum, G. mosseae*
Oryza sativa L. var. Asgo	*Acaulospora delicata, A. mellea, Glomus fasciculatum*
Oryza sativa L. var. Jyoti	*Acaulospora mellea, Glomus formosanum*
Oryza sativa L.var. Korgut	*Glomus claroideum*
Paspalum scrobiculatum L.	*Acaulospora nicolsonii, Glomus fasciculatum*
Phyllanthus niruri L.	*Acaulospora laevis, Glomus claroideum, G. fasciculatum*
Salvinia molesta Mitch.	*Acaulospora mellea, A. spinosa,*
Scirpus lateriflorus Gmel	*Glomus claroideum, G. fasciculatum, G. multicaule*
Sesuvium portulacastrum L.	*Acaulospora delicata, A. dilatata, A. nicolsonii, Glomus aggregatum, Scutellospora weresubiae*
Sphaeranthus africanus L.	*Acaulospora delicata, A. laevis, A. myriocarpa, Glomus claroideum, G. mosseae*
Sphenoclea zeylanica Goert. Frud.	*Glomus formosanum, G. geosporum*

Many of the known non-mycotrophic plant families, except Cyperaceae, also showed AM association, with intracellular hyphae and vesicles. Intensity of AM colonization varied both with the species and situations of their occurrence, being more intense and extensive in less saline dry ridge mangroves than in more saline formative and developed swamp mangroves. Spores from seven species of AM fungi were isolated from root-free rhizosphere soils of the mangroves. *Glomus mosseae*, *G. fasciculatum* and *Gigaspora margarita* appeared to be the predominant species as these were present in all successional stages. These species, individually and in combination, colonized roots of salinity tolerant herbs and trees in both locational silt and upstream alluvial soil with improved biomass yield and phosphorus nutrition. The studies showed that AM endophytes are widespread in the roots of mangrove and associate plants of the estuarine habitats.

Gupta et al. (2002) reported mycorrhizal association in three out of 12 mangrove species growing in the intertidal regions of the Bhitarkanika wildlife sanctuary of Orissa. Jaiswal et al. (2003) surveyed arbuscular mycorrhizal status of seven dominating mangrove plant species from Goa. Only two of the seven plant species selected for the study exhibited the presence of AM fungal colonization. Ten species of AM fungi belonging to three genera viz. *Acaulospora*, *Glomus* and *Scutellospora* were recovered. *Glomus macrocarpum* was the most frequently occurring species of AM fungi associated with the rhizosphere soil of the mangrove species.

Edaphic and physiographic differences between estuarine and maritime mangrove habitats explain the common absence of mycorrhiza in mangrove habitats (Kannan and Laskminaryan 1989). Mycorrhizas are sensitive to salinity and inundation stress and ecologically variable soil physical chemical factors appeared to be the primary determinant of mycorrhizal association in the ecosystem. Data from the studies also proved that within the mangrove ecosystem considerable adjustment exists between natural plant colonizers which could adversely affect the mycorrhizal relationships.

CONCLUSION

The importance and functioning of AM in aquatic and semi-aquatic habitats is only now beginning to be realized. These fungi are known to have the potential to alter at least some aspects of plant morphology under wetland conditions but further research is needed to understand the mechanisms. As the demand grows for preserving and re-establishing wetland plant communities for the many beneficial functions they provide, it will be increasingly necessary to understand the role AM fungi play in these areas.

ACKNOWLEDGMENTS

The authors gratefully acknowledge the financial support provided by the University Grants Commission (UGC) New Delhi and the Planning Commission, Government of India, New Delhi.

REFERENCES

Anderson, R.C., Liberta, A.E., Dickman, L.A. 1984. Interaction of vascular plants and vesicular-arbuscular mycorrhizal fungi across a soil moisture gradient. *Oecologia* 64: 111-117.

Azi, T., Sylvia, D.M., Dore, R.F. 1995. Activity and species composition of arbuscular mycorrhizal fungi following soil removal. *Ecol. Appl.* 5: 776-784.

Bagyaraj, D.J., Munjunath, A., Patil, R.B. 1979. Occurrence of vesicular arbuscular mycorrhizae in tropical aquatic plants. *Trans. Br. Mycol. Soc.*72: 164-167.

Beck-Nielsen, D., Vindaek, M. 2001. Occurrence of vesicular-arbuscular mycorrhiza in aquatic macrophytes from lakes and streams. *Aquat. Bot.* 71: 141-148.

Brown, A.M., Bledsoe, C. 1996. Spatial and temporal dynamics of mycorrhizas in *Jaumea carnosa*, a tidal salt marsh halophyte. *J. Ecol.* 84: 703-715.

Chaubal, R., Sharma, G.D., Mishra, R.R. 1982. Vesicular-arbuscular mycorrhiza in subtropical aquatic and marshy plant communities. *Proc. Ind. Acad. Sci. (Plant Sci.)* 91: 69-77.

Clayton, J.S., Bagyaraj, D.J. 1984. Vesicular-arbuscular mycorrhizas in submerged aquatic plants of New Zealand. *Aquatic Plants* 19: 251-262.

Cooke, J.C., Lefor, M.W. 1998. Comparison of vesicular-arbuscular mycorrhizae in plants from disturbed and adjacent undisturbed regions of a costal salt marsh in Clinton, Connecticut, USA. *J. Environ. Manage.* 14: 131-137.

Facelli, E., Facelli, J.M. 2002. Soil phosphorus heterogeneity and mycorrhizal symbiosis regulate plant intra-specific competition and size distribution. *Oceologia* 133: 54-61.

Jackson, R.M., Mason, P.A. 1984. Mycorrhiza. Edward Arnold Ltd., London, UK.

Jaiswal, V., Bukhari, M.J., Khade, S.W., Gaonkar, U.C. 2003. Preliminary survey of arbuscular mycorrhizal association in mangrove vegetation of Goa. *Plant Archives* 3: 73-76.

Kannan, K., Lakshminarayan, C. 1989. Survey of VAM of maritime stand plants of Point Calimere. In: Proceedings of the First Asian Conference on Mycorrhiza. Mahadevan, A., Raman, N., Natarajan, K. (eds.). Mycorrhizae for Green Asia. University of Madras, Madras, pp. 53-55.

Keeley, J.E. 1980. Endomycorrhizae influence growth of black gum seedlings in flooded soils. *Am. J. Bot.* 67: 6-9.

Koske, R.E., Gemma, J.N., Flynn, T. 1992. Mycorrhizae in Hawaiian angiosperms: A survey with implications for the origin of the native flora. *Am. J. Bot.* 79: 853-862.

Lodge, D.J. 1989. The influence of soil moisture and flooding on formation of VA- endo- and ectomycorrhizae in Populus and Salix. *Plant Soil* 117: 43-253.

Louis I. 1990. A mycorrhizal survey of plant species colonizing coastal reclaimed land in Singapore. *Mycologia* 82: 772-778.

Lugo, A.E., Snedaker, S.C. 1974. The ecology of mangroves. *Annu. Rev. Ecol. Syst.* 5:39-64.

Mejstrik, V. 1984. Ecology of vesicular arbuscular mycorrhizae of the Schoenetum nigricantis bohemicum community in the Grabanovsky Swamps Reserve. *Sov. J. Ecol.* 15: 18-23.

Mohankumar, V., Mahadevan, A. 1986. Survey of vesicular-arbuscular mycorrhizae in mangrove vegetation. *Curr. Sci.* 55:936.

Mosse, B., Stribley, D.P., LeTacon, F. 1981. Ecology of mycorrhizas and mycorrhizal fungi. In: Advances in Microbial Ecology. Alexander, M. (ed). Plenum Press, New York. pp. 137–210.

Nadarajah P, Nawawi A. 1988. VAM fungi associated with *Theobroma cocao* in Malaysia. In: Proceedings of the First Asian Conference on Mycorrhiza. Mahadevan, A., Raman, N., Natarajan, K., (eds.). Mycorrhizae for Green Asia. University of Madras, Madras, India. pp. 4-6.

Nibha Gupta, Basak U.G., Das, P. 2002. Arbuscular mycorrhizal association of mangroves in saline and non-saline soils. *Mycorrhiza News* 13: 16-19.

Radhika, K.P., Rodrigues, B.F. 2007. Arbuscular Mycorrhizae in association with aquatic and marshy plant species in Goa, India. *Aquat. Bot.* 86(3): 291-294.

Rodrigues, B.F., Anuradha N. Naik 2009. Arbuscular mycorrhizal fungi of the 'Khazan Land' agro-ecosystem. In: Frontiers in Fungal Ecology, diversity and Metabolites. Sridhar K.R. (ed.) I. K. International Publishing House Pvt. Ltd. New Delhi. pp. 141-150.

Rozema, J., Arp, W., van Diggelen, J., van Esbroek, M., Broekman, R., Punte, H. 1986. Occurrence and ecological significance of vesicular-arbuscular in the salt marsh environment. *Acta Bot. Neerlandica* 35: 457-467.

Ragupathy, S., Mohankumar, V., Mahadevan, A. 1990. Occurrence of vesicular arbuscular mycorrhizae in tropical hydrophytes. *Aquat. Bot.* 36(3):287-291.

Saif, R., Ali, I., Zaidi, A.A. 1977. Arbuscular mycorrhizae in plants and endogonaceous spore in the soil of northern areas of Pakistan. *Pak. J. Bot.* 9(2): 129-148.

Sengupta, A., Chaudhuri, S. 1990. Vesicular-arbuscular mycorrhiza (VAM) in pioneer salt marsh plants of the Ganges river delta in West Bengal (India). *Plant Soil* 122:111–113.

Sengupta, A., Chaudhuri, S. 1994. Atypical root endophytic fungi of mangrove plant community of Sundarban and their possible significance as mycorrhiza. *J. Mycopathol. Res.* 32:29-39.

Sengupta, A., Chaudhuri, S. 2002. Arbuscular mycorrhizal relations of mangrove plant community at the Ganges river estuary in India. *Mycorrhiza* 12:169-174.

Smith, S.E., Read, D.J. 1997. Mycorrhizal Symbiosis, Second ed. Academic Press, Inc., San Diego.

Sondegaard, M., Laegaard, S., 1977. Vesicular-arbuscular mycorrhiza in some vascular aquatic plants. *Nature* 268: 232-233.

Muthukumar, T., Udaiyan, K., Manian, S. 1996. Vesicular arbuscular mycorrhizae in tropical sedges of southern India. *Biol. Fertil. Soils* 22: 96-100.

Tanner, C.C., Clayton, J.S. 1985. Effects of vesicular- arbuscular mycorrhizas on growth and nutrition of a submerged aquatic plant. *Aquat. Bot.* 22: 377-386.

Turner, S.D., Friese, C.F. 1998. Plant mycorrhizal community dynamics associated with a moisture gradient within a rehabilitated prairie fen. *Restor. Ecol.* 6: 44-51.

Valiela, I., Teal, J. M. 1979. The nitrogen budget of a salt marsh ecosystem. *Nature* 280: 652-656.

Van Duin, W.E., Rozema, J., Ernst, W.H.O. 1990. Seasonal and spatial variation in the occurrence of vesicular-arbuscular (VA) mycorrhiza in salt marsh plants. *Agric. Ecosyst. Environ.* 29: 107-110.

Wetzel, P.R., van der Valk, A.G. 1996. Vesicular-arbuscular mycorrhizae in prairie pothole wetland vegetation in Iowa and North Dakota. *Can. J. Bot.* 74: 883-890.

In: Mycorrhiza: Occurrence in Natural and Restored... ISBN: 978-1-61209-226-3
Editor: Marcela Pagano © 2012 Nova Science Publishers, Inc.

Chapter 13

IMPORTANCE OF ARBUSCULAR MYCORRHIZAL FUNGI FOR RECOVERY OF RIPARIAN SITES IN SOUTHERN BRAZIL

Sidney Luiz Stürmer[1], Andressa Franzoi Sgrott[2], Felipe Luiz Braghirolli[2], Alexandre Uhlmann[3] and Rosete Pescador[4]

[1]Universidade Regional de Blumenau (FURB), Departamento de Ciências Naturais. Cx.P. 1507. 89012-900 Blumenau, SC, Brazil
[2]Graduation Program in Environmental Engineering,
Universidade Regional de Blumenau. Cx.P. 1507. 89012-900 Blumenau, SC, Brazil
[3]EMBRAPA Florestas, Estrada da Ribeira, Cx. P. 319. 83411-000 Colombo, PR, Brazil
4 Universidade Federal de Santa Catarina (UFSC), Departamento de Fitotecnia,
Cx.P. 476, 88040-900 - Florianópolis, SC, Brazil

ABSTRACT

The endomycorrhizal association established between arbuscular mycorrhizal fungi (Glomeromycetes) and plant roots are known to have a pivotal role on plant survival and nutrition. In this chapter, we convey that the triad riparian forest-carbon sequestration-arbuscular mycorrhizal fungi should be considered as environmental friendly strategy to recovery riparian sites. Greenhouse and field experiments carried out in soils occurring at the Itajai valley basin have demonstrated the importance of this association for growth and carbon allocation on plants species used to revegetation of riparian sites. Inoculation of pioneer species with selected fungal isolates have shown that more carbon is allocated to roots when compared to non-inoculated plants or plants inoculated with indigenous fungi. For *Schinus terebinthifolius* we have evidence at field conditions that inoculation with *Scutellospora heterogama*, ,*Glomus clarum* and *Acaulospora koskei* increased significantly the quantity of carbon stored in individual plants, calculated using allometric equations. Riparian sites are habitats that harbors a great diversity of glomeromycetes and 42 species has been registered in works developed in the USA and Brazil, most of them in the genera *Glomus* and *Acaulospora*. It is also noticeable that all members of

Paraglomus (*P. occultum, P. brasilianum, P. laccatum*) and *Archaeospora trappei* are commonly found in riparian sites. Considering the role of arbuscular mycorrhizal fungi on growth and carbon accumulation of pioneer plants and the high diversity of glomeromycetes in riparian soils, the inoculation of plants to be used in revegetation process of riparian sites should be viewed as a environmental friendly strategy to recovery these areas and improve carbon sequestration.

INTRODUCTION

Since the beginning of the Industrial Revolution, the composition of planet Earth atmosphere has been changed due to increases on the concentration of greenhouse gases such as carbon dioxide (CO_2) and metane (CH_4) and the impact of the elevation of carbon concentration has been of concern by the scientific community (Desjardins et al. 2005). To sequestrate terrestrial carbon (C) in living plant biomass is one strategy to decrease carbon concentration in the atmosphere and provide a pathway to increase carbon storage in soils (Rees et al. 2005). Increasing of plant biomass is directly dependent on the process of photosynthesis that converts atmosphere CO_2 into organic carbon, and also is influenced by internal plant factors such as leaf age, leaf structure and position, and chlorophyll content (Marabesi 2007). In this scenario is paramount to consider the role of soil organisms establishing mutualist symbiosis with plant species which influence above and belowground plant growth rates (Stulen and der Hertog 1993, Jakobsen and Rosendahl 1990).

Forest ecosystems are considered one of the most precious natural resources that can be used by humankind to provide several goods (*e.g.*, timber, food, wood, etc.) and environmental services, and they are important at the global level in the terrestrial carbon cycle (Caldeira et al. 2002). Deforestation contributes to the greenhouse effect by either releasing C in the atmosphere due to burning and decreasing living plant biomass that function as a sink of atmosphere C (Marcene et al. 2006). In this context, recovery of areas occupied by riparian forest is recognized as important and necessary to preserve one of the most important natural resources for human society: water (Rodrigues and Leitão Filho 2000). WWF (2003) points riparian forests as strategic partners to maintain and to regulate water supply to urban centers and recovery of riparian forests starts with revegetation using woody tree species considering their physioecological characteristics, their relations with soil classes and their association with soil mutualists.

The Itajai river watershed is considered one of the most important basins in Brazil. It is located in the state of Santa Catarina, southern Brazil, within an economically very wealthy region. Despite the relatively recent human colonization in this region by Germans and Italians immigrants, riparian forests along the Itajai river are found in advanced stage of environmental degradation. Since 2001, several research activities have been developed to provide the basic knowledge to recover riparian forest within the Itajai river watershed. In this chapter, we summarize results of this research agenda concerning arbuscular mycorrhizal fungi (AMF) and the symbiosis they establish with plant species used to recover riparian forests. We first comment on the importance to recover these forests. We also provide information on species richness occurring in riparian soils and include results of plant growth experiments carried out under controlled and field conditions. Throughout this chapter, we convey the idea that the triad riparian forest-carbon sequestration-arbuscular mycorrhizal fungi should be considered as environmental friendly strategy to recovery riparian forests.

RIPARIAN FORESTS AND SOIL CARBON SEQUESTRATION

Riparian forests are considered vegetation types associated with stream dynamics found along river banks, lakes and watersprings (Ab'Saber 2000) which experience the influence of geology, geomorphology, soils and general climate of the watershed basin (Naiman and Decamps 1997, Ab'Saber 2000, Lima and Zakia 2000). Riparian forests are inundated for part or most of the growing season (Kozlowski 2002) and therefore, the establishment of plant species depends upon their morphological, physiological and reproductive strategies and their interaction with soil organisms to face stresses imposed by the soil flooding and the process of river dynamic (Rodrigues 1992, Naiman and Decamps 1997, Lobo and Joly 2000).

Riparian forests are important not only because of plant biomass production and hence, forest products, but also because they provide a wide range of ecological services (Kozlowski 2002). These forests influence soil hydrology by improving infiltration and percolation processes, decreasing surface runoff and dissipating energy from rain drops upon the soil surface preventing erosion (Bigarella 2003). The typology of riparian forests allows the formation of extensive and continuous areas of vegetation that creates favorable conditions to survival and genetic flow among populations of animals and plants, functioning therefore as ecological corridors (Macedo 1993, Durigan 1994). They also provide shelter for birds and mammals species due to low fire incidence, proximity with agricultural fields and abundant water supply (Kozlowski 2002). From an anthropological view, riparian forests are important to improve water quality, lighten the effects of air and water pollution (Kozlowski 2002). Because of their fertile alluvial soils, their areas are exploited for agriculture and pasture for cattle ranching which causes loss of original vegetation of riparian forests.

The regeneration process of riparian forest brings up a new environmental service: the potential for carbon sequestration. Recovery of areas occupied by riparian forests starts by transplanting seedlings of woody tree species, a process that must consider particularities of each plant species and their relation to edaphic factors. To maximize the input of C in soils it is necessary to maximize plant productivity (Rees et al. 2005) as primary productivity aboveground is the major source of organic matter to most soils (van de Geijn and van Veen 1993). Carbon enters the pool of soil organic matter (SOM) through litter production, incorporation of crop plants after harvesting and via root system (van de Gejin and van Veen 1993). Therefore riparian forest regeneration processes using fast growth species have the potential to promote not only the recovery of the area by revegetation but also to function as atmospheric carbon sink in the plant biomass and soil.

The role of riparian forest revegetation to act as carbon sinks is better understood considering that the carbon allocated to roots is also utilized to maintain symbiosis established by plant roots with soil microorganisms, particularly nitrogen fixing bacteria and arbuscular mycorrhizal fungi (AMF). The latter are keystone component of terrestrial ecosystems forming the arbuscular mycorrhizal association with plant roots. They have been viewed not only as plant symbionts but also as a critical link in the plant-soil continuum (Wilson et al. 2009) as they improve plant nutrition and nutrient uptake (specially phosphorus) and might be an important component of soil organic carbon in addition to facilitate carbon sequestration through stabilizing soil aggregates (Zhu and Miller 2003). In terms of global carbon cycle, it has been estimated that plants can allocate 10-20% of their photosynthate to AMF growth (Jakobsen and Rosendahl 1990) which is a significant amount considering that AMF establishes association with 80% of plant species in nearly every terrestrial ecosystem in the world (Smith and Read 1997, Wang and Qiu 2006). Mycorrhizal fungi can enhance carbon sequestration by distinct mechanisms. AMF external hyphae extending into the bulk soils are expected to translocate C away from the respiratory activity around the roots (Treseder and Allen 2000). Growth and turnover of extra-radical hyphae (ERH) within the bulk soil can influence directly soil carbon dynamics as the residence time of ERH is usually short (Zhu and Miller 2003). ERH also plays an important role in the formation and maintenance of soil aggregates (Wilson et al. 2009); soil aggregation is an important mechanism to protect C-rich compounds from microbial degradation and therefore to improve C sequestration. Glomalin, a recalcitrant protein produced solely by AMF, can be responsible for the production of 4-5% of carbon and nitrogen in some soils (Rillig et al. 2001). Besides these processes, the amount of chitin in the mycelia cell walls represent C allocated to mycorrhizal tissues that could be long-lived in soils (Treseder and Allen 2000). On the other hand, the contribution of asexual spores produced by AMF as carbon sink has not been readily measured.

The role of mycorrhizal fungi in contributing to increase carbon sequestration is evidenced when the successional model for recovery of riparian forests is applied. In this model, pioneer species are pivotal as they will provide conditions for development of non-pioneer species (Kageyama et al. 1990). Pioneer species are characterized by fast growth, shade intolerant and small seeded species (Barbosa 2000). Evidences have been built in the literature that pioneer species are highly responsive to inoculation of AMF. Siqueira et al. (1998) studied the mycotrophic growth and response to superphosphate of 28 woody species belonging to different successional stages and observed that pioneers were markedly more responsive to mycorrhizal and P treatments that climax one. Considering shoot dry biomass, mycorrhizal response of pioneers and climax species over controls were 813% and 0%, respectively. Zangaro et al. (2000) calculated that the mycorrhizal dependency of pioneer species was 90% compared to 48%, 12% and 14% for early secondary, late secondary and climax species, all respectively. They also showed that greenhouse and field mycorrhizal colonization tended to be higher in pioneer species compared to climax species. Based on both studies it is clear that recovery of riparian areas using seedlings of pioneer species inoculated with mycorrhizal fungi represents an important strategy to increase plant biomass accumulation and, consequently, carbon sequestration.

In the last decade, works have evaluated the relation of arbuscular mycorrhizal colonization on plant biomass production and nutrient uptake of several tropical woody species. Carneiro et al. (1996) evaluated root colonization and mycorrhizal dependency of 31 species and observed that the latter was affected by phosphorus addition. Nevertheless, inoculation with AMF and addition of superphosphate acted synergistically on initial growth of 90% of the species studied. Using an assemblage of AMF as inoculum, Pouyú-Rojas and Siqueira (2000) observed that inoculation of seven woody species at nursery or at transplanting increased growth response and nutrition after transplant into the field. Carneiro et al. (2004) found that inoculated *Cecropia pachystachya* had higher survival and growth after transplanting compared to non-mycorrhizal seedlings. Conversely, Vandresen et al.

(2007) observed that AMF inoculation and fertilization had no effect on seedling survival after transplanting, a result that could be partially explained by using an assemblage of fungal species multiplied under trap cultures conditions and stored for two years in refrigerator before inoculation. Compatibility studies of AMF isolates and tropical woody species has demonstrated that isolates of *Glomus clarum, Acaulospora colombiana, Scutellospora pellucida* and *Glomus etunicatum* benefited 80% of the 16 plant species studied by Pouyú-Rojas et al. (2006). Relationship between root morphology, soil fertility and sucessional stages with mycorrhizal colonization and mycotrophic growth has been extensively studied by Zangaro and coworkers (Zangaro et al. 2003, 2005, 2007).

In Santa Catarina state, studies with AMF in riparian forests have emphasized their role in improving plant biomass accumulation and carbon sequestration considering different woody species and soil classes. In the Itajai river watershed, permanent flooded areas are occupied by *Organossolos* and *Gleissolos* while higher, well or moderated drained terrains are dominated by *Neossolos* and *Cambissolos*. At greenhouse conditions, Pasqualini et al. (2007) studied plant growth response of pioneer (*Schinus terebinthifolius, Alchornea glandulosa, Cedrella fissilis, Cytharexyllum myrianthum*) and late secondary (*Annona cacans, Cabralea canjerana, Marlierea tomentosa, Magnolia ovata*) woody species to mycorrhizal inoculation and phosphorus addition. Plants were inoculated with assemblage of AMF originating from a *Cambissolo* and *Neossolo*, which correspond to Inceptisols and Entisols according to the US Soil Survey, respectively. The authors categorized mycorrhizal dependency of pioneer species in highly responsive and very highly responsive while late secondary species were independent, marginally or moderately responsive, regardless of soil classes that plant were developed. They also observed that mycorrhizal dependency was inversely correlated with plant seed mass: small-seeded species tended to be more dependent on mycorrhizal than large-seeded species, a pattern also detected by Siqueira et al. (1998) and Zangaro et al. (2000).

Quantities of carbon sequestration by woody plants can be estimated using different approaches: allocation of biomass to distinct root organs (stems, roots, and leaves), physiological measurements of C-rich compounds like carbohydrates and starch, and estimates based on equations considering height and stem diameter. Under greenhouse conditions, AMF inoculation altered the allocation of C compounds measured as soluble carbohydrates according to the host and soil (Figure 1). Pioneer species *Schinus terebinthifolius* Raddi and *Cytharexyllum myrianthum* Cham. were grown in a *Cambissolo* or *Organossolo* and submitted to the following treatments: sterilized soil to obtain a free-mycorrhizal treatment (Control), non-sterilized soils (Native), and non-sterilized soil added with an assemblage of AMF (Inoculated). Plants under Native or Mycorrhizal treatments tended to have larger amounts of shoot soluble carbohydrates compared to Control plants. Inoculation with an assemblage of AMF resulted in large amounts of soluble carbohydrates to the roots in *C. myrianthum* and *S. terebinthifolius* grown in *Cambissolo*. The main conclusion that inoculation significantly changes carbon allocated to shoots and roots is noticeable. More important is that fungal development in Inoculated plants was sufficient to function as carbon sink to roots and certainly provides a pathway for this carbon enter the soil. *Schinus terebinthifolius* is a pioneer species highly recommended to initiate the revegetation process during recovery of riparian forests. We established in greenhouse for 5 months *S. terebinthifolius* seedlings in a non-sterilized *Cambissolo* and inoculated with isolates of

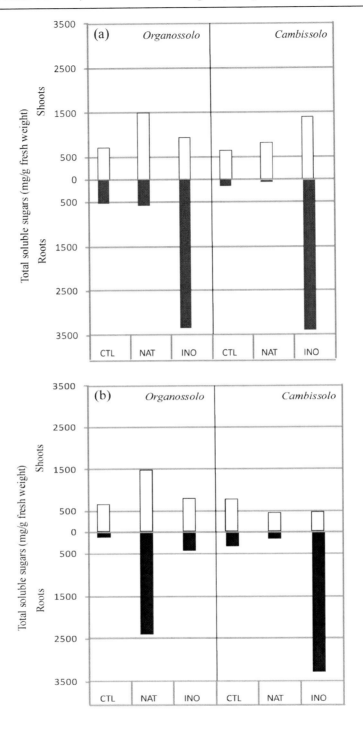

Figure 1. Total soluble sugars (mg/g fresh weight) in (a) *Cytharexyllum mirianthum* and (b) *Schinus terebinthifolius* grown in *Organossolo* and *Cambissolo* and left non-inoculated (CTL), inoculated with native AMF (NAT), and inoculated with an assemblage of AMF isolates (INO).

Glomus clarum, Scutellospora heterogama, and *Acaulospora koskei.* Plants were transplanted to a riparian area under recuperation and after one year at field conditions, plants inoculated with *G. clarum* and *A. koskei* averaged 0.88 Kg C tree^{-1} and 0.87 Kg C tree^{-1}, respectively, compared to plants grown on non-sterile *Cambissolo* that averaged 0.55 Kg C tree^{-1} (Figure 2). This data shows that inoculation of seedlings with selected fungal isolates can influence plant growth and carbon fixation on plant biomass on a long-term basis and therefore, inoculation should be considered a key process on the seedling establishment for revegetation of riparian areas.

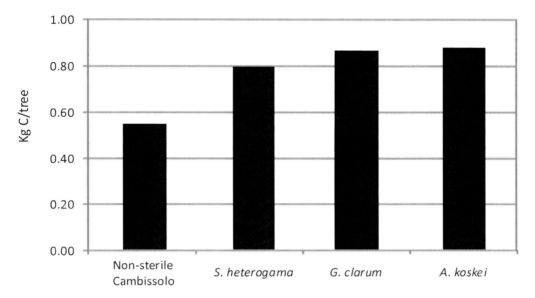

Figure 2. Carbon (Kg C/tree) of *Schinus terebinthifolius* after one year at field conditions and inoculated with native fungi (non-sterile Cambissolo), *Scutellospora heterogama, Acaulospora koskei,* and *Glomus clarum.*

AMF SPECIES RICHNESS IN RIPARIAN ENVIRONMENT

Several edaphic factors are known to influence AMF species distribution and abundance including pH (Siqueira et al. 1989, Porter et al. 1987), nutrients (Johnson et al. 1991) and moisture (Anderson et al. 1984, Miller 2000). In riparian ecosystems, environmental heterogeneity is created by a moisture gradient occurring from the active river channel to inland that is directly related with some factors like the river width, alluvial deposition, and frequency and duration of flooding (Durigan et al. 2000). This gradient impact upon plant community composition and therefore, has the potential to influence mycorrhizal fungal communities in riparian soils. There are few studies in the literature on AMF or ectomycorrhizal fungi in riparian environment and we provide a summary of the main results of these studies (Table 1) and the list of AMF species found in riparian sites (Table 2).

Table 1. Published works of arbuscular mycorrhizal fungi and ectomycorrhizae in riparian systems

Reference	Country/ state	Main results
Carrenho et al. (2001)	Brazil – São Paulo	- AMF species richness was higher under the rhizosphere of *Croton urucurana* and *Inga striata* than *Genipa americana*. - AMF spore number tended to increase with sucessional stages of the plant host. - Mycorrhiza community was dominated by species of *Glomus* and *Acaulospora*. - Diversity indices, richness and evenness for AMF community tended to decrease in the rizosphere of climax species compared to pioneer species. - Most frequent species: *Glomus macrocarpum, G. etunicatum, G. claroideum, Entrophospoa kentinensis,* and *Acaulospora tuberculata*
Kennedy et al. (2002)	USA - Arizona	- Presence of 15 AMF species associated with *Sporobulus wrightii* in four riparian sites. - *Glomus* was the dominant genus associated with *S. wrightii*. - Root colonization was significantly different in plants of *S. wrightii* occupying upper terraces and lower floodplains, but changes in total colonization were not correlated with percentage relative moisture. - Most frequent species: *Glomus mosseae, G. intraradices, G. spurcum,* and *Paraglomus occultum*
Beauchamp et al. (2006)	USA – Arizona	- AMF species richness declined with stand age and distance from and elevation above the channel - The following factors played a role in structuring AMF community: distance from and elevation above the active channel, stand age, annual species cover, perennial species richness, and exchangeable potassium concentration - Despite that most AMF species were found across a wide range of soil conditions, a subset of AMF species tended to be more often found in hydric areas. - Most frequent species: *G. intraradices, G. microaggregatum, G. spurcum, G. eburneum* and *G. mosseae*).
Parádi and Baar (2006)	The Netherlands	- Twelve distinct types of ectomycorrhizal fungi were detected in roots of *Salix alba* and ectomycorrhizal fungal community was dominated by *Tuber* sp. - Basidiomycete fungi pertaining to *Hebeloma* sp. and telephoroid spp. were found mainly in 20-year-old forest with lowest soil nutrient content. - Only a limited number of ectomycorrhizal fungi can resist environmental conditions caused by flooding and drought.

Table 2. Occurrence of AMF fungal species in riparian soils

AM Fungal Species	Carrenho et al. (1997)	Carrenho et al. (2001)	Kennedy et al. (2002)	Beauchamp et al. (2006)	Stürmer et al. (unpublished)
Acaulosporaceae					
Acaulospora capsicula Blaszkowski					X
A. delicata Walker, Pfeiffer & Bloss			X	X	X
A. foveata Trappe & Janos					X
A. koskei Blaszkowski					X
A. longula Spain & Schenck	X	X			
A. mellea Spain & Schenck		X			X
A. morrowiae Spain & Schenck	X		X	X	X
A. scrobiculata Trappe		X	X		X
A. spinosa Walker & Trappe		X		X	
A. tuberculata Janos & Trappe	X	X			
A. undulata Sieverding					X
A. colombiana (Spain & Schenck) Kaonongbua, Morton & Bever					X
A.. kentinensis (Wu & Liu) Kaonongbua, Morton & Bever		X			
Glomeraceae					
Glomus claroideum Schenck & Smith		X		X	
G. clarum Nicol. & Schenck		X			X
G. deserticola Trappe, Bloss & Menge				X	
G. eburneum Kennedy, Stutz & Morton			X	X	
G. etunicatum Becker & Gerd.	X	X		X	
G. fasciculatum (Thaxter) Gerd. & Trappe emend. Walker & Koske				X	
G. globiferum Koske & Walker		X			
G. intraradices Schenck & Smith			X	X	
G. invermaium Hall		X			
G. luteum Kennedy, Stutz & Morton			X	X	
G. macrocarpum Tul & Tul.	X	X	X	X	
G. microaggregatum Koske, Gemma & Oleixa		X	X	X	
G. microcarpum Gerd. & Trappe		X			
G. mosseae (Nicol. & Gerd.) Gerd. & Trappe			X	X	
G. spurcum Pfeiffer, Walker & Bloss emend. Kennedy, Stutz & Morton			X	X	
G. tortuosum Schenck & Smith				X	
Gigasporaceae					
Gigaspora decipiens Hall & Abbott	X				X
G. margarita Becker & Hall		X			
G. ramisporophora Spain, Sieverding & Schenck	X				
Scutellospora calospora (Nicol & Gerd.) Walker & Sanders		X			
S. erythropa (Koske & Walker) Walker & Sanders		X			
Racocetra. fulgida (Koske & Walker) Oehl, Souza & Sieverding		X			
S. heterogama Nicol. & Gerd.		X			

Table 2 (Continued).

AM Fungal Species	Carrenho et al. (1997)	Carrenho et al. (2001)	Kennedy et al. (2002)	Beauchamp et al. (2006)	Stürmer et al. (unpublished)
Racocetra persica (Koske & Walker) Oehl, Souza & Sieverding	X				
Paraglomeraceae					
Paraglomus brasilianum (Spain & Miranda) Morton & Redecker					X
P. occultum Walker		X	X	X	
P. laccatum Renker, Błaszk. & Buscot					X
Archaeosporaceae					
Archaeospora trappei (Ames & Linderman) Morton & Redecker			X	X	X
Ambisporaceae					
Ambispora gerdemanni (Rose, Daniels & Trappe) Walker, Vestberg & Schuessler		X			
Identified species	8	1	13	17	14
Non-identified species	0	21	2	10	13
Total Species Richness	8	22	15	27	27

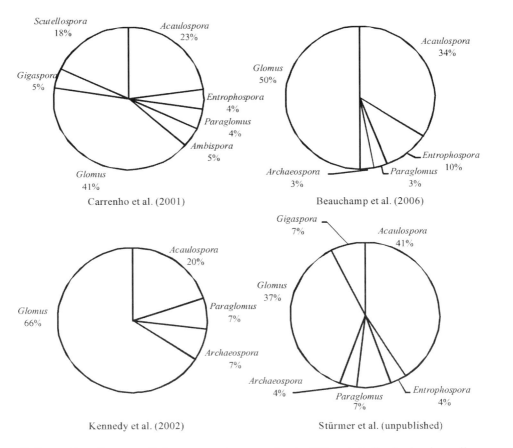

Figure 3. Proportion of genera of glomeromycetes occurring in different studies in riparian soils.

In the United States, AMF occurrence and species diversity in riparian habitats have been studied in the Arizona state. Kennedy et al. (2002) observed the seasonal dynamics of AMF associated with *Sporobulus wrightii* in four riparian sites. They detected fifteen AMF species associated with the host and species richness ranging from 9 to 13 according to the site. Most species detected were *Glomus* followed by *Acaulospora*. Beauchamp et al. (2006) studied the occurrence of AMF species and determined environmental variables related to AMF species richness associated with *Populus* and *Salix* stands. They found 27 AMF species in trap cultures and fourteen of them pertained to *Glomus* and eight to *Acaulospora*. Other genera detected were *Entrophospora, Paraglomus* and *Archaeospora*. In this study, AMF species richness declined with plant species stand age and distance from the river channel. Interesting is that a subset of all species found tended to occur with high frequency in hydric areas. Parádi and Baar (2006) in the Netherlands studied the ectomycorrhizal fungal community associated with three homogenous *Salix alba* riparian forest that experienced flooding for up to 144 days during the last ten years. A total of 12 types of ectomycorrhizal fungi were detected using molecular methods; *Tuber* spp. were the most abundant representative in all sites and species of *Hebeloma* and *Cenococcum* were also detected.

In Brazil, Carrenho et al. (1997) and Carrenho et al. (2001) surveyed the occurrence of AMF species associated with rhizosphere of *Croton urucurana, Inga striata* and *Genipa americana* used in a revegetation process to recuperate riparian areas. In the work of 1997, they found 8 AMF species associated with the host surveyed and no differences on AMF species richness was detected between young and mature plants. Carrenho et al. (2001) found 22 species belonging to *Acaulospora, Glomus, Entrophospora, Gigaspora, Scutellospora,* and *Paraglomus*. Overall *Glomus macrocarpum, G. etunicatum, G. claroideum, E. kentinensis,* and *A. tuberculata* were the most often species recovered. Spore abundance of AMF was correlated with the successional stage: spore numbers were low in the pioneer *C. urucurana* and tended to increase in the climax *G. americana*. In riparian soils in the Itajai river valley we have investigated AMF species richness in soil classes occurring in distinct land use systems. Results from surveys in *Organossolos* and *Gleissolos*, both characterized by being hydromorphic soils, under pasture, forest and agricultural systems have recorded 27 species (Table 2). *Paraglomus laccatum* and *Archaeospora trappei* were the most abundant and frequent species recovered, especially in *Gleissolos*. Both soils shared a high percentage of AMF species detected but land use systems under the same type of soil influenced occurrence of some fungal species. For instance, *Paraglomus laccatum* and *Acaulospora morrowiae* were the most abundant species in agriculture while *A. mellea* and *Acaulospora* sp. were the most often recovered in forest, both under *Organossolo*. Some species were detected exclusively under a particular combination of soil and land use system: *Entrophospora colombiana* was detected in both soils only in pasture lands, and *Gigaspora decipiens* was found only in pasture in *Gleissolos*.

Analysis of fungal species richness from studies of Carrenho et al. (1997), Carrenho et al. (2001), Kennedy et al. (2002), Beauchamp et al. (2006), and surveys of riparian ecosystems in the Itajai river valley demonstrate some pattern of AMF communities in riparian soils. All species of *Paraglomus* (*P. occultum, P. laccatum,* and *P. brasilianum*) have been detected in riparian soils and *Archaeospora trappei* is also a common species found in these soils.

CONCLUDING REMARKS

In this chapter we provided some evidences that inoculate seedlings of pioneer woody species with arbuscular mycorrhizal fungi is an environmental friendly strategy to pursue fast plant growth at field conditions and speed up the recovery process of riparian areas. Models for recovery of riparian areas consider the pivotal role of pioneer plants which are highly responsive to mycorrhizal fungi and therefore inoculation of pioneer seedlings should turn a common practice in nurseries. Besides impacting initial plant growth, AMF function also as an important way to introduce and maintain carbon in the soil improving carbon sequestration in riparian areas through hyphal turnover, glomalin production, increasing soil aggregation and source of chitin in spores and hyphae.

Studies of occurrence of arbuscular mycorrhizal species reveal that riparian soils and the vegetation established herein support a relatively high species richness of glomeromycetes dominated by the genera *Glomus* and *Acaulospora*. Common species in these environments pertain to *Paraglomus* and *Archaeospora*, two genera that are basal from a phylogenetically within the phylum Glomeromycota. Further works in riparian areas should include the role of external mycelium on increasing soil aggregation and the assessment of mycorrhizal inoculum potential of these soils.

REFERENCES

Ab'Saber, AN. O suporte Geoecológico das Florestas Beiradeiras (Ciliares). In: Rodrigues RR, Leitão-Filho HF (editors). *Matas ciliares: conservação e recuperação.* São Paulo: Editora USP- Fapesp; 2000; 15-25.

Anderson RC; Liberta AE; Dickman LA. Interaction of vascular plants and vesicular-arbuscular mycorrhizal fungi across a soil moisture–nutrient gradient. *Oecologia,* 1984, 64, 111–117.

Barbosa, LM. Considerações gerais e modelos de recuperação de formações cilitares. In: Rodrigues RR, Leitão-Filho HF (editors). *Matas ciliares: conservação e recuperação.* São Paulo: Editora USP- Fapesp; 2000; 289-312.

Beauchmap, VB; Stromberg, JC; Stutz, JC. Arbuscular mycorrhizal fungi associated with *Populus-Salix* stands in a semiarid riparian ecosystem. *New Phytologist,* 2006, 170, 369-380.

Bigarella, JL. *Estrutura e origem das paisagens tropicais e subtropicais.* Florianópolis: UFSC, 2003.

Caldeira, MVW, LF Watzlawick, MV Schumacher, R Balbinot, CR Sanquetta. Carbono orgânico em solos florestais. In: Sanquetta CR, Watzlawich LF, Balbinot R, Ziliotto MAB, Gomes FS (Orgs.). *As florestas e o carbono.* Curitiba:Imprensa Universitária da UFPR; 2002; 191-213.

Carneiro, MAC; Siqueira, JO; Davide, AC; Gomes, LJ; Curi, N; Vale, FR. Fungo micorrízico e superfosfato no crescimento de espécies arbóreas tropicais. *Scientia Forestalis,* 1996, (50), 21-36.

Carneiro, MAC; Siqueira, JO; Davide, AC. Fósforo e inoculação com fungos micorrízicos arbusculares no estabelecimento de mudas de embaúba (*Cecropia pachystachya* Trec). *Pesquisa Agropecuária Tropical*, 2004, 34, 119-125.

Carrenho, R; Bononi, VLR; Barbosa, LM. Glomales em áreas de recomposição de mata ciliar de Moji-Guaçu, SP, Brasil. *Hoehnea*, 1997, 24, 107-113.

Carrenho, R; Trufem, SFB; Bononi, VLR. Fungos micorrízicos arbusculares em rizosferas de três espécies de fitobiontes instaladas em área de mata ciliar revegetada. *Acta botânica brasílica*, 2001, 15, 115-124.

Desjardins, RL; Smith, W; Grant, B; Campbell, C; Riznek, R. Management strategies to sequester carbon in agricultural soils and to mitigate greenhouse gas emissions. *Climatic Change*, 2005, 70, 283-297.

Durigan, G. *Floristica, fitossociologia e produção de folhedo em matas ciliares da região oeste do Estado de São Paulo*. Tese de Doutorado. Instituto de Biologia - Universidade Estadual de Campinas. Campinas, SP, 1994. 149p.

Durigan, G; Rodrigues, RR; Schiavini, I. A heterogeneidade ambiental definindo a metodologia de amostragem da floresta ciliar. In: Rodrigues RR, Leitão-Filho HF (editors). *Matas ciliares: conservação e recuperação*. São Paulo: Editora USP- Fapesp; 2000; 159-167.

Jakobsen, I; Rosendahl, L. Carbon flow into soil and external hyphae from roots of mycorrhizal cucumber plants. *New Phytologist*, 1990, 115, 77-83.

Johnson, NC; Zak, DR; Tilman, D; Pfleger, FL. Dynamics of vesicular arbuscular mycorrhizae during old field succession. *Oecologia*, 1991, 86, 349-358.

Kageyama, PY; Biella, LC; Palermo Jr., A. Plantações mistas com espécies nativas com fins de proteção a reservatório. In: Anais...Congresso Florestal Brasileiro, 6, Campos do Jordão. Sao Paulo:Sociedade Brasileira de Silvicultura, 1990, 1, 109-112.

Kennedy, LJ; Tiller, RL; Stutz, JC. Associations between arbuscular mycorrhizal fungi and *Sporobolus wrightii* in riparian habitats in arid South-western North America. *Journal of Arid Environments*, 2002, 50, 459-475.

Kozlowski, TT. Physiological-ecological impacts of flooding on riparian forest ecosystems. *Wetlands*, 2002, 22, 550-561.

Lima, WP; Zakia, MJB. Hidrologia de Matas Ciliares. In: Rodrigues RR, Leitão-Filho HF (editors). *Matas ciliares: conservação e recuperação*. São Paulo: Editora USP- Fapesp; 2000; 33-44.

Lobo, PC; Joly, CA. Aspectos ecofisiológicos da vegetação de Mata ciliar do Sudeste do Brasil. In: Rodrigues RR, Leitão-Filho HF (editors). *Matas ciliares: conservação e recuperação*. São Paulo: Editora USP- Fapesp; 2000; 143-157.

Macedo, AC. *Revegetação: matas ciliares e de proteção ambiental*. São Paulo: Fundação Florestal; 1993.

Marabesi, MA. Efeito do alto CO_2 no crescimento inicial e na fisiologia da fotossíntese em plântulas *Senna alata* (L.) Roxb.. Dissertação (Mestrado) - Instituto de Botânica da Secretaria de Estado do Meio Ambiente, São Paulo, 2007.

Marcene EA, Dalla Corte AP, Sanquetta, CR, Schneider, CR. Variações nos teores e estoques individuais de carbono fixado com o crescimento de *Gmelina arborea* Rosb. na região litorânea do Paraná, Brasil. *Revista Scientia Florestalis*, 2006, 71, 55-63.

Miller, SP. Arbuscular mycorrhizal colonization of semi-aquatic grasses along a wide hydrologic gradient. *New Phytologist*, 2000, 145, 145–155.

Naiman, RJ; Décamps, H. The ecology of interfaces: Riparian Zones. *Annual Review of Ecology and Systematics*, 1997, 28, 621–58.

Parádi, I; Baar, J. Mycorrhizal fungal diversity in willow forests of different age along the river Waal, The Netherlands. *Forest Ecology and Management*, 2006, 237, 366-372.

Pasqualini, D; Uhlmann, A; Stürmer, SL. Arbuscular mycorrhizal fungal communities influence growth and phosphorus concentration of woody plants species from the Atlantic rain forest in South Brazil. *Forest Ecology and Management*, 2007, 245, 148-155.

Porter, WM; Robson, AD; Abbott, LK. Factors controlling the distribution of vesicular arbuscular mycorrhizal fungi in relation to soil pH. *Journal of Applied Ecology*, 1987, 24, 663–672.

Pouyú-Rojas, E; Siqueira, JO. Micorriza arbuscular e fertilização do solo no desenvolvimento pós-transplante de mudas de sete espécies florestais. *Pesquisa Agropecuária Brasileira*, 2000, 35, 103-114.

Pouyú-Rojas, E; Siqueira, JO; Santos, JGD. Compatibilidade simbiótica de fungos micorrízicos arbusculares com espécies arbóreas tropicais. *Revista Brasileira de Ciência do Solo*, 2006, 30, 413-424.

Rees, RM; Bingham, IJ; Baddeley, JA; Watson, CA. 2005. The role of plants and land management in sequestring soil carbon in temperate arable and grassland ecosystems. *Geoderma*, 2005, 128, 130-154.

Rillig, MC; Wright, SF; Nichols, KA; Schmidt, WF; Torn, MS. Large contributions of arbuscular mycorrhizal fungi to soil carbon pools in tropical forest soils. *Plant Soil*, 2001, 233,167-177.

Rodrigues, RR. *Análise de um Remanescente de vegetação natural às margens do Rio Passa-Cinco, Ipeúna, SP*. Tese (Doutorado) - Instituto de Biologia. Campinas, UNICAMP. 1992. 373p.

Rodrigues RR; Filho, HF. *Matas ciliares: conservação e recuperação*. São Paulo:Edusp-Fapesp; 2000.

Siqueira, JO; Colozzi-Filho, A; Oliveira, E; Schenck, NC. Ocorrência de micorrizas vesículo-arbusculares em agro e ecossistemas naturais do Estado de Minas Gerais. *Pesquisa Agropecuária Brasileira*, 1989, 24, 1499-1506.

Siqueira, JO; Carneiro, MAC; Curi, N; Rosado, SCS; Davide, AC. Mycorrhizal colonization and mycotrophic growth of native woody species as related to successional groups in Southeastern Brazil. *Forest Ecology and Management*, 1998, 107, 241–252

Smith, SE; Read, DJ. *Mycorrhizal Symbiosis*. 2nd editions, San Diego: Academic Press; 1997.

Stulen, I; den Hertog, J. 1993. Root growth and functioning under atmospheric CO_2 enrichment. *Vegetatio*, 1993, 104/105, 99-115.

Treseder, KK; Allen, MF. Mycorrhizal fungi have a potential role in soil carbon storage under elevated CO_2 and nitrogen deposition. *New Phytologist*, 2000, 147,189-200.

Van de Geijn, SC; van Veen, JA. Implications of increased carbon dioxide levels for carbon input and turnover in soils. *Vegetatio*, 1993, 104/105, 283-292.

Vandresen, J; Nishidate, FR; Torezan, JMD; Zangaro, W. Inoculação de fungos micorrízicos arbsuculares e adubação na formação e pós-transplante de mudas de cinco espécies arbóreas nativas do sul do Brasil. *Acta botânica brasílica*, 2007, 21, 753-765.

Wilson, GWT; Rice, CW; Rillig, MC; Springer, A; Hartnett, DC. Soil aggregation and carbon sequestration are tightly correlated with the abundance of arbuscular mycorrhizal fungi: results from long-term field experiments. *Ecology Letters*, 2009, 12,452-461.

Zangaro, W; Bononi, VLR; Trufem, SB. Mycorrhizal dependency, inoculum potential and habitat preference of native woody species in South Brazil. *Journal of Tropical Ecology*, 2000, 16, 603-622.

Zangaro, W; Nisizaki, SMA; Domingos, JCB; Nakano, EM. Mycorrhizal response and successional status in 80 woody species from south Brazil. *Journal of Tropical Ecology*, 2003, 19:315-324.

Zangaro, W.; Nishidate, FR; Camargo, FRS; Romagnoli, GG; Vandresen, J. Relationships among arbuscular mycorrhizas, root morphology and seedling growth of tropical native woody species in southern Brazil. *Journal of Tropical Ecology*, 2005, 21, 529-540.

Zangaro, W; Nishidate, FR; Vandresen, J; Andrade, G; Nogueira, MA. Root mycorrhizal colonization and plant responsiveness are related to root plasticity, soil fertility and successional status of native woody species in southern Brazil. *Journal of Tropical Ecology*, 2007, 23, 53-62.

Zhu, Y-G; Miller, RM. Carbon cycling by arbuscular mycorrhizal fungi in soil-plant systems. *TRENDS in Plant Science*, 2003, 8,407-409.

Wang, B; Qiu Y-L. Phylogenetic distribution and evolution of mycorrhizas in land plants. *Mycorrhiza,* 2006, 16, 299-363.

WWF. 2003. A água, as cidades e as florestas. Disponível em: <*http://www.wwf.org.br*>

In: Mycorrhiza: Occurrence in Natural and Restored…
Editor: Marcela Pagano

ISBN: 978-1-61209-226-3
© 2012 Nova Science Publishers, Inc.

Chapter 14

MYCORRHIZAS IN NATURAL AND RESTORED RIPARIAN ZONES

Marcela C. Pagano[1] and Marta Noemi Cabello[2]

[1]Federal University of Ceará, Fortaleza, Ceará, Brazil
[2]Instituto Spegazzini, Researcher from CIC, National University of La Plata, Argentina

ABSTRACT

The riparian vegetation, used for cattle and agriculture practices, has long been negatively influenced by human activities, especially domestic sewage and mining activity. Only recently the river basins are considered as part of the landscape and have begun to be studied for their environmental management. Moreover, a study of interactions among the functional groups of microorganisms in riparian regions is essential to a successful restoration, and to improve restoration programs. The purpose of this review is to explore the current information on the occurrence of mycorrhizas in riparian ecosystems in Brazil, and to speculate about the role of symbiosis. This chapter discusses arbuscular mycorrhizal (AM) root colonization, drawing on results of research of native and invasive plant species in Brazil. As expected, the studies revealed that most native plant species in riparian areas show mycotrophy. AM spore diversity found in rhizospheric soils is illustrated, and relevant findings are emphasized. As generally found, legumes present a higher AM colonization, being the *Arum*-type the most commonly observed. A high AMF spore richness in the rhizospheric soils of most plant species in riparian zones was observed. In general, AM fungal species belongs to eight genera, which are related to the sites and regions studied; however, *Glomus* and *Scutellospora* spores were common. Some terrestrial pteridophytes showed presence of AM structures; however, some ferns show dominant extraradical hyphae whereas their mycorrizal status remains uncertain. The restored areas presented higher AM species richness than the degraded areas. The AM symbiosis is a common and important component in the riparian vegetation, and should be included in future restoration programs. The benefits and problems encountered are discussed in this chapter.

Keywords: Arbuscular mycorrhizal fungi, Colonization pattern, Riparian forest, Restoration

INTRODUCTION

Because riparian zones link the stream with its terrestrial catchment, they can modify, incorporate, dilute, or concentrate substances before they enter a lotic system. In small to mid-size streams forested riparian zones can moderate temperatures, reduce sediment inputs, provide important sources of organic matter, and stabilize stream banks (Osborne and Kovacic 1993). Riparian forest buffers are natural or re-established streamside forests made up of tree, shrub, and grass plantings, which buffer non-point source pollution of waterways from adjacent land. Thus, riparian forests reduce bank erosion, protect aquatic environments, and enhance wildlife, increasing biodiversity. Riparian zones have been shown to contribute to beta diversity, but there is no guarantee that the width of the riparian zone is similar for groups with different life forms (Sabo et al. 2005). Donaldson et al. (2007) have showed that some waterbird species are sensitive to anthropogenic disturbance in urban riparian environments. Allan (2004) presented a review of the principal mechanisms by which land use influences stream ecosystems.

Variations in topography, landform and soils have strong effects on species composition, distribution and structure of riparian zones in tropical regions (Campbell et al. 1992, Oliveira-Filho et al. 1994), and the river dynamics can also determine patterns of succession and distribution of species (Schnitzler 1997).

In floodplains, inundations are an important disturbing factor, and the presence of plant species, which colonize different environment, dispersion features, tolerance to inundation and shade tolerance, reflects the timing and frequency of floods (Junk 1989). However, there is little information about the effects of exceptional floods on riparian forests (Vervuren et al. 2003.)

Floodplain forests have been greatly reduced by agriculture and river control has altered the natural flooding and disturbance regime, influencing the nutrient cycling, the plant growth and soil structure, thus decreasing the water retention, the resistance to erosion, the root development, and the microbial activity. When the higher microbial activity (occurring in the upper layer of soils) is reduced by erosive processes, soil biota numbers and diversity decrease (Alvarenga et al. 1999).

In Brazil, there is an expanding occupation of the riparian areas, converting them into agrosystems, pastures, and forests of exotic species such as *Pinus* and *Eucalyptus* (Schäffer and Prochnow 2002). Thus, riparian areas are fragmented as a function of forest exploitation and agricultural expansion.

Restoration of riparian areas along rivers and adjacent to water springs, designated for permanent preservation, is urgently needed (Pasqualini et al. 2007). Ecosystem services of riparian forests are the absorption of precipitation mostly as rain into the soil, promoting continuous and moderated release of water into rivers and moderate flow across landscapes reducing soil erosion and flooding (Barbosa 2000), and the conservation of the diversity of animal and plant communities as ecological corridors allowing gene flow (Kageyama and Gandara 2000).

Seasonal inundation, commonly in Brazil, can produce many effects on plant communities: decrease the growth rate of trees (Worbes 1985), change the metabolism of woody species (Joly 1994) and influence the morphology of individuals and the richness, structure, and distribution of plant species and communities (Junk 1996). Trees with deep roots and higher transpiration rates help the water stability of the stream, and their litter, with high lignin content and nitrogen (in legumes), increase the humified soil organic matter. On the other hand, grasses and forbs cover slow surface runoff of water and sediments infiltrating more water and nutrients (Shultz et al. 2004).

The forest cover of rivers, in Brazil, is highly reduced and fragmented as a function of forest exploitation and agricultural expansion. Elevated costs and weed invasion are constrains to forest restoration in the tropics, and grasses (*Brachiaria, Pennisetum, Cynodon, Sorghum*) are weeds with the great invasive capacity, which seems to prevent forest re-establishment (Hüller et al. 2009)

In Brazil most urban sewage is discharged without treatment into rivers, which are canalized in the large cities. Together with mining these are important threats to the Brazilian aquatic fauna (Pompeu et al. 2004).

Restoration of the natural cover is an important aspect of environmental management. The soil disturb by human activities decrease nitrogen-fixing bacteria and mycorrhizal fungi usually associated with roots (Cooke and Lefor 1990).

In recent years there has been an increasing interest in restoration programs with native species; however, the mycorrhizal status of them was few studied. The riparian vegetation, used for cattle and agriculture practices, has long been negatively influenced by human activities, especially domestic sewage and mining activity. Only recently were the river basins considered as part of the landscape and began to be studied for their environmental management (Osborne and Kovacic 1993). Moreover, a study of interactions among the functional groups of microorganisms in riparian regions of Brazil is fundamental to improve restoration programs.

This chapter presents an overview of field studies conducted in Brazil, especially in the Velhas River basin, in Minas Gerais. Data showing mycorrhizal symbioses in natural and restored sites are here compiled.

MYCORRHIZAS IN RIPARIAN ZONES

Mycorrhizal fungi have been studied extensively in upland ecosystems, but we know few about their ecology at riparian areas (Harner et al. 2009). Most common reports on arbuscular mycorrhizal fungi (AMF) in riparian vegetation are the following: Hashimoto and Higuchi (2003), Beauchamp et al. (2007), Piotrowski et al. (2007).

AMF and ectomycorrhizas (EM) occur in riparian areas (Jacobson 2004, Beauchamp et al. 2006, Piotrowski et al. 2008), and some riparian species (*Populus*) form associations with both types of mycorrhizas (Jacobson 2004). Harner et al. (2009) hypothesized that an analogous process to plant dispersion may distribute propagules of mycorrhizal fungi to early-successional riparian sites, enhancing primary succession. They found viable propagules (mostly small spores, probably of *Glomus*) and hyphae in freshly deposited sediments along river banks; however their abundance was low in relation to many upland habitats, and their distribution was heterogeneous among sites. Moreover, they hypothesized a similar viability of AMF propagules (20 days) during transport in river to in sea water.

MYCORRHIZAS IN RIPARIAN ZONES FROM BRAZIL

In Brazil, there is little information about patterns of occurrence of mycorrhizas in riparian forests. Studies by Melloni et al. (2001) on Minas Gerais's riparian forests of the Camargos dam, in Itutinga (917 m altitude) informed the number of phosphate solubilizing microorganisms and AMF spores, and the microbial activity in this woodland site; however, they did not identify the isolated AMF species.

I performed a literature survey and found 4 studies where AMF were studied in riparian areas from Brazil (Table 1).

Table 1. Summary of evidence on AMF in riparian zones, Brazil

Source	Location/ forest type	Experiment Type	Spore number	AMF Species Richness	Mycorrhizal colonization#
This study	Riparian forest, Sabará and Velhas Rivers, Minas Gerais	Field (natural and restored sites)	51*	27	65.92%
Pasqualini et al. (2007)	Soil from riparian forest, Itajaí river, Santa Catarina	Inoculation experiment in glasshouse (woody pioneer or late secondary species native to the Atlantic Rain Forest, utilized to recover riparian forests)	nd	nd	36%
Patreze and Cordeiro (2005)	Soil from riparian forest, Corumbataí, São Paulo	Inoculation experiment in glasshouse	nd	nd	45.25%
Carrenho et al. (2001)	Riparian forests of Moji-Guaçu River, São Paulo	Field (revegetated site)	511	22	nd
Melloni et al. (2001)	Riparian forests of the Camargos dam, in Itutinga, Minas Gerais	Field (natural site)	150	nd	nd

nd = not determined in the study, # Maximal AM colonization reported, *Spore number 100 g^{-1} soil.

Furthermore, in South Brazil, studies on revegetated riparian forests of Moji-Guaçu River (Carrenho et al. 2001) showed that each plant species favored a different community of AMF, and that *Glomus macrocarpum* was dominant. Other studies in seedlings showed that AMF inoculation influence plant growth and phosphorus (P) uptake of woody pioneer or late secondary species native to the Atlantic Rain Forest, utilized to recover riparian forests (Pasqualini et al. 2007).

Despite the necessity to increase the systematic research of AM in pteridophytes (Zhang et al. 2004), which will be important to indicate AM inoculum for fern or for other plant species, also for restoration of riparian areas, little information is available on the association between AM and ferns in Brazil, especially in riparian areas.

There are few reports on arbuscular mycorrhizal fungi (AMF) in pteridophytes (Zhao 2000, Zhang et al. 2004, Wang and Qiu 2006, Becerra et al. 2007, Menoyo et al. 2007). Most reports of AM symbioses in ferns refer to colonization, and few reports showed the spore number in their rhizospheres (Zhang et al. 2004). In Brazil, Marins et al. (2009) reported no arbuscular mycorrhizal (AM) colonization in 3 species of *Salvinia*, and to my knowledge there is no other report on AMF colonizing ferns in Brazil.

Ferns are important in restoration of degraded lands. Pteridophytes evolved, presenting adaptations to extremes environments, some of them presenting tolerance to dissecation, to salinity, and to heavy metals, and more knowledge is necessary for well understood these characteristics (Rathinasabapathi 2006). Furthermore, reports for the inoculum obtained from fern rooting-soil increasing the colonization of a leguminous nitrogen-fixing tree species (Asbjorsen and Montagnini 1994); as well as reports of revegetation with a fern *Dicranopteris linearis* (Burm. f.) Gleicheniaceae associated with Hawaiian Rain Forests (Follett et al. 2003), and *Dicranopteris curranii* Copel (Negishi et al. 2006), pioneer species, showed ecological roles in recovery mitigating erosion processes or plant invasions. Ferns as *Nephrolepsis biserrata*, which had high mycorrhizal colonization may have important implications for the restoration and management of degraded lands (Asbjorsen and Montagnini 1994).

MYCORRHIZAE IN RIPARIAN ZONES FROM SOUTHERN BRAZIL

The present chapter also presents information on root colonization and investigates which AM fungi are associated with species of natural or restored riparian ecosystems in Brazil, which was part of my postdoctoral research carried out between 2007 and 2009, which objective was to describe the diversity and potential of arbuscular mycorrhizal fungi vegetation of riparian forests along the Velhas River basin, Minas Gerais State. The benefit of this improved knowledge of mycorrhizal biology may involve the reduction of agricultural and forest surpluses, nature conservation, and the consequent relations to patterns of evolution within the plant families.

The silvicultural development and ecological contribution of native tree species associated with a riparian system, employed in the recovery of riparian forest in the Velhas River basin, Brazil, were previously reported in local (Pagano et al. 2008a, Pagano et al. 2008b) as well as in International Conferences (Pagano and Scotti 2008, Pagano et al. 2008c, Pagano et al. 2008d, Pagano et al. 2009). The principal variables addressed were initial growth, effect of fertilization, rhizospheric soil microorganisms (Pagano and Scotti 2008), and glomalin content of soil (Pagano et al. 2009). In this chapter, the results are discussed in relation to restoration practices and according to the literature.

The Velhas River is the main tributary of the São Francisco, one of the largest Brazilian rivers, and is subject to flooding. A possible restoration of the Velhas River is expected, however the continued mining in the headwaters, the rising number of exotic species, and contamination by agricultural pesticides remain important challenges (Pompeu et al. 2004). The Velhas River has social and economic importance, because the capital of Minas Gerais is located about 100 km from its headwaters, which provides most of the water supply. Only a 27.5% of sewage receives secondary treatment. In addition, the industrial waste, other impact, has produced the most polluted large river of the State. However, well-preserved tributaries persist in the basin (Pompeu et al. 2005).

The mixed native tree planting can be a good method for forest rehabilitation (Parrota and Knowles 1999). Nonetheless, little is known about silviculture and ecological responses of Neotropical tree species, and this difficult planning of forest restoration (Hüller et al. 2009).

Experimental riparian plots (~1 ha) established in different rivers (Sabará and Velhas) and surveys in other streams and tributaries of Velhas River have showed that artificial restoration with native species can be successful in southern Brazil. The native trees (7-10 plant species) were planted in 2006, with alternate lines of two species groups: pioneer and non-pioneer species, with a spacing of 2.5 m between lines and 2.0 m between plants. The species sequence in each group was established at random. The woody species were selected by their occurrence and adaptation at local riparian forest conditions. The pioneer species used were: *Inga edulis* Mart., *Mimosa bimucromata* (D.C.) O. Kuntze, *Anadenanthera colubrina* (Vell.) Brenan and *Plathymenia reticulata* Benth. The non-pioneers species planted were *Peltophorum dubium* (Spreng.) Taubert, *Centrolobium tomentosum*. Guill. Ex Benth and *Erythrina speciosa* Tod., *Samanea inopinata* (Harms) Ducke, considered early secondary (Cruz et al. 2006), and *Enterolobium contortisiliquum* (Vellozo) Morong. (Table 2, Figure 1). An 80% of Legumes were used. Legume trees usually facilitate the growth of non-legumes, since Leguminosae can support rhizobia and mycorrhizae; the arbuscular mycorrhiza being the most frequent (Frioni et al. 1999). In legumes, AMF may conceivably enhance plant performance by promoting plant vigor and hence biomass production and nitrogen (N) uptake, and it is also known that legumes are generally more mycotrophic than other plants (Plenchette et al. 2005), and they can increase the concentration of AMF spores in the soil (Colozzi and Cardoso 2000).

We hypothesized that using dual inoculated legumes in riparian zones, can increase plant growth, facilitating the riparian buffer function and propitiating future establishment of more exigent plant species. Thus, plants were inoculated with AMF. The noduliferous legumes were inoculated with specific rhizobia and AMF: spores mixed with AMF propagules. The experimental area also presented grasses as *Sporolobus indicus*, *Pennisetum setosum* (Poaceae) and herbs such as *Triumfetta* sp., *Wissadula* cf *contracta* (Malvaceae), *Marsypianthes chamaedrys* (Vahl) Kuntze (Lamiaceae) and *Aeschynomene* sp. (Fabaceae) (Table 2).

The disturbed site, presenting vegetation dominated by herbs, and a preserved site upper the river was sampled. Representative species from the degraded site were: *Megathyrsus maximus*, *Andropogon* sp., *Cynodon dactylon*, *Urochloa decumbens* (Poaceae), and *Calyptocarpus biaristatus*, *Tagetes minuta*, *Vernonanthura brasiliana*, *Tithonia speciosa* (Asteraceae), *Melochia villosa*, *Sida micrantha* (Malvaceae) (Table 2, Figure 1).

Mycorrhizas in Natural and Restored Riparian Zones

Table 2. Some characteristics of the Velhas River and tributaries riparian zones under study for AMF [#] (m.a.s.l.)

River	Coordinates	Locality	Elevation[#]	Soil type	Vegetation	River conditions	Studied plant species
Itabirito River	20°13'21.2" S 43°48' 9.3" W	Itabirito	823	Sandy	-	Disturbed	-
Paraúna River	18°37'54.2" S 44°03' 48.2 W	Presidente Juscelino	596		Graminous herbaceous cover	Receives waste city	-
Paraúna River	18°40' 56.6" S 43°35' 57.7"W	Presidente Kubitschek	1109	Sandy	Graminous herbaceous	Free-flowing river /stones	-
Peixe River	20° 10' 58"S 43°54' 35.7"W	Itabirito (Estoril farm)	-		-	Natural river	-
Gaia stream	19° 52' S 43° 47'W	Sabará (Reserve forest)	735	Clay loam	Native herbaceous and woody species (Atlantic Forest and the Cerrado savannas)	Natural river	-
Sabará River	19°53'32"S 43°48'31"W	Sabará (urban site)	637	Loamy sand	Revegetated with native trees	Urban river	*Anadenanthera colubrina, Centrolobium tomentosum, Erythrina speciosa, Inga edulis, Mimosa bimucronata, Plathymenia reticulata, Samanea inopinata, Peltophorum dubium, Aeschynomene* sp., *Croton urucurana,* Weeds: *Urochloa plantaginea*
Sabará River	19°53'32"S 43°48'31"W	Sabará (urban site)	637	Loamy sand	Graminous herbaceous cover	Disturbed	*Urochloa decumbens, Digitaria ciliaris, Calyptocarpus biaristatus*
Velhas River	20°18' 44"S 43° 34' 46.3"	São Bartolomeu	1124	Sandy loam	-	Disturbed	-
Velhas River	43° 51' 58" S 19° 50' 51" W	Sabará (Farm)	662	Sandy loam	Revegetated with native trees	Disturbed	Weeds: *Urochloa plantaginea, Sporobolus indicus Pennisetum setosum*
Velhas River	43° 51' 58" S 19° 50' 51" W	Sabará (Farm)	662	Sandy loam	Graminous herbaceous cover	Disturbed	*Urochloa decumbens, Andropogon* sp., *Megathyrsus maximus, Cynodon dactylon, Triumfetta* sp., *Wissadula contracta, Calyptocarpus biaristatus, Tagetes minuta, Vernonanthura brasiliana*

Figure 1. Disturbed sites that are in need of riparian plant cover (A and B). Graminous cover at Sabará River (A); weeds at Sabará River (B); bulk soil (C) at Sabará River; native trees planted at Sabará River (D, E); riparian pristine site (F) in Brazil.

Mycorrhizas in Natural and Restored Riparian Zones

Table 3. Analysis of the soil from some sampling sites at Velhas River and tributaries, Brazil

Soil property [a]	Gaia stream (Reserve forest)	Sabará River (Revegetated)	Paranaúba River	Itabirito River	Velhas River (São Bartolomeu)
pH (H_2O) 1:1	4.9	6.7	5.7	6.6	6.2
Soil organic matter (%)	5.74	0.99	1.58	1.58	5.01
C (%)	3.32	0.57	0.92	0.92	2.9
N			0.09	0.09	0.23
C/N			10.22	10.22	12.60
Avail. P (mg dm^3)	4.02	16.46	2.6	5.8	8.1
Avail. K (mg dm^3)	100	264.6	13	25	169
Exchang. Ca^{2+} (cmol (+) kg^{-1})	1.70	3.06	0.22	1.52	3.4
Exchang. Mg^{2+} (cmol (+) kg^{-1})	0.73	0.63	0.11	0.41	1.73
CEC (cmol (+) kg^{-1})	9.26	5.48	1.95	3.02	7.96
Base saturation (%)	29.03	78.9	18.68	66.25	69.87
Texture (%)[b]					
Coarse sand	17	24.8	63.3	14.5	22.3
Fine sand	36	40.2	33.1	72.16	46.42
Clay	37	13.8	0.4	8.66	17.36
Silt	10	21.2	3.2	4.68	13.92
Macroaggregates	22.48	22.26	-	23.07	22.57
Microaggregates	43.98	19.37	-	17.64	28.91
Aggregate stability (%)	88.8	52.8	-	-	-

[a] Mean of two measures from one composite sample. [b] Particle size distribution: coarse sand 2-0.2 mm, fine sand 0.2-0.02 mm, silt 0.02-0.002 mm and clay < 0.002 mm. mg L^{-1} = milligram per liter, CEC = cation exchange capacity. (-) Not determined.

Restored riparian sites presented a steep vertical stream bank of ≥ 2 m (Velhas River) and a lower bank (Sabará River), and were supplied of chemical fertilization (N-P-K, 04:30:10), 50 Kg ha^{-1}, organic matter addition, 400 g/hole, and P fertilization (19 g/hole). At Sabará River, a nearby degraded site, presenting vegetation dominated by herbs: *Urochloa plantaginea*, *Urochloa decumbens*, *Digitaria ciliaris* (Poaceae) and *Calyptocarpus biaristatus* (Asteraceae), was studied, and a preserved site upper the river was used as reference area. Details of the original experimental design and sampling are provided by (Pagano et al. 2008a,b, 2009); I present here an overview necessary to place in perspective the findings from the present chapter.

Abiotic Conditions

The total rainfall ranges from 1.300–1.400 mm. Rainfall is unimodal, most coming between November and March, followed by a prolonged dry season. In this study, wet season refers to the November to April and dry season refers to the May to October months (Figure 2).

Soils, which belong to the sand textural class, were slightly acid or neutral in some disturbed sites, and the organic matter (OM) content increased along the time. Commonly, Base saturation was high, and P content, moderated. In general, natural riparian forest had relatively higher soil OM content than restored sites (except Velhas River). The C (carbon) soil contents also were dramatically reduced once under disturbs. Available P (phosphorus) in the disturbed sites, however, was similar to forest sites, probably as a result of fertilizer application (Table 2).

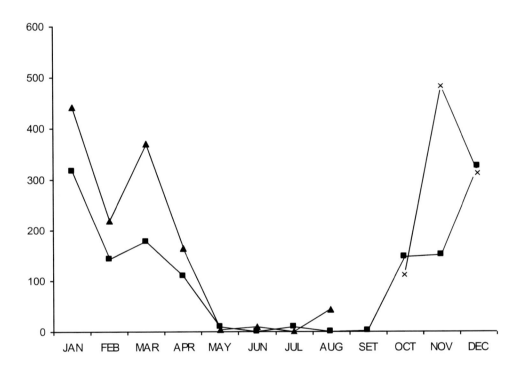

Figure 2. Regional precipitation for the period 2006-2008. Meteorological data from Estação Climatológica Automática da Pampulha/Belo Horizonte/MG. Lat: 19°53'00" S Long: 043°58'00" W Alt: 869 m a.s.l. Instituto Nacional de Meteorologia – INMET. (×) 2006, (■) 2007, (▲) 2008. Jan=January, Feb= February, Mar= March, Apr= April, May= May, Jun= June, Jul= July, Aug= August, Sep= September, Oct= October, Nov= November, Dec= December.

Plant Growth in Riparian Environments

The results obtained corroborate the important contribution of ecologically complementary species in mixed systems of tree planting for riparian forest restoration. *S. inopinata* and *M. bimucronata* presented the best performance in relation to the highest maximum height and diameter (Table 4).

Peltophorum dubium also showed high growth, but a low spore number in their rhizosphere. *Peltophorum dubium* is considered an initial secondary tree species (Duringan and Nogueira 1990) with fast growth, or climax (Siqueira et al. 1998). This caducifolious full-sun plant tree attaining 12-20 m height, and is used in agroforestry systems (Carvalho 2003) and has a potential use in restoration of degraded lands, used in mixed plantations (Lorenzi 1992). Also, *P. dubium* tolerates flooding periods (Medri et al. 1998). Their nutritional requirements are few know and the AM symbioses have been showed (Faria et al. 1995) in greenhouse conditions. However, Carneiro et al. (1998) and Zangaro et al. (2002) have reported *P. dubium* as a non-mycorrhizal plant. Frioni et al. (1999) reported 48% colonization in this species. The great majority of Caesalpinoideae are non-nodulating (Sprent and Sprent 1990); however, Frioni et al. (2001) reported isolates of rhizobia wich fast-growing from *P. dubium* in Uruguay.

Erythrina speciosa and *Inga edulis* showed a high growth and spore number especially the last plant species. *C. tomentosum* showed the lowest growth. Notwithstanding its low growth can be explained at least partially by its successional characteristics since it is not a pioneer species. *C. tomentosum* is an important contribution to shading species, and a noduliferous legume which fix nitrogen (Pagano 2009).

The mortality of individual plants from 22% (Velhas River) to 23% (Sabará River) in the beginning of the restoration process was expected due to water stress during the dry period, and also occurrence of wildfires (crown fires). At Sabará River, *C. tomentosum* presented the lower mortality (15%), followed by *E. speciosa* (26%) and *I. edulis* (40%). The mortality of individuals occurring in higher positions, where flooding is an unusual event, is expected to be higher than that of individuals occurring in lower positions submitted to annual flooding where the species are theoretically more tolerant to inundation and anoxia (Damasceno-Junior et al. 2004); however it depends of the plant species identity, as for *Inga vera* ssp. *affinis*, the most abundant species in riparian forest of the Paraguay River, mortality was 3.15% per year (Damasceno-Junior et al. 2004).

The set of plant species studied at the Velhas River basin showed a significant growth in the rainy period, which could favor a high AMF colonization in roots.

Spore Numbers and Phosphate Solubilizing Microorganisms

The number of solubilizing bacteria at restored and degraded riparian zones (Sabará River) were higher than those of pristine sites (Table 4), and this can be favored by the planting of trees fertilized with rock phosphate, or by previous fertilization in these areas. The quality and quantity of OM, the vegetal cover and the mycorrhizal associations affect the capability of solubilizing inorganic phosphate by microbes on soils (Eira 1992).

Native ferns favoured different AMF species, particularly *Glomus*. This suggests that it may be important to consider differences in the mycorrhization in ferns, and its influence on plant establishment, growth, and competitive interactions, in designing restoration and management strategies for degraded riparian areas. Further studies are needed to determine whether AMF associated with pteridophytes are functionally involved in riparian areas dynamics.

Table 4. AMF spore number, richness and phosphate solubilizing culturable microorganism at Sabará River riparian zone, Minas Gerais, Brazil

Site	AMF spore number	AMF species richness	PSM
Restored	36*	10	41.44[#ab]
Preserved	49	7	11,67[b]
Degraded	18	5	49,67[a]

* Spore number x 100 g^{-1} dry soil, [#] N° CFU x 10^5 g^{-1} dry soil. Letters indicate significant differences by the Tukey's test ($P < 0.05$). CFU (colony forming units).

Table 5. Average values of height, diameter and AMF spore number and richness, 24 months after planting, in Velhas River and tributaries riparian zones, Minas Gerais, Brazil

Vegetal Species	Height (cm)	Diameter (mm)	Spore number	AMF species richness
M. bimucronata	406	42.08	51[#]	6.5
E. speciosa	64.44	39.46	46.5	8
C. tomentosum	42	9.14	38	7
I. edulis	64.4	8.65	68.33	10
P. dubium	327.5	34.64	8.33	5
P. reticulata	173.75	26.15	ND	ND
S. inopinata	466	64.56	ND	ND

Table 6. AMF status in roots from plants of the different studied areas at Sabará River riparian zone, Minas Gerais, Brazil

Host	Length	NS	Coils	V	Type	PC[‡]	EH	Previous report
C. tomentosum	34.52	+	+	17.14	Arum	h, ar, ov	-	AM[2,7]
I. edulis [a]	30	+	-	0	Arum	h, ar, ov	+	
E. speciosa [a]	50	+	+	10	Arum	h, ar, ov	+	
M. bimucronata	50	+	-	8.57	Arum	h, ov		AM[3]
P. dubium	65.92	+	-	48.4	Arum	h, ov	-	AM[1], NM[4,5]
U. plantaginea [a]	38.88	+	-	-	ND	h	-	
Degraded área	44.44	+	-	2.2	Arum	h, ov	-	
C. biaristatus [a]	20	+	-	10	Arum	h	-	
Preserved forest site	66.21	+	-	10.8	Arum	h, ar, iv, ov, rh	25.5	

[‡]PC: Patterns of AM colonization; *h*: intra- or intercellular aseptate hyphae, *ar*: arbuscules, *ov*: oval vesicles, *iv*: irregular vesicles, *c*: coils, *rh*: root hairs. Relative development of structures shown as: ++ always present in significant numbers, + always present, − not detected, ND not determined, Length AM root length %, NS non-septate hyphae, coils hyphal coils, V vesicles, type *Arum* or *Paris*, EH extra-radical hyphae, [a]New records of AM type. [1]Frioni et al. (1999), [2]Marques et al. (2001), [3] Patreze and Cordeiro (2004), [4]Siqueira et al. (1998), [5]Zangaro et al. (2003).

AMF spore numbers were generally higher at the restored and the natural sites than at the degraded ones (Table 4). With a few exceptions, spore numbers from rhizospheric soil of noduliferous legumes (restored sites) were consistently higher compared with non noduliferous legumes (Table 5, Figure 3). The highest number of spores occurred in the rhizosphere of *I. edulis*.

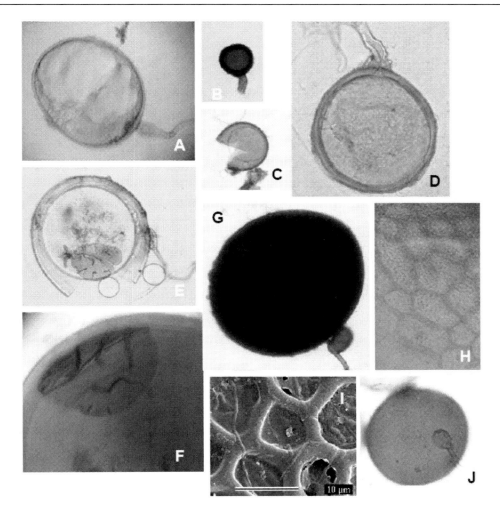

Figure 3. Spores of AMF species found in riparian sites in Brazil: spores of *Scutellospora* (A,C,E-J), and *Glomus* (B,D). Ornamentation on the surface (H) and scanning electron micrograph (SEM) (I) of *S. reticulata* (G) in PVLG.

Root Colonization In Riparian Areas

Arbuscular mycorrhizal colonization was evident in tree species, herbaceous dicotyledons, and 3 herbaceous monocotyledons. Aseptate intra and intercellular hyphae, vesicles, were observed in the majority of the samples. Arbuscules or hyphal coils were less frequent. In general, the extent of root colonization varied from about 34% to 50% in the revegetated site, 20 to 44% in the degraded site and 66% in the preserved environment. Moreover, colonization by AMF in degraded site was higher than colonization of herbs within the experimental site.

Although the colonization pattern varied among the species, intracellular aseptate hyphae and vesicles were the most frequent AM structures present in the studied species. Arbuscular mycorrhizal colonization varied among the species studied and among the sampled sites (Table 3).

In the preserved site, the vegetation cover showed the highest colonization (66%) and the highest percent of arbuscules (28%), whereas the percent of vesicles was low (14%).

In this study, the mycorrhizal status of some herbs, which belong to the Asteraceae and Poaceae families, on the riparian vegetation of Sabará River is reported for the first time.

Table 7. Mycorrhizal status of the terrestrial ferns studied in riparian sites from southeast of Brazil

Order	Family	Plant species	NS	PC	Type	AMF %	Previous report
Polypodiales	Blechnaceae	*Blechnum occidentale* L.	-	eh, rh	-	0	AM[1]
		Blechnum polypoides (Sw.) Kuhn[a]	+		*Paris*	IV	
	Gleicheniaceae	*Dicranopteris flexuosa* Und.[a]	-	eh, ac, rh	-	0	
Pteridales	Pteridaceae	*Pityrogramma trifoliata* (L.) R.M.Tryon[a]	+	ar, c, eh, ov, h	Intermediate	V	
Ophioglossales	Thelypteridaceae	*Thelypteris dentata* (Forssk.) E.P.St.John[a]	+	ac, c, eh, h, iv, rh	*Arum* and *Paris*	IV	
		Thelypteris serrata Alston [a]	-	eh, rh	-	0	

[a]New records of AM type. Note: Indicated are the families. NS non-septate hyphae, PC: Patterns of AM colonization; ar: arbuscules, ac: auxiliary cells, c: hyphal coils, eh: extraradical hyphae, h: intra- or intercellular aseptate hyphae, iv: irregular vesicles, ov: oval vesicles, rh: root hairs. Structures shown as: + always present, – not detected. AM type: colonization type: *Arum* type, *Paris* type, I= intermediate. Arbuscular mycorrhizal (AM) colonization, class: I, 1-5%; II, 6-25%; III, 26-50%; IV, 51-75% and V, 76-100%. Rh = root hairs. Ac = auxiliary cells. (+) = presence, (-) = absence. [1]Gemma et al. (1992).

The results obtained showed that all the species present colonization of the *Arum*-type. This type was seen to be dominant in the herbaceous understory plants of the revegetated area, in the degraded area, and in preserved area. It is known that the *Arum*-type is formed in most plants that usually grow in sunlight and that the spreading rate of colonization is faster than the *Paris*-type. The slower colonization of the *Paris*-type is usually found in plants of slow growth in woodland environments (Brundrett and Kendrick 1990). Nevertheless, these results must be considered preliminary, since they cover only a small proportion of the plant diversity of these forests.

As regards nutritional demand plants species differ due to their ecophysiological characteristics and their capacity to form symbioses with soil microorganisms, especially with mycorrhizal fungi. The need to know the nutrient requirements of the plant species and their mycorrhizal dependency is thus crucial. All native tree species studied presented *Arum*-type colonization in their roots, and the significant AM morphological structures were documented

(Figure 4). Variations in occurrence of fungal structures provide information about the fungi in relation to nutrient transfer and plant growth (Jakobsen et al. 2003), as the external hyphae produced by a mycorrhizal fungus can indicate its relative ability to uptake phosphorus (Jones et al. 1990). Moreover, hyphae connected by "h"-shape anastomosis pattern, often observed in riparian roots (Figure 4), is a Glomineae-type colonization.

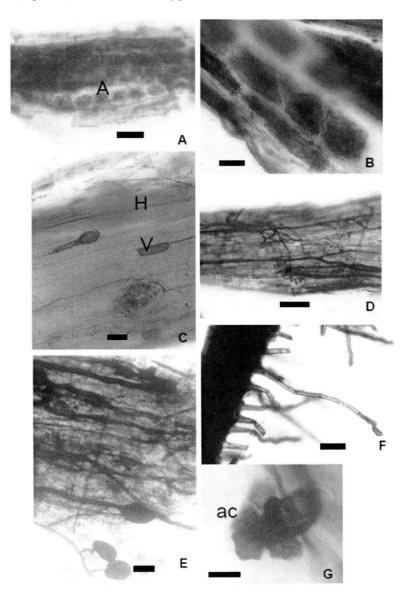

Figure 4. Differences in colonization pattern of AM fungi in roots of riparian plants. Arbuscules in *I. edulis* a root segment (A); coils in *B. polypodioides* (B); intra radical hyphae bearing vesicles in *C. tomentosum* (C); intra radical hyphae in *M. bimucronata* showing "y" branching pattern (D); intra radical hyphae bearing vesicles in *P. dubium* (E); root hairs in *D. flexuosa*, external auxiliary cell in *D. flexuosa* root (G). A = arbuscule; AC = auxilliary cells, H = hyphae, V = vesicle. Scales: (a, b, c, d) = 100 μm, (e, f) = 50 μm, (g) = 25 μm.

Table 8. AM spore diversity in some pristine, restored and degraded riparian sites in Brazil

AMF Species (authority)	1	2	3	4	5	6	7	8	9
Acaulosporaceae									
Acaulospora bireticulata Rothwell & Trappe								X	
A. excavata Inglebly & Walker								X	
A. foveata Trappe & Janos				X				X	
A. mellea Spain & Schenck								X	
A. paulineae Blazkowski				X					
A. scrobiculata Trappe				X	X	X	X	X	
A spinosa Walker & Trappe								X	
Acaulospora sp. 1					X	X		X	X
Acaulospora laevis Gerdemann & Trappe							X	X	
A. rhemii Sieverding & Toro			X						
Entrophosporaceae									
Entrophospora infrequens (Hall) Ames & Schneider								X	
Gigasporaceae									
Gigaspora sp. 1					X	X		X	
Scutellosporaceae									
Scutellospora aurigloba (Hall) Walker & Sanders					X	X		X	
Scutellospora sp. 1					X	X		X	
S. reticulata (Koske, Miller & Walker) Walker & Sanders								X	
S. biornata (Spain, Sieverd. & S. Toro) Sieverd., F.A.Souza & Oehl		X							
Scutellospora cf *cerradensis*							X		
S. rubra (Stürmer & J.B Morton) Oehl, F.A.Souza & Sieverd								X	
Scutellospora sp. 2		X							
Scutellospora sp. 3				X					
Racocetraceae									
Racocetra fulgida (Koske & Walker) Oehl, F.A.Souza & Sieverd.	X		X	X	X			X	X
Racocetra gregaria (N.C. Schenck & T.H. Nicolson) Oehl, F.A.Souza & Sieverd.		X						X	
Glomeraceae									
Glomus brohultii Sieverd. & Herrera								X	X
Glomus constrictum Trappe						X		X	
Glomus sp.1	X	X	X		X	X	X		
Glomus etunicatum Becker & Gerdemann						X			
Glomus macrocarpum Tulasne & Tulasne					X				

1 Paraúna river (Presidente Juscelino), 2 Paraúna river (Presidente Kubitschek), 3 Peixe river, 4 Gaia stream - Reserve forest, 5 Sabará river – restored site, 6 Sabará river – disturbed site, 7 Velhas River – São Bartolomeu, 8 Velhas River Restored, 9 Velhas River Degraded.

In general, aseptate intra and intercellular hyphae, vesicles, were the most frequent AM structures present in the studied tree species. Arbuscules or hyphal coils were less frequent (Table 6 and 7), being observed only in *I. edulis* and *B. polypoides* (Figure 4).

The mycorrhizal status of five terrestrial ferns studied in riparian sites from southeast of Brazil is report for the first time (Table 7). The *Arum*-type of colonization was the most common observed. Three ferns showed dominant extraradical hyphae, whereas their mycorrizal status remained uncertain. Two fern species formed *Arum*-type and one *Paris*-type AM. AMF spore richness was higher in the rhizospheric soils of *P. trifoliata*, which occurred in the pristine site.

The Diversity of AMF Fungi in Riparian Areas

A total of 27 AMF species were detected in soils sampled from the riparian sites (Table 8). Five species belonged to the genus *Glomus* in the family Glomeraceae, 10 species to *Acaulospora*, in the Acaulosporaceae. There were eight *Scutellospora*, epresentants, two of Racocetraceae, and one *Gigaspora* species in the Gigasporaceae. One species belonged to the families Entrophosporaceae. This is similar to Carrenho's (2001) identification of 22 species for conventional revegetated riparian sites.

AMF genetic diversity, evaluated by denaturing gradient gel electrophoresis (PCR-DGGE) procedure (Pagano et al. 2008a), showed that the highest diversity was found in the preserved area. DGGE bands revealed 24 AMF species (assuming each band represented a different isolate), that were related to plant host and vegetation cover. This survey by molecular techniques found 0–15 species per site. This study is one of the first using PCR–DGGE to characterize AMF communities in riparian environments, and is the first molecular survey of AMF in a these ecosystems in Brazil.

Denaturing gradient gel electrophoresis (DGGE) is a molecular fingerprinting method that separates polymerase chain reaction (PCR)-generated DNA products. The sampling strategy used to assess the AMF community composition in soils may dilute the number of spores per gram of soil because of patchy occurrence of AMF spores (Smith and Read 2008); however the detection of either spores or hyphal fragments showed a higher diversity, especially in the pristine site.

A high species richness of AMF suggests a high level of functional diversity in these environments.

AMF Species Richness and Land Use

Land degradation and ruderal species invasion in Sabará, State of Minas Gerais, reduces AMF species richness, thus AMF can be used as indicators of riparian land degradation. Independent of the river location, land disturbance negatively affected the AMF species richness, particularly species of Gigasporaceae, Scutellosporaceae, and Racocetraceae. Acaulosporaceae was also reduced, while *Acaulospora scrobiculata* were less affected, and recorded from five of the sites under investigation. Only a species of *Scutellospora* and one of *Gigaspora* was found in the degraded sites, thus contrasting with the preserved site. In the preserved areas Glomeraceae and Acaulosporaceae spores were dominant.

In general soil disturbance seems to select for mycorrhizal fungi (*Acaulospora* and *Glomus* spp.) that could differ in their strategies to exploit limiting resources (Miranda and Reader 2002). These results demonstrate that native trees and herbaceous species favoured different AMF species richness, and that soil disturbance (cattle and human impacts) decrease AMF species.

The evaluation of the AMF genetic diversity also showed that the variations in AMF populations were related to the level of degradation of the riparian areas, and none of Gigasporaceae species was found in the degraded area.

Glomalin-Related Soil Protein Content and Land Use

Glomalin is a glycoprotein produced by arbuscular mycorrhizal fungi, which concentration was typically highly correlated with soil aggregate water stability (Wright and Upadhyaya 1998). These fungi produce glomalin within their hyphal walls (Driver et al. 2005, Wright and Upadhyaya 1996), which, is deposited within the soil, As the hyphae senesce, accumulating until it represents as much as 5% of soil C (Rillig et al. 2003, 2001) and N (Lovelock et al. 2004). Recently, Purin and Rillig (2007) have proposed a primary physiological function relating to wall-location of the protein and effects on palatability of the mycelium and the secondary environmental function in the soil in the context of soil aggregation.

The extraction from soil of 'Bradford-reactive soil protein (BRSP) was higher for the pristine (Forest reserve) and restored sites than for the disturbed site at Sabará River (Figure 5), as was expected due to the fact that the availability of plant C and composition of the plant community appears to be an important determinant of glomalin stocks (Treseder and Turner 2007, Rillig et al. 2002). Moreover, the restored riparian zones studied (Sabará and Velhas Rivers), which are flooded every year; receive a substantial input of OM, which potentially makes the soils in this zone richer than those in disturbed areas.

THE IMPORTANCE OF RIPARIAN AREAS CONSERVATION AND RESTORATION

The spontaneous regeneration of riparian forests is insufficient to meet the increasing human impact, thus management protocols to accelerate their restoration are urgently needed.

As the long-term treatment of vegetation is the simplest solution for land restoration, allowing to the natural or artificial succession (Bradshaw 2002), a theoretical base is needed to develop technologies on management of native species, for riparian land restoration and fragments of natural ecosystems conservation. It was pointed out that the principles and practices of conservation from temperate, developed-world regions can generally be applied to tropical, developing regions, but the specific solutions are likely to be determined by regional ecological and socio-economic factors (Moulton and Wantzen 2006). In Brazil, some analysis of the possibilities of carbon credit attainment by low-income community, as part of incentive programs for the restoration of these areas, can help contributing for local restoration of the areas and also for carbon capture by the atmosphere, which this is a global subject (Crisci 2007).

Thus, specific studies and monitoring are necessary, as well as the conservation process; in parallel to the management actions and refining the conservation planning (Moulton and

Wantzen 2006). Moreover, Osborne and Kovacic (1993) stressed that several questions on the utility and efficiency of vegetated buffer strips for stream restoration still remain unanswered, including: the most efficient types (grass or forest); the nutrient saturation; the function as temporary sinks; the species composition influence; and, the optimal width of buffer to facilitate nutrient reduction under different conditions. Statements about buffer effectiveness over a wide range of landscapes must be made carefully and further research must be conducted to establish the range of process rates in other ecoregions around the world (see Schultz et al. 2004).

Alternative restoration practices, most restoration efforts and the imitation of the geomorphology or of the riparian vegetation of a quasi-natural or natural reference channel may prove to be a more effective means of controlling non-point-source agricultural inputs of nutrients and need more attention (Osborne and Kovacic 1993).

Successful stream restoration requires a multidisciplinary approach monitoring the outcome of past, existing and future stream-restoration projects for information on the feasibility of alternative techniques and approaches. For that, it was recommended that systems in pristine condition serve as a point of reference and not as a goal for most stream restoration projects, and that all restoration programs should consider geomorphic, hydrological, biological, aesthetic, and water quality aspects of the system (Osborne et al. 1991).

To create a system with a stable channel, or a channel in dynamic equilibrium that supports a self-sustaining and functionally diverse community assemblage; with more than one species or group must be the goal of restoration programs (Osborne et al. 1991).

The environmental benefits of mycorrhizas on riparian environments in Brazil have been scarcely studied, and require rigorous analyses. It also remains to be tested if the dominant fungal species are the most functionally relevant to ecosystem and the minor species are functionally equivalents of the dominant ones, then the minor species contributing to ecosystem resilience as hypothesized by Allen et al. (2003).

CONCLUSIONS AND FUTURE GOALS

In the introduction to this chapter, I briefly describe the function of riparian zones, as well as the mycorrhizal fungi significant benefits to their plant hosts, and that many efforts have been made in recent years to accrue benefits from mycorrhizae in riparian environments, however a few works were carried out in Brazil.

Throughout the chapter, the importance of mycorrrhizae as an essential component for riparian plant communities was highlighted. Nonetheless, further studies are required to achieve maximum benefits from these microorganisms and their associations. All these results show that mycothropic native tree species are indicated for mixture in riparian zones and that restoration programs should take mycorrhizae into account. Moreover, the presence and amounts of these symbionts can affect successional trajectories of riparian plant communities.

Studies incorporating a larger number of sites and seasonal sampling will help clarify these relationships with greater confidence. Investigations such as these can shed light on the complex patterns in AMF communities, interactions between sediment texture, and plant diversity on flood plains.

Finally, a more complete monitoring of the community composition of AMF is a first step towards understanding their ecology, and requires not only the development of methodologies able to detect the whole range of AMF groups, but also more ambitious sampling strategies, both in terms of space and time.

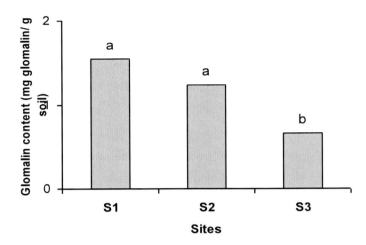

Figure 5. Concentration of easily-extractable Bradford-reactive soil protein in relation to soil disturbance. For each histogram bar, $n = 5$. Histogram bars labeled with the same lower case letter are not significantly different at $p<0.05$. S1 Gaia stream - Reserve forest, S2 Sabará River – restored site, S3 Sabará River – disturbed site.

In Brazil, the occurrence of AMF in riparian trees is not yet well documented; however, this study provides the first detailed report ever published on the mycorrhizal status of some of the species examined. These results emphasize the need to consider the symbiotic fungi in riparian restoration practices, which should select highly dependent tree hosts over mycorrhizal- independent. The choice of tree species would therefore have great implication in the persistence of AMF species.

ACKNOWLEDGMENTS

This research was supported by FAPEMIG (Minas Gerais Research Funding Agency - Process 311/07). Marcela Pagano is grateful to CAPES (Council for the Development of Higher Education at Graduate Level, Brazil) for Post-doctoral scholarships granted from 2009 to 2010. Marta Noemí Cabello is researcher from Comisión de Investigaciones Científicas de la Provincia de Buenos Aires (CIC), Argentina. Marcela Pagano is grateful to Fritz Oehl for help with spore identification.

REFERENCES

Allan, JD. Landscapes and Riverscapes: The Influence of Land Use on Stream Ecosystems. *Annu. Rev. Ecol. Evol. Syst.* 2004, 35, 257-84.

Allen, MF; Swenson, W; Querejeta JI; Egerton-Warburton, LM; Treseder, KK. Ecology of Mycorrhizae: A conceptual framework for complex interactions among plants and fungi. *Annu. Rev. Phytopathol.*, 2003, 41, 271-303.

Alvarenga, MIN; Siqueira, JO; Davide, AC. Teor de carbono, biomassa microbiana, agregação e micorriza em solos de cerrado com diferentes usos. *Ciênc. agrotec.* 1999, 23, 617-625. (in Portuguese).

Asbjornsen, H; Montagnini, F. Vesicular-arbuscular mycorrhizal inoculum potential affects the growth of *Stryphnodendron microstachyum* seedlings in a Costa Rican human tropical lowland. Mycorrhiza. 1994, 5, 45-51.

Barbosa, LM. Considerações Gerais e Modelos de Recuperação de Formações Ciliares. In: Rodrigues, R.R., Leitao Filho, H.F., editors. *Matas Ciliares. Conservação e Recuperação*, São Paulo: Edusp; 2000; 289-312 (in Portuguese).

Beauchamp, VB; Stromberg, JC; Stutz JC. Arbuscular mycorrhizal fungi associated with Populus-Salix stands in a semiarid riparian ecosystem. *New Phytologist*, 2006, 170, 369-380.

Beauchamp, VB; Stromberg, JC; Stutz, JC. Flow regulation has minimal influence on mycorrhizal fungi of a semi-arid floodplain ecosystem despite changes in hydrology, soils, and vegetation. *J. Arid Environ.*, 2007, 68, 188-205.

Becerra, A; Cabello, M; Chiarini, F. Arbuscular mycorrhizal colonization of vascular plants from the Yungas forests, Argentina. *Ann. For. Sci.*, 2007, 64, 765-772.

Bradshaw, AD. Introduction and philosophy. In: Perrow MR, Davy AJ, editors. *Handbook of ecological restoration*, Cambridge University Press, Cambridge; 2002; 1–9.

Brundrett, M; Kendrick, B. The roots and mycorrhizas of herbaceous woodland plants. I. Quantitative aspects of morphology. *New Phytol.*, 1990, 114, 457-468.

Campbell, DG; Stone, JL; Rosas JrA. A comparison of the phytosociology and dynamics of three floodplain (Varzea) forests of known ages, rio Juruá, western Brazilian Amazon. *Bot. J. Linn. Soc.* 1992, 108, 213–237.

Carneiro, MAC; Siqueira, JO; Moreira, FMS; Carvalho, D; Botelho, AS; Junior, OJS. Micorriza arbuscular em espécies arbóreas e arbustivas nativas de ocorrência no Sudeste do Brasil. CERNE, 1998, 4, 129-145. (in Portuguese).

Carrenho, R; Trufem, SFB; Bononi, VLR. Arbuscular mycorrhizal fungi in rhizospheres of three phytobionts established in a riparian area. *Acta bot. bras.,* 2001, 15, 115-124.

Carvalho, PER. *Espécies florestais brasileiras. Recomendações Silviculturais, potencialidades e uso da madeira*. Brasília: EMBRAPA-CNPF; 2003 (in Portuguese).

Cooke, JC; Lefor, MW. Comparison of Vesicular-Arbuscular Mycorrhizae in Plants from Disturbed and Adjacent Undisturbed Regions of a Coastal Salt Marsh in Clinton, Connecticut, *USA Environmental Management,*1990, 14, 131-137.

Colozzi, A; Cardoso, EJBN. Detection of arbuscular mycorrhizal fungi in roots of coffee plants and Crotalaria cultivated between rows. Pesqui. Agropecu. Bras., 2000, 35,2033-2042.

Crisci, MC. *Gallery Forest Restoration by the Attainment of Carbon Credit: a proposal social-environmental for low-income community Coelho*. Master thesis, 2007 (in Portuguese).

Cruz, CAF; Paiva, HN; Guerrero, CRA. Efeito da adubação nitrogenada na produção de mudas de sete-cascas (Samanea *inopinata* (Harms) Ducke). *Rev. Árvore*, 2006, 30, 537-546, (in Portuguese).

Damasceno-Junior, GA; Semir, J; Santos FAM; Leitão-Filho, HF. Tree mortality in a riparian forest at Rio Paraguai, Pantanal, Brazil, after an extreme flooding. *Acta Bot. Bras.*, 2004, 18, 839-846.

Donaldson, MR; Henein, KM; Runtz, MW. Assessing the effect of developed habitat on waterbird behaviour in an urban riparian system in Ottawa, Canadá. *Urban Ecosyst.*, 2007, 10:139-151.

Driver, JD; Holben, WE; Rillig, MC. Characterization of glomalin as a hyphal wall component of arbuscular mycorrhizal fungi. *Soil Biol. Biochem.*, 2005, 37,101-106.

Durigan, G; Nogueira, JCB. *Recomposição de matas ciliares: orientações básicas*. São Paulo: IF, 4; 1990 (in Portuguese).

Eira, AF. Solubilização microbiana de fosfatos. In: Cardoso, EJBN; Tsai, SM; Neves, MCP. *Microbiologia do Solo*. Campinas: Sociedade Brasileira de Ciências do Solo; 1992; 243-255 (in Portuguese).

Faria, MP; Vale, FR; Siqueira, JO; Curi, N. Crescimento de leguminosas arbóreas em resposta a fósforo, fungo micorrízico e rizóbio. II. *Peltophorum dubium* (Spreng.) Taub. *R. Árv.*, 1995, 19, 4, 433-446 (in Portuguese).

Follett, PA; Puanani, AW; Johnson, M; Jones, T; Vincent, P. Revegetation in Dead *Dicranopteris* (Gleicheniaceae) Fern Patches Associated with Hawaiian Rain Forests. *Pacific Science*, 2003, 57, 347-357.

Frioni, L; Minasian, H; Volfovicz, R. Arbuscular mycorrhizae and ectomycorrhizae in native tree legumes in Uruguay. *Forest Ecology and Management*, 1999, 115, 41-47.

Frioni, L; Rodríguez, A; Meerhoff, M. Differentiation of rhizobia isolated from native legume trees in Uruguay. *Applied Soil Ecology*, 2001, 16, 275–282.

Harner, MJ; Piotrowski, JS; Lekberg, Y; Stanford, JA; Rillig, MC. Heterogeneity in mycorrhizal inoculum potential of flood-deposited sediments. *Aquat. Sci.*, 2009, 71, 331-337.

Hashimoto, Y; Higuchi, R. Ectomycorrhizal and arbuscular mycorrhizal colonization of two species of floodplain willows. *Mycoscience*, 2003, 44,339-343.

Hüller, A; Coelho, GC; Lucchese, OA; Schirmer, JA. Comparative Study Of Four Tree Species Used In Riparian Forest Restoration Along Uruguay River, Brazil. *R. Árvore*, 2009, 33, 297-304.

Jakobsen, I; Smith, SE; Smith, FA. Function and diversity of Arbuscular mycorrhizae in carbon and mineral nutrition. In: van der Heijden MGA and Sanders IR, editors. *Mycorrhizal Ecology*. Berlin: Springer; 2003; 75-92.

Jacobson, KM. The effects of flooding regimes on mycorrhizal associations of *Populus fremontii* in dryland riparian forests. In: Cripps C. editor, *Fungi in Forest Ecosystems*: Diversity, Systematics, and Ecology, New York Botanical Gardens, New York; 2004; 275-280.

Joly, CA. Flooding tolerance: a reinterpretation of Crawford's metabolic theory. *Proc. R. Soc. Edinburgh*, 1994, 102, 343-354.

Jones, MD; Durall, DM; Tinker, PB. Phosphorus relationships and production of extramatrical hyphae by two types of willow ectomycorrhizas at different soil phosphorus levels. *New Phytologist*, 1990, 115, 259-267.

Junk, W. Ecology of floodplains—a challenge for tropical limnology. In: Shiemer F, Boland, KT, editors. *Perspectives in Tropical Limnology*. SPB Academic Publishing, Amsterdam; 1996; 255-265.

Kageyama, P; Gandara, FB. Considerações Gerais e Modelos de Recuperação de Formações Ciliares. In: Rodrigues, RR, Leitão Filho, HF, editors. *Matas Ciliares. Conservação e Recuperação*, São Paulo: Edusp; 2000; 249-269 (in Portuguese).

Lorenzi, H. *Árvores brasileiras: manual de identificação e cultivo de plantas arbóreas nativas do Brasil*. Nova Odessa: Plantarum, 1992. (in Portuguese).

Lovelock, CE; Wright, SF; Nichols, KA. Using glomalin as an indicator for arbuscular mycorrhizal hyphal growth: an example from a tropical rain forest soil. *Soil Biology and Biochemistry*, 2004, 36, 1009-1012.

Marques, MS; Pagano, MC; Alvarenga, A; Lages, M; Raposeiras, R; Scotti, MR. Arbuscular mycorrhizal communities in revegetated riparian areas in Brazil. Krakow Microbial Plant Interaction, 2-6 July 2008, Poland.

Marins, JF; Carrenho, R; Thomaz, SM. Occurrence and coexistence of arbuscular mycorrhizal fungi and dark septate fungi in aquatic macrophytes in a tropical river–floodplain system. *Aquatic Botany*, 2009, 91, 13-19.

Medri, ME; Bianchini, E; Pimenta, JA; Delgado, MF; Correa, GT. Morpho-anatomic and physiological aspects of *Peltophorum dubium* (Spr.) Taub. submitted to flooding and ethrel application. *Rev. Brasil. Bot.*, 1998, 21, 3.

Melloni, REG; Pereira, I; Trannin, CB; Dos Santos, DR; Moreira, FMS; Siqueira, JO. Características biológicas de solos sob mata ciliar e campo cerrado no sul de Minas Gerais. *Ciênc. agrotec.*, 2001, 25, 7-13 (in Portuguese).

Menoyo, E; Becerra, AG; Renison D. Mycorrhizal associations in *Polylepis* woodlands of Central Argentina. *Can. J. Bot.* 2007, 85, 526-531.

Miranda, MH; Reader, RJ. Taxonomic basis for variation in the colonization strategy of arbuscular mycorrhizal fungi. New Phytol., 2002, 153, 335-344.

Moulton, TP; Wantzen, KM. Conservation of tropical streams - special questions or conventional paradigms? *Aquatic Conservation: Marine and Freshwater Ecosystems*, 2006, 16, 659-663.

Negishi, JN; Sidle, RC; Noguchi, S; Nik, AR; Stanforth, R. Ecological roles of roadside fern (*Dicranopteris curranii*) on logging road recovery in Peninsular Malaysia: Preliminary results. *For. Ecol. Manage.*, 2006, 224,176-186.

Oliveira-Filho, AT; Vilela, EA; Gavilanes, ML; Carvalho, DA;. Effect of flooding regime and understorey bamboos on the physiognomy and tree species composition of a tropical semideciduous forest in southeastern Brazil. *Vegetatio*, 1994, 113, 99-124.

Osborne, LL; Kovacic, DA. Riparian vegetated buffer strips in water-quality restoration and stream management. *Freshwater Biology,* 29, 243-258.

Osborne, LL; Bayley, PB; Higler LWGB; Statzner, F; Triska, T; Iversen, M. Restoration of lowland streams: an introduction. *Freshwater Biology*, 1993, 29, 187-194.

Pagano, MC. Rhizobia associated with neotropical tree *Centrolobium tomentosum* used in riparian restoration. *Plant Soil Environ.*, 2008, 54, 498-508.

Pagano, MC; Raposeiras, R; Scotti, MR. Arbuscular Mycorrhizal Diversity In A Revegetated Riparian Area In Brazil, A DGGE Analysis. In: FERTBIO 2008- Londrina, PR, Brasil. 15-19 Setembro, 2008a (in Portuguese).

Pagano, MC; Marques, MS; Cabello, MN; Scotti, MR. Mycorrhizal Associations In Native Species For The Restoration Of Velhas River Riparian Forest, Brazil. In: FERTBIO 2008- Londrina, PR, Brasil. 15-19 Setembro, 2008b (in Portuguese).

Pagano, MC; Passos RV; Viana P; Cabello MN; Scotti MR Riparian forest restoration: arbuscular mycorrhizae in disturbed and undisturbed soils. In: VI Congreso Latinoamericano de Micología, Mar del Plata, Argentina, 10-13 Novembro de 2008c.

Pagano, MC; Marques, MS; Sobral, M; Scotti, MR. Screening for arbuscular mycorrhizal fungi for the revegetation of eroded riparian soils in Brazil. In: VI Congreso Latinoamericano de Micología, Mar del Plata, Argentina. 10-13 Novembro de 2008d.

Pagano, MC; Scotti, MR. Recuperação De Mata Ciliar Degradada Do Rio Sabará - Minas Gerais. VII Simpósio Nacional sobre Recuperação de Áreas Degradadas. Curitiba-PR, Brazil, 9 a 11 de outubro de 2008, (in Portuguese).

Pagano, MC; Cabello, MN; Bellote, AF; As, NM; Scotti, MR. Intercropping system of tropical leguminous species and *Eucalyptus camaldulensis*, inoculated with rhizobia and/or mycorrhizal fungi in semiarid Brazil. *Agrofor. Syst.*, 2008, 74: 231-242.

Pagano, MC; Persiano, AIC; Cabello, MN; Scotti, MR. Survey of arbuscular mycorrhizas in preserved and impacted riparian environments. In: 6th International Conference on Mycorrhiza ICOM6, 2009a, Belo Horizonte. Abstracts ICOM6. Viçosa: editors, 2009a, 1, 59-60.

Pagano, MC; Scotti, MR; Cabello, MN. Effect of the inoculation and distribution of mycorrhizae in *Plathymenia reticulata* Benth under monoculture and mixed plantation in Brazil. *New Forests*, 2009b, 38,197-214.

Parrota, JA; Knowles, OH. Restoration of tropical moist forests on bauxite-mined lands in the Brazilian Amazon. *Restoration Ecology*, 1999, 7, 103-116.

Pasqualini, D; Uhlmann, A; Stürmer, SL. Arbuscular mycorrhizal fungal communities influence growth and phosphorus concentration of woody plants species from the Atlantic rain forest in South Brazil. *For. Ecol. Manage.*, 2007, 245, 148-155.

Patreze, CM; Cordeiro, L. Nodulation, arbuscular mycorrhizal colonization and growth of some legumes native from Brazil. *Acta Bot. Bras.*, 2005, 19, 527-537.

Piotrowski, JS; Lekberg Y; Harner, MJ; Ramsey, PW; Rillig, MC. Dynamics of mycorrhizae during development of riparian forests along an unregulated river. *Ecography*, 2008. 31, 245-253.

Plenchette, C; Clermont-Dauphin, C; Meynard, JM; Fortin, JA. Managing arbuscular mycorrhizal fungi in cropping systems. *Can. J. Plant Sci.*, 2005, 85,31-40.

Pompeu, PS; Alves, CBM; Hughes, R. Restoration of the das Velhas River basin, Brazil: challenges and potential. In: Lastra DGJ and Martinez PV, editors. *Aquatic habitats: analysis & restoration*. Proceedings of the Fifth International Symposium on Ecohydraulics, September 2004, Madrid, Spain, Volume 1. Inter- national Association of Hydraulic Engineering & Research, Madrid; 2004; 589-594.

Pompeu, PS; Alves, CBM. The Effects of Urbanization on Biodiversity and Water Quality in the Rio das Velhas Basin, Brazil. *American Fisheries Society Symposium*, 2005, 47,11-22.

Purin, S; Rillig, MC. The arbuscular mycorrhizal fungal protein glomalin: Limitations, progress, and a new hypothesis for its function. *Pedobiologia*, 2007, 51, 123-130.

Rathinasabapathi, B. Ferns represent an untapped biodiversity for improving crops for environmental stress tolerance. New Phytol., 2006, 172, 385-390.

Rillig, MC; Wright, SF, Eviner, VT. The role of arbuscular mycorrhizal fungi and glomalin in soil aggregation: Comparing effects of five plant species. *Plant Soil*, 2002, 238,325-333.

Rillig, MC; Ramsey, PW; Morris, S; Paul, EA. Glomalin, na arbuscular-mycorrhizal fungal soil protein, responds to land-use change. *Plant Soil*, 2003, 253,293-299.

Rillig, MC; Wright, SF; Nichols, KA; Schmidt, WF; Torn, MS. Large contribution of arbuscular mycorrhizal fungi to soil carbon pools in tropical forest soils. *Plant Soil*, 2001, 233,167-177.

Sabo, JL; Sponseller, R; Dixon, M; Gade, K; Harms, T; Heffernan, J; Jani, A; Katz, G; Soykan, C; Watts, J; Welter, J.. Riparian zones increase regional species richness by harboring different, not more, species. *Ecology*, 2005, 86,56-62.

Schäffer, WB; Prochnow, M. Mata Atlântica. In: Schäffer, W.B., Prochnow, M., editors. A Mata Atlântica e Você: como preservar, recuperar e se beneficiar da mais ameaçada floresta brasileira. Brasília, Apremavi, 2002; 12-45.

Schnitzler, A. River dynamics as a forest process: Interaction between fluvial systems and alluvial forests in large European River Plains. *Bot. Rev.*, 1997, 63, 40-64.

Schultz, RC; Isenhart, TM; Simpkins, WW; Colletti, JP. Riparian forest buffers in agroecosystems – lessons learned from the Bear Creek Watershed, central Iowa, USA *Agroforestry Systems,* 2004, 61, 35-50.

Siqueira, JO; Saggin-Júnior, OJ; Flores-Ayles, WW; Guimaraes, PTG. Arbuscular mycorrhizal inoculation and superphosphate application influence plant development and yield of coffee in Brazil. *Mycorrhiza,* 1998, 7, 293-300.

Smith, SE; Read, DJ. Mycorrhizal Symbiosis. New York: Elsevier; 2008.

Sprent JI, Sprent P. 1990. *Nitrogen fixing organisms. Pure and Applied Aspects*. London: Chapman & Hall.

Treseder, KK; Turner, KM. Glomalin in Ecosystems. *SSSA*, 2007, 71, 1257-1266.

Venturin, N; Duboc, E; Vale, FR; Davide, AC. Adubação Mineral Do Angico-Amarelo (Peltophorum Dubium (Spreng.) Taub.). *Pesq. agropec. bras.*, 1999, 34, 441-448.

Vervuren, PJA; Blom, CWPM; Kroon, H. Extreme flooding events on the Rhine and the survival and distribution of riparian plant species. *Journal of Ecology*, 2003, 91, 135-146.

Wang, B; Qiu, YL. Phylogenetic distribution and evolution of mycorrhizas in land plants. *Mycorrhiza,* 2006, 16, 299-363.

Worbes, M. Structural and other adaptations to longterm flooding by trees in Central Amazonia. *Amazoniana*, 1985, 9, 459-484.

Wright, SF; Upadhyaya, A. Extraction of an abundant and unusual protein from soil and comparison with hyphal protein of arbuscular mycorrhizal fungi. *Soil Sci.,* 1996, 161, 575-586.

Wright, SF; Upadhyaya, A. A survey of soils for aggregate stability and glomalin, a glycoprotein produced by hyphae of arbuscular mycorrhizal fungi. *Plant Soil*, 1998, 198, 97-107.

Zangaro, W; Nisizaki, SMA; Domingos, JCB; Nakano, EM. Micorriza Arbuscular Em Espécies Arbóreas Nativas Da Bacia Do Rio Tibagi, Paraná. *Cerne*, 2002, 8, 77-87 (in Portuguese).

Zhao, X; Yu, T; Wang, Y; Yan, X. Effect of arbuscular mycorrhiza on the growth of *Camptotheca acuminata* seedlings *Journal of Forestry Research,* 2006, 17,121-123.

Zhang, Y; Guo, LD; Liu, RJ. Arbuscular mycorrhizal fungi associated with common pteridophytes in Dujiangyan, southwest China. *Mycorrhiza*, 2004, 14, 25-30.

Zhao, ZW. The arbuscular mycorrhizas of pteridophytes in Yunnan, southwest China: evolutionary interpretations. *Mycorrhiza*, 2000, 10,145-149.

INDEX

A

access, 88, 128, 204
accounting, 124
acetic acid, 106
acid, 24, 88, 89, 106, 131, 204, 215, 219, 220, 300
acidic, 76, 106, 204, 236
acidity, 89
adaptability, xii, 265
adaptation, 24, 63, 83, 109, 138, 236, 249, 253, 255, 296
adaptations, xi, 37, 62, 88, 223, 225, 226, 231, 232, 240, 249, 255, 260, 269, 295, 315
adjustment, 272
adverse conditions, 269
adverse effects, 155
aesthetic, 309
Africa, 37, 42, 46, 98, 137, 138, 146, 166, 195, 202
age, xi, 5, 166, 201, 205, 214, 218, 221, 232, 233, 238, 276, 282, 285, 288
aggregation, xi, 20, 21, 201, 278, 286, 289, 308, 315
aggressiveness, 64
agriculture, vii, xii, 5, 29, 38, 42, 43, 83, 128, 236, 243, 277, 285, 291, 292, 293
Alaska, 219
alimentation, 97, 144
ALT, 252
aluminium, 22, 76
amino, 204, 205
amino acid, 204, 205
amino acids, 204, 205
ammonium, 69
amplitude, 231, 232, 238, 239
anastomosis, 305
anatomy, 202, 220
angiosperm, 166
annuals, 88
ANOVA, 107, 113, 152, 155
anoxia, 301

B

antagonism, 161
aquatic habitats, 10, 12, 17, 18, 26, 266, 272
Asia, 273, 274
assessment, 123, 129, 165, 167, 286
atmosphere, 276, 308
attachment, 136
Australasia, 102, 202
authority, 306
avoidance, 246

bacteria, 3, 23, 97, 277, 301
Bangladesh, 16, 22
banks, 277, 292, 293
barriers, 103, 237
base, x, 67, 89, 131, 165, 167, 246, 308
bauxite, 314
beneficial effect, 72, 124, 158
benefits, 3, 4, 21, 24, 29, 42, 69, 87, 96, 142, 145, 148, 156, 161, 201, 215, 222, 291, 309
bias, 259
biodiversity, 26, 43, 46, 47, 76, 124, 158, 161, 162, 223, 224, 236, 244, 292, 315
biogeography, 30, 43, 44, 88, 197, 257, 258, 261
bioindicators, 103
biomass, 22, 32, 54, 55, 60, 63, 65, 70, 152, 153, 154, 155, 163, 203, 219, 225, 252, 254, 262, 265, 272, 276, 277, 278, 279, 296
biotechnology, 73
biotic, 20, 30, 37, 38, 45, 60, 64, 93, 129, 156, 203
biotic factor, 64
birds, 109, 122, 277
Bolivia, 103, 232
boreal forest, 166, 202, 204, 205, 221
Botswana, 259
branching, 62, 305
breakdown, 205

C

Britain, 46, 47, 222
burn, 38, 83

cacao, 37, 85, 235
calcium, 22, 131
CAM, 10, 17, 18
Cameroon, 198
canals, 245
carbohydrates, 54, 62, 64, 65, 279
carbon, 20, 21, 23, 25, 30, 38, 44, 54, 55, 60, 62, 63, 64, 65, 69, 70, 71, 72, 76, 90, 95, 131, 148, 155, 158, 194, 205, 220, 275, 276, 277, 278, 279, 286, 287, 288, 289, 300, 308, 312, 315
carbon dioxide, 276, 288
Caribbean, 249
Caribbean Islands, 249
case study, 27, 97
catalysis, 204
cation, 69, 75, 76, 299
cattle, 83, 138, 277, 291, 293, 308
CEC, 299
cell biology, 221
cellulose, 205
Central Europe, 45, 145, 162
challenges, 121, 159, 225, 296, 314
changing environment, 69
chemical, 45, 65, 75, 201, 205, 231, 238, 265, 272, 299
chemical degradation, 65
chemical properties, 45
Chile, 9, 10, 11, 12, 13, 15, 16, 17, 18, 19, 22, 102, 103, 105, 108, 120, 122, 123, 192
China, 9, 13, 15, 18, 23, 93, 124, 125, 137, 143, 202, 316
chitin, 205, 278, 286
chlorine, 35
chlorophyll, 276
cities, 293
City, 35, 260
classes, 59, 72, 123, 260, 276, 279, 285
classification, 4, 50, 68, 106, 121, 124, 131, 144, 167, 188, 195, 196, 217, 218, 228, 233, 246, 255
cleaning, 203
climate, 29, 30, 31, 33, 46, 76, 89, 116, 120, 131, 223, 224, 227, 228, 229, 231, 237, 244, 249, 250, 253, 257, 277
climate change, 31, 89, 223, 224, 231, 249, 250, 253
climates, 93, 98, 139, 194, 250
climatic factors, 237
clone, 199
cluster analysis, 132, 140

clusters, 192
CO2, 21, 31, 46, 195, 203, 204, 225, 276, 287, 288
coastal region, 24
cocoa, 260
coffee, 311, 315
Colombia, 75, 77, 78, 79, 80, 85, 232, 241
colonisation, 26, 148
color, 215
commercial, vii, 1, 97, 101
compaction, 31, 43, 45
competition, 3, 22, 65, 66, 151, 152, 159, 204, 232, 267, 273
competitiveness, 140
compilation, 2
complex interactions, 37, 311
complexity, 30, 203
composition, 2, 20, 23, 24, 26, 30, 31, 32, 37, 38, 42, 43, 44, 45, 46, 75, 76, 77, 93, 99, 100, 127, 128, 129, 132, 140, 144, 146, 148, 151, 155, 156, 157, 196, 197, 202, 204, 213, 224, 225, 226, 230, 231, 232, 233, 236, 238, 243, 245, 252, 254, 259, 262, 273, 276, 281, 292, 307, 308, 309, 310, 313
compost, 67
compounds, 63, 194, 205, 225, 226, 278, 279
compression, 237
condensation, 245
conditioning, 119
conductivity, 214
conference, 21
conifer, 35, 169, 197
connectivity, 223, 228, 250
consensus, 215
conservation, 30, 42, 63, 78, 88, 89, 95, 96, 103, 120, 143, 201, 222, 244, 255, 257, 258, 259, 260, 261, 269, 292, 308
construction, 54, 63, 227
consumption, 225
contaminated soil, 163
contaminated soils, 163
contamination, 203, 296
controversial, 4, 151, 203, 206, 243
copper, 145
correlation, 117, 206, 258
correlation analysis, 258
correlations, 60, 73, 207, 213
cortex, 100, 109, 114, 116
cosmetics, 94
cost, 4, 60, 63, 64, 65, 102, 109, 156
Costa Rica, 4, 20, 37, 55, 68, 78, 97, 98, 122, 123, 124, 311
covering, 2
crop, 35, 37, 43, 158, 159, 160, 255, 277
crops, 1, 38, 42, 65, 100, 158, 315

Index 319

crown, 188, 232, 301
crown fires, 301
crowns, 232, 258
crystalline, 237
Cuba, 137
cultivation, 83, 84
culture, x, 165, 167
cycles, 70, 77, 116, 144, 148, 233, 249
cycling, 65, 71, 72, 158, 162, 265, 289, 292
Cyprus, 270, 271

D

data set, 32, 38, 210, 211, 212
database, vii, 29, 33, 78, 215
decomposition, 65, 225
defence, 68
deficiencies, 110
deficiency, 225, 226, 237, 241
deforestation, 224
degradation, 65, 278, 307, 308
degraded area, 49, 65, 66, 138, 291, 304, 308
Delta, 259
Denmark, 266
Department of Agriculture, 199
deposition, 31, 43, 46, 76, 281, 288
deposits, 89, 225, 238
depression, 237
deprivation, 252
depth, 56, 57, 58, 81, 89, 206, 221, 245
destruction, 95, 243
detection, 144, 307
detoxification, 194
deviation, 111
dichotomy, 101
diet, 197
diffusion, 225
discrimination, 77
diseases, 162
dispersion, 292, 293
displacement, 102
distinctness, 233, 253
diversification, 30, 83, 124, 195
DNA, 77, 78, 83, 85, 86, 195, 215, 307
DOI, 161
dominance, 76, 78, 101, 138, 140, 151, 156, 159, 228, 238, 250
drainage, 236
drawing, 1, 291
drought, 38, 203, 213, 217, 226, 233, 238, 239, 241, 246, 250, 251, 256, 258, 269, 282
dry matter, 50
drying, 63

dynamic systems, 229, 232
dynamism, 232

E

ECM, 165, 166, 167, 168, 169, 187, 188, 189, 192, 193, 194, 201, 202, 203, 204, 205, 206, 207, 208, 209, 210, 211, 212, 213, 214, 215, 216, 217, 221
ecological restoration, 147, 159, 267, 311
ecological roles, 266, 295
ecological systems, 254
ecology, 2, 4, 5, 25, 26, 72, 84, 85, 88, 97, 102, 120, 122, 159, 160, 195, 196, 197, 199, 203, 204, 218, 228, 248, 252, 253, 256, 261, 269, 273, 288, 293, 310
ecosystem, 2, 25, 26, 29, 30, 31, 42, 44, 45, 46, 47, 66, 68, 75, 78, 83, 87, 93, 96, 97, 100, 103, 117, 124, 140, 158, 159, 160, 161, 163, 196, 199, 218, 222, 246, 249, 263, 265, 267, 269, 272, 274, 278, 286, 309, 311
ecosystem restoration, vii, 29, 42, 159
Ecuador, 122, 232
editors, 23, 24, 25, 26, 43, 44, 45, 68, 70, 71, 72, 73, 97, 144, 146, 217, 219, 220, 222, 253, 254, 286, 287, 311, 312, 313, 314, 315
electrical conductivity, 214
electron, 121, 124, 136, 303
electron microscopy, 124
electrophoresis, 307
elongation, 262
encoding, 195
endangered species, 21
energy, 63, 100, 109, 116, 277
energy supply, 100, 116
England, 84, 156
environment, 3, 10, 24, 37, 38, 60, 63, 66, 69, 96, 117, 158, 205, 206, 213, 219, 225, 226, 249, 250, 269, 274, 281, 292, 303
environmental change, 117, 121
environmental characteristics, 205
environmental conditions, ix, xii, 38, 45, 100, 103, 116, 119, 127, 216, 217, 224, 226, 238, 249, 260, 263, 282
environmental degradation, 276
environmental factors, 3, 52, 124, 203, 206
environmental management, xii, 3, 291, 293
environmental stress, 72, 256, 315
environmental stresses, 256
environmental variables, 3, 70, 285
enzymatic activity, 205
enzyme, 194, 204
enzymes, 194, 195, 196, 205, 219, 225
epidermis, 109, 114

equilibrium, 309
erosion, 97, 224, 232, 277, 292, 295
estuarine systems, 250
eucalyptus, 131
Europe, 10, 12, 17, 18, 45, 145, 162, 202, 260
evaporation, 131, 229, 237
Everglades, 266
evidence, 52, 76, 96, 113, 121, 122, 124, 130, 143, 147, 148, 151, 152, 153, 156, 157, 158, 159, 165, 167, 169, 187, 192, 193, 194, 195, 197, 205, 218, 227, 250, 253, 266, 275, 294
evolution, 3, 26, 46, 76, 84, 88, 99, 101, 118, 119, 123, 124, 146, 166, 195, 197, 199, 226, 231, 248, 250, 251, 289, 295, 315
evolutional processes, xii, 223, 250
exclusion, 37, 169
experimental design, 299
exploitation, 292, 293
exposure, 88, 224
extinction, 30, 105, 224
extraction, 76, 77, 85, 128, 141, 308

F

FAA, 106
families, 2, 6, 76, 87, 90, 94, 95, 97, 99, 105, 107, 109, 110, 113, 116, 118, 127, 128, 142, 144, 145, 148, 151, 166, 195, 206, 234, 235, 241, 242, 243, 265, 269, 272, 295, 304, 307
farmland, 93
fauna, 78, 89, 156, 160, 223, 224, 293
fertility, 1, 3, 4, 21, 27, 50, 52, 55, 60, 62, 64, 65, 71, 73, 76, 81, 127, 279, 289
fertilization, 31, 41, 43, 69, 81, 84, 203, 218, 279, 295, 299, 301
fertilizers, 65, 95, 158, 199
financial, 42, 95, 142, 272
financial support, 42, 95, 142, 272
fire event, 102
fires, 36, 38, 150, 226, 249, 250, 258
fitness, 2, 43, 123
fixation, 281
flooding, 22, 82, 220, 223, 225, 226, 228, 230, 231, 232, 236, 239, 240, 241, 245, 249, 250, 254, 255, 258, 262, 269, 273, 277, 281, 282, 285, 287, 292, 296, 301, 312, 313, 315
floods, 226, 227, 231, 239, 292
flora, 68, 76, 78, 88, 89, 95, 102, 105, 107, 117, 133, 160, 219, 224, 225, 227, 228, 229, 231, 232, 233, 237, 243, 246, 250, 251, 256, 261, 269, 273
flora and fauna, 78
fluctuations, 202, 213, 227, 231, 254, 262
food, xi, 65, 158, 223, 224, 276

food web, 158, 223
forbs, 70, 122, 152, 155, 293
force, 41
forest ecosystem, 69, 75, 100, 196, 201, 263, 287
forest fragments, 66, 257, 258
forest habitats, 228
forest management, 94
forest restoration, 25, 293, 296, 300, 314
formation, 30, 52, 62, 76, 82, 84, 94, 141, 199, 200, 206, 218, 219, 225, 231, 236, 237, 250, 255, 267, 269, 273, 277, 278
fragments, 66, 90, 152, 244, 257, 258, 307, 308
freshwater, 223, 224, 227, 231, 248, 250, 260, 265
frost, 237
FSB, 24, 145
fungus, 25, 32, 45, 49, 50, 52, 54, 55, 59, 60, 62, 63, 64, 65, 66, 71, 100, 101, 109, 116, 119, 124, 128, 148, 162, 167, 169, 193, 194, 197, 198, 204, 213, 305
fungus spores, 25, 45, 71

G

gametophyte, 121
gel, 307
gene expression, 222
general knowledge, 103, 119
genes, 195, 197
genetic background, 117
genetic diversity, 26, 72, 86, 307, 308
genetics, 69, 119
genome, 52
genotype, 100, 117
genus, 6, 8, 10, 12, 14, 16, 18, 75, 76, 77, 91, 92, 95, 102, 105, 110, 129, 140, 167, 168, 169, 187, 189, 192, 193, 195, 196, 197, 198, 202, 215, 282, 307
geography, 168
geology, 243, 245, 277
Germany, 209, 220, 221, 223
germination, 88, 92, 136, 225, 254, 262
Gigaspora margarita, 10, 12, 14, 81, 83, 91, 93, 138, 139, 141, 272
global climate change, 231
global scale, 166, 195
Glomus intraradices, 80
glycerol, 34, 106
goods and services, 96
gracilis, 34, 35
grants, 119
grass, 9, 10, 37, 55, 56, 83, 86, 162, 220, 258, 292, 309
grasses, 55, 64, 124, 152, 160, 220, 287, 293, 296

grasslands, 4, 5, 20, 34, 43, 57, 88, 93, 138, 189, 236, 237, 238, 244
grazing, 138, 150
Great Britain, 46, 47
greenhouse, 54, 64, 95, 122, 152, 153, 163, 276, 278, 279, 287, 301
greenhouse gases, 276
274, 275, 277, 278, 286, 288, 289, 292, 295, 296, 301, 304, 311, 313, 314, 316
growth dynamics, 252
growth rate, 3, 4, 23, 25, 54, 61, 62, 63, 68, 116, 218, 292
Guyana, 197, 231, 236, 254
gymnosperm, 166

H

habitat, 46, 60, 73, 89, 98, 111, 119, 168, 169, 192, 193, 197, 201, 203, 205, 223, 224, 232, 233, 239, 243, 244, 248, 249, 250, 258, 289, 312
habitats, 1, 2, 6, 10, 12, 17, 18, 22, 26, 62, 63, 64, 87, 88, 91, 93, 96, 99, 105, 107, 109, 111, 113, 116, 117, 118, 127, 200, 201, 203, 205, 209, 210, 213, 215, 219, 223, 226, 227, 228, 231, 232, 233, 238, 239, 240, 241, 246, 247, 248, 249, 맴250, 256, 258, 260, 265, 266, 267, 272, 275, 285, 287, 293, 314
hair, 4, 53, 55, 60, 69, 72, 112, 114, 124, 267
hairless, 69
halophyte, 273
harbors, 275
hardwood forest, 252
harvesting, 26, 277
health, 128, 224
heavy metals, 204, 295
height, 95, 106, 214, 225, 229, 230, 232, 233, 238, 240, 241, 279, 300, 301, 302
height growth, 95
hemisphere, 148, 158, 166, 249
heterogeneity, 3, 22, 205, 273, 281
highlands, 87, 91, 93, 94, 96, 254
Highlands, 87, 88
highways, 95, 142
histogram, 310
history, 23, 29, 30, 33, 41, 70, 249
hormones, 262
host, 1, 2, 5, 21, 30, 31, 38, 50, 52, 62, 64, 66, 70, 71, 76, 77, 78, 82, 83, 87, 89, 95, 100, 101, 117, 119, 122, 124, 128, 148, 151, 157, 160, 169, 195, 199, 201, 202, 203, 204, 205, 215, 216, 220, 279, 282, 285, 307
hot spots, 37, 88, 224
hotspots, 107, 257

House, 73, 274
human, 3, 68, 89, 127, 150, 244, 276, 291, 293, 308, 311
humidity, 215
humus, 237
Hungary, 6
hydrogen, 204
hypothesis, 3, 21, 54, 116, 117, 118, 119, 165, 194, 315

I

identification, 5, 19, 32, 42, 46, 76, 77, 90, 95, 119, 129, 137, 144, 165, 167, 168, 169, 194, 195, 307, 310
identity, 2, 24, 100, 121, 124, 160, 195, 301
IMA, 89
images, 106, 107
imitation, 63, 309
immigrants, 223, 224, 276
improvements, 5
in transition, 236
in vitro, 84, 189, 200, 220
incidence, 4, 53, 55, 60, 62, 118, 267, 277
income, 308, 312
incompatibility, 219
India, 25, 26, 123, 196, 202, 265, 266, 267, 269, 272, 274
indirect effect, 204
individuals, 225, 240, 245, 292, 301
Indonesia, 6, 19, 70
induction, 225
industries, 224
inertia, 195
infection, 4, 27, 42, 43, 62, 68, 69, 70, 86, 123, 203, 204, 215, 218, 219, 222
inferences, 32, 197
inoculation, 3, 24, 25, 49, 50, 51, 52, 59, 61, 66, 92, 93, 95, 96, 141, 145, 275, 278, 279, 286, 295, 314, 315
inoculum, 2, 59, 60, 66, 68, 69, 73, 77, 93, 95, 140, 148, 152, 278, 286, 289, 295, 311, 312
insects, 143, 156
integration, 163
interaction effect, 215
interface, 54, 72, 119, 167, 206, 269
intervention, 75
invasions, 31, 46, 295
invertebrates, 72, 151
investment, 26, 54, 55, 61, 63
ions, 204
Iowa, 274, 315
iron, 76, 89, 97

322 Index

J

Japan, 120, 169, 193, 202
Jordan, 72, 158, 161, 163, 236, 255

K

kill, 226
Korea, 122

L

lakes, 2, 22, 23, 65, 202, 227, 229, 233, 273, 277
landscape, 22, 26, 150, 223, 224, 229, 231, 235, 241, 243, 244, 250, 259, 261, 262, 291, 293
landscapes, 88, 94, 96, 102, 224, 226, 241, 248, 292, 309
Latin America, 251
leaching, 76
lead, 5, 64, 217, 225, 229, 237
legend, 192
legume, 77, 91, 96, 128, 133, 301, 312
levees, 229
life cycle, 101, 116, 233, 249
light, 38, 60, 63, 64, 65, 69, 72, 116, 121, 188, 209, 233, 309
lignin, 194, 205, 293
livestock, 150
locus, 198
logging, 141, 150, 313
longevity, 50, 63, 65, 69
longleaf pine, 258
low temperatures, 102
LTD, 161
lying, 105

irrigation, 128
islands, 102
isolation, 88, 105, 133, 143, 154
isotope, x, 165, 167, 187
issues, 159
Italy, 165

M

macropores, 131
magnesium, 22
magnitude, 60, 148
majority, 2, 5, 101, 128, 151, 271, 301, 303
Malaysia, 6, 19, 193, 196, 274, 313
mammal, 197

mammals, 109, 122, 277
management, 1, 3, 31, 35, 38, 42, 43, 65, 71, 75, 94, 95, 96, 103, 147, 148, 158, 159, 161, 257, 262, 288, 295, 301, 308, 313
manganese, 194
mangroves, 244, 265, 269, 272, 273, 274
manipulation, 96, 222
mantle, 167, 208, 214, 215, 216
manufacturing, 94
marsh, 266, 269, 273, 274
mass, 53, 60, 72, 255, 279
materials, 78
matrix, 96
matter, 24, 50, 62, 65, 81, 89, 90, 123, 131, 196, 199, 214, 219, 225, 231, 245, 277
MCP, 312
measurement, 207, 220
measurements, 279
mediation, 159
Mediterranean, 68
MES, 122
meta-analysis, 31, 46
metabolism, 54, 60, 204, 222, 225, 253, 255, 292
metal ion, 204
metal ions, 204
metals, 72, 204
methodology, 77, 78, 105
Mexico, 22, 23, 29, 33, 35, 36, 37, 41, 55, 68, 120, 168, 252
microbial communities, 69
microbiota, 95
microclimate, 245
microcosmos, 3
microhabitats, 88, 206, 217, 229
microorganism, 302
microorganisms, 21, 31, 44, 95, 96, 97, 142, 158, 205, 225, 277, 291, 293, 294, 295, 304, 309
microscope, 34, 35, 106, 121, 215
microscopy, 116, 124
Middle East, 202
migration, 223, 251
mineralization, 245
Miocene, 250
Missouri, 125, 145, 163, 195, 246, 254
model system, 161
models, 3, 147, 148, 151, 159
moisture, 110, 202, 203, 207, 215, 217, 220, 266, 273, 274, 281, 282, 286
molecules, 205
Montana, 24, 219
Montenegro, 85
morphogenesis, 69

morphology, 3, 4, 22, 26, 53, 54, 60, 62, 70, 72, 73, 95, 100, 109, 110, 113, 116, 117, 119, 120, 123, 124, 127, 129, 133, 170, 172, 174, 176, 178, 180, 182, 184, 186, 190, 192, 201, 202, 253, 261, 272, 279, 289, 292, 311

mortality, 162, 245, 254, 301, 312

mortality rate, 245

mosaic, 150, 205

multiplication, 77, 144

mutant, 69

mycelium, 4, 25, 66, 71, 83, 286, 308

mycorrhiza, 3, 22, 23, 26, 27, 43, 52, 68, 71, 75, 76, 83, 84, 86, 97, 98, 100, 112, 114, 120, 123, 144, 145, 146, 147, 165, 167, 197, 202, 219, 267, 272, 273, 274, 296, 316

N

Namibia, 46, 146

national parks, 88

native population, 52

native species, 4, 5, 20, 24, 83, 87, 89, 95, 96, 107, 128, 142, 203, 213, 293, 296, 308

natural habitats, vii, 1, 2, 6

natural resources, 224, 233, 276

nature conservation, 295

negative effects, 156, 158

negative relation, 31, 36

nested PCR, 86

Netherlands, 25, 45, 161, 165, 207, 221, 254, 282, 285, 288

neutral, 119, 151, 214, 300

New South Wales, 220

New Zealand, 102, 105, 121, 273

Nicaragua, 20, 37, 55

nitrogen, 3, 21, 37, 43, 46, 69, 71, 195, 217, 218, 220, 221, 274, 277, 288, 293, 295, 296, 301

nitrogen-fixing bacteria, 3, 293

North America, 29, 30, 31, 46, 102, 138, 146, 169, 197, 199, 202, 260, 287

Norway, 121, 219

Norway spruce, 219

nucleic acid, 76

nutrient, 1, 3, 4, 5, 20, 23, 26, 43, 50, 52, 53, 54, 55, 56, 60, 62, 63, 64, 65, 66, 68, 69, 70, 71, 72, 73, 75, 82, 88, 95, 97, 100, 109, 122, 124, 131, 148, 156, 158, 162, 202, 204, 214, 215, 227, 228, 229, 231, 233, 236, 245, 249, 256, 265, 267, 277, 278, 282, 286, 292, 304, 309

nutrient transfer, 5, 305

nutrients, 22, 44, 52, 54, 55, 62, 64, 65, 66, 70, 72, 88, 95, 110, 128, 148, 201, 203, 204, 229, 231, 236, 243, 245, 266, 281, 293, 309

nutrition, 4, 23, 54, 68, 69, 71, 86, 89, 99, 100, 160, 194, 195, 204, 266, 272, 274, 275, 277, 278, 312

nutritional status, 204

O

Oceania, 202

OH, 314

oil, 27, 30, 52, 55, 73, 94, 97, 123, 158, 201, 203, 207, 214, 215, 238, 266, 313

orchid, 200, 202

organic matter, 24, 62, 65, 81, 89, 90, 131, 196, 199, 214, 219, 225, 245, 277, 292, 293, 299, 300

organic soils, 206

organism, 76

organize, 21

organs, 63, 223, 239, 243, 279

overlap, 137

oxidative damage, 252

oxidative stress, 194

oxygen, xi, xii, 20, 206, 223, 225, 226, 236, 237, 252, 253, 265, 266, 267

oxygen consumption, 225

P

Pacific, 9, 19, 22, 45, 102, 105, 219, 312

Pakistan, 274

Panama, 246

Paraguay, 103, 226, 227, 228, 237, 238, 239, 240, 253, 301

parallel, 3, 166, 243, 308

pasture, 25, 33, 37, 45, 55, 71, 72, 83, 124, 277, 285

pastures, 37, 65, 66, 83, 292

pathogens, 65, 66, 101, 147, 151, 156, 162

PCR, 77, 84, 85, 86, 220, 307

peat, 105, 106

percolation, 277

periodicity, 229, 260

permeability, 204

permit, 54, 60, 65, 77

personal communication, 266

Peru, 84, 103, 122, 232

pesticide, 158

pharmaceuticals, 94

phenolic compounds, 63, 194

phosphate, 67, 97, 128, 294, 301, 302

phosphates, 204, 219

phosphorous, 76

phosphorus, 3, 22, 31, 37, 45, 46, 50, 69, 70, 71, 75, 82, 195, 197, 218, 266, 272, 273, 278, 279, 288, 295, 300, 305, 313, 314

photosynthesis, 109, 253, 276
phylum, 46, 76, 129, 187, 286
Physiological, 123, 218, 256, 260, 287
physiological mechanisms, 54
physiology, 2, 203, 256
phytoplankton, 37, 42
phytoremediation, x, 147, 159, 162, 217
pioneer species, x, xii, 4, 147, 156, 159, 233, 275, 278, 279, 282, 295, 296, 301
plant establishment, 45, 160, 301
plant growth, 5, 23, 24, 45, 52, 53, 61, 64, 70, 101, 128, 145, 148, 151, 152, 153, 159, 161, 204, 218, 226, 228, 245, 255, 262, 265, 269, 276, 279, 281, 286, 292, 295, 296, 305
plasticity, 27, 73, 216, 289
plastid, 197
Pliocene, 238
PM, 24, 26, 72
Poland, 207, 209, 210, 313
pollutants, xi, 201, 203
pollution, 65, 224, 277, 292
polymerase, 307
polymerase chain reaction, 307
polymorphisms, 86
ponds, 202, 265, 266
pools, 25, 72, 288, 315
population, 32, 66, 128, 196, 206, 218, 224, 226, 227, 233, 249, 250, 254, 256
population growth, 218
population size, 250
population structure, 256
positive correlation, 60
positive feedback, 157
positive relationship, 37
potassium, 282
poverty, 96
precipitation, 20, 93, 102, 105, 131, 149, 229, 233, 244, 292, 300
predation, 63, 65
predators, 63
preparation, 5, 31, 77, 94
preservation, 292
primary function, 53
principles, 43, 146, 308
probability, 239, 246
project, 42, 83
proliferation, 54, 62
propagation, 77
protection, 224, 260, 261
Puerto Rico, 70

Q

quantification, 82, 260
quantitative estimation, 43
Queensland, 16

R

radiation, 38
radius, 69
rain forest, 6, 16, 18, 19, 20, 22, 23, 26, 35, 71, 72, 75, 78, 86, 120, 122, 161, 198, 252, 288, 313, 314
rainfall, 73, 89, 93, 105, 139, 149, 221, 229, 235, 237, 243, 259, 261, 299
rainforest, 99, 100, 102, 103, 104, 105, 109, 113, 117, 119, 249, 250, 251, 255, 256, 260
random walk, 99, 105
recommendations, 45
recovery, 43, 72, 138, 159, 203, 275, 276, 277, 278, 279, 286, 295, 313
recovery process, 286
regenerate, 250
regeneration, 30, 63, 87, 91, 93, 96, 97, 127, 277, 308
regions of the world, 2, 110
rehabilitation, 43, 50, 66, 143, 269, 296
reintroduction, 66
relatives, 105, 122, 169
remediation, 203, 220
remote sensing, 163, 262
reproduction, 221, 233, 249
requirements, 60, 65, 203, 301, 304
researchers, 4, 33
reserves, 55, 69, 88
resilience, 41, 163, 196, 309
resistance, 63, 64, 65, 256, 292
resource availability, 26
resources, 30, 31, 38, 41, 53, 155, 224, 233, 255, 276, 307
respiration, 63
response, 3, 21, 23, 24, 26, 49, 50, 51, 52, 59, 61, 66, 70, 71, 73, 84, 95, 103, 162, 163, 200, 203, 220, 226, 269, 278, 279, 289
responsiveness, 3, 26, 27, 49, 50, 52, 54, 64, 72, 73, 98, 148, 152, 157, 289
restoration, 3, 5, 21, 24, 25, 29, 42, 87, 89, 91, 93, 94, 95, 96, 127, 129, 140, 141, 142, 147, 159, 162, 203, 217, 291, 293, 295, 296, 301, 308, 309, 310, 313, 314
restoration programs, 87, 91, 93, 96, 140, 142, 291, 293, 309
restrictions, 5, 249

reticulum, 136
risks, 260
river basins, 291, 293
river systems, 201, 206, 223, 227, 229
root hair, 3, 4, 50, 54, 55, 60, 62, 69, 70, 90, 109, 110, 112, 114, 115, 116, 302, 304, 305
root system, 1, 54, 55, 60, 62, 68, 81, 105, 109, 110, 120, 199, 206, 214, 236, 277
routes, 223, 251
rowing, 1, 3, 68, 81, 88, 89, 92, 100, 147, 162
Royal Society, 162, 259
rubber, 82
rules, 229
runoff, 277, 293

S

salinity, 138, 226, 266, 269, 272, 295
salts, 204
saltwater, 265
samplings, 90
saturation, 89, 131, 224, 226, 243, 245, 299, 300, 309
savannah, 34, 138, 144, 222
scarcity, 88, 216, 236, 245
science, 20
Scots pine, 195
sea level, 93, 131, 250
sea-level, 231, 244
seasonality, 37, 229
secretion, 196, 219
sediment, 34, 207, 227, 228, 229, 231, 233, 238, 243, 292, 310
sedimentation, 225, 232, 233
sediments, 76, 206, 230, 231, 267, 293, 312
seed, 55, 72, 88, 91, 93, 98, 101, 105, 118, 119, 153, 225, 279
seeding, 203
seedlings, 4, 22, 23, 27, 50, 55, 59, 63, 64, 67, 68, 69, 70, 72, 95, 142, 195, 201, 205, 225, 251, 258, 273, 277, 278, 279, 286, 294, 311, 316
segregation, 250
senses, 213
sensing, 163, 257, 258, 262
services, 96, 158, 159, 160, 161, 276, 277, 292
sewage, 291, 293, 296
sex, 254
Seychelles, 199
shade, 63, 66, 70, 72, 116, 278, 292
shape, 117, 226, 238, 305
shelter, 224, 277
shoot, 50, 72, 124, 152, 153, 155, 225, 262, 278, 279
shoots, 225, 279

shortage, 224, 244, 245
showing, 3, 37, 60, 129, 136, 137, 141, 158, 159, 205, 215, 293, 305
shrubland, 33, 150, 152
shrubs, 68, 101, 128, 131, 152, 196, 239, 246, 258
signals, 194
Singapore, 273
society, 276
software, 107
soil erosion, 292
soil particles, 107
soil seed bank, 91, 93
soil type, 219
solubility, 204, 225
solution, 3, 35, 308
South Africa, 42
South America, 24, 56, 97, 102, 103, 104, 105, 120, 137, 145, 148, 166, 201, 202, 203, 206, 254, 255, 257, 259
Southeast Asia, 23, 122
Soviet Union, 202
Spain, 79, 91, 93, 133, 283, 284, 306, 314
specialists, 138, 213
speciation, 30, 255
species richness, 29, 31, 32, 36, 37, 38, 39, 40, 41, 45, 81, 87, 91, 96, 127, 128, 133, 138, 141, 165, 167, 193, 203, 230, 231, 232, 233, 239, 241, 243, 244, 245, 254, 259, 276, 282, 285, 286, 291, 302, 307, 308, 315
specific knowledge, 224, 228
specific tax, 187
spore, 4, 5, 20, 29, 31, 32, 36, 37, 38, 39, 40, 42, 44, 75, 77, 78, 79, 83, 87, 89, 91, 92, 93, 95, 96, 101, 118, 127, 133, 136, 138, 140, 142, 158, 195, 267, 270, 274, 282, 285, 291, 295, 301, 302, 306, 307, 310
sporophyte, 23, 111, 113, 121
stability, 26, 46, 66, 293, 299, 308, 315
stabilization, 97
standard deviation, 111
standard error, 41
starch, 279
state, 55, 83, 95, 98, 142, 146, 243, 276, 279, 282, 285
states, 117, 128, 129, 245
stele, 116
sterile, 281
stoichiometry, 44
storage, 276, 288
strategy use, 307
stress, 72, 92, 100, 116, 128, 129, 138, 194, 223, 238, 249, 252, 255, 258, 265, 266, 269, 272, 301
stress factors, 249

stressors, 225, 226
strong force, 41
structural characteristics, 254
structure, 1, 21, 25, 26, 30, 32, 37, 42, 44, 45, 64, 66,
    78, 84, 99, 100, 103, 116, 123, 129, 140, 148,
    156, 158, 160, 161, 166, 197, 203, 204, 205, 218,
    219, 220, 222, 232, 236, 237, 243, 245, 250, 252,
    253, 256, 262, 265, 276, 292
structuring, 41, 49, 50, 64, 65, 66, 147, 148, 151,
    156, 159, 282
style, 169, 193
Styles, 193
substrate, 67, 89, 105, 109, 110, 113, 117
substrates, 102, 105, 233, 266
succession, 4, 23, 30, 44, 49, 50, 55, 57, 59, 60, 63,
    64, 65, 66, 69, 72, 73, 128, 147, 148, 151, 156,
    157, 159, 160, 161, 162, 220, 222, 245, 255, 262,
    267, 269, 287, 292, 293, 308
sucrose, 34, 35
supplier, 76
suppression, 156
surface area, 53, 66
surplus, 62
survival, viii, xii, 4, 26, 49, 50, 62, 63, 65, 66, 67, 95,
    128, 253, 265, 269, 275, 277, 278, 315
susceptibility, 4, 52
sustainability, 21, 96, 142, 162
Sweden, 169, 220
symbiosis, 1, 2, 4, 5, 23, 26, 30, 44, 46, 50, 52, 55,
    60, 64, 65, 70, 72, 75, 76, 82, 83, 84, 98, 101,
    103, 109, 110, 116, 117, 122, 123, 124, 127, 145,
    147, 159, 161, 162, 166, 169, 194, 199, 201, 203,
    215, 216, 217, 273, 276, 277, 291
synthesis, 47, 165, 167, 188, 193, 200, 258

T

taxa, 5, 29, 30, 32, 33, 38, 41, 90, 91, 92, 118, 122,
    129, 144, 169, 187, 188, 192, 193, 199, 209, 215,
    221, 227, 269
taxonomy, 46, 78, 111, 119, 189, 199
technical support, 95, 142
techniques, 33, 76, 77, 219, 258, 262, 307, 309
technologies, 5, 94, 308
temperate rain forest, 120, 122
temperature, 20, 36, 45, 89, 93, 105, 109, 131, 149,
    203, 206, 215, 221, 237
terraces, 282
terrestrial ecosystems, 30, 204, 265, 267, 277
territorial, 85
testing, 159, 168
texture, 37, 89, 131, 214, 310
Thailand, 193, 196

thinning, 45, 141
threats, 293
Tibet, 97, 143
tissue, 60, 63
tobacco, 71
tooth, 189
toxicity, 159
traits, 27, 32, 54, 55, 60, 62, 68, 70, 73, 94, 127, 162,
    196, 249
transformation, 83, 107
transformations, 65, 81
transpiration, 293
transplant, 278
transport, 3, 69, 201, 245, 293
treatment, 38, 45, 72, 154, 155, 156, 168, 279, 293,
    296, 308
tropical dry forest, 33, 70, 127, 128
tropical forests, 4, 5, 49, 50, 56, 65, 70, 72, 144, 226,
    244, 248
tropical rain forests, 20, 198
tropical savannas, 258
turnover, 62, 64, 157, 278, 286, 288

U

underlying mechanisms, 146
United, 29, 33, 42, 46, 123, 285
United Kingdom, 46
United Nations, 42
United States, 29, 33, 123, 285
urban, 276, 292, 293, 297, 312
Uruguay, 103, 301, 312
USA, 6, 22, 29, 33, 34, 35, 36, 37, 38, 39, 40, 41, 43,
    46, 120, 121, 122, 124, 161, 196, 199, 217, 220,
    273, 275, 282, 311, 315
USDA, 42

V

variables, 3, 55, 64, 70, 210, 226, 295
variations, 5, 23, 230, 231, 237, 308
vegetation, 4, 20, 29, 33, 38, 60, 65, 70, 75, 76, 83,
    88, 96, 98, 102, 104, 121, 127, 128, 129, 130,
    132, 133, 137, 138, 139, 140, 142, 149, 160, 161,
    162, 226, 227, 228, 230, 233, 235, 236, 237, 238,
    243, 244, 245, 252, 253, 255, 256, 258, 259, 260,
    261, 263, 265, 266, 269, 273, 274, 277, 286, 291,
    293, 295, 296, 299, 304, 307, 308, 309, 311
vegetative reproduction, 233
Venezuela, 44, 55, 77, 79, 80, 84, 85, 138, 140, 144,
    162, 252, 255, 260
vesicle, 91, 305

vision, 189
vocabulary, 116

W

Wales, 220
warts, 136
waste, 95, 203, 296, 297
water, 20, 26, 34, 35, 46, 65, 66, 72, 76, 88, 97, 105, 106, 109, 127, 128, 138, 194, 201, 202, 203, 206, 207, 210, 213, 214, 215, 216, 217, 224, 225, 226, 227, 228, 229, 230, 231, 232, 233, 234, 238, 239, 240, 241, 243, 244, 245, 246, 247, 249, 250, 254, 258, 260, 262, 266, 267, 276, 277, 292, 293, 296, 301, 308, 309, 313
water fluctuations, 202
water quality, 277, 309
watershed, 260, 276, 277, 279

waterways, 292
web, 44, 144, 158, 165, 167, 222
West Africa, 98, 146
wetlands, 220, 224, 227, 233, 248, 253, 255, 257, 266, 267
wildlife, 272, 292
Wisconsin, 43, 253
wood, 44, 94, 97, 128, 131, 133, 144, 189, 222, 225, 262, 276
woodland, 6, 7, 19, 22, 23, 44, 120, 149, 152, 198, 199, 228, 252, 294, 304, 311
workers, 267
worldwide, vii, 1, 2, 21, 93, 138, 167, 202, 229, 231
worms, 122

Y

yield, 97, 156, 158, 272, 315